HAZARDOUS DECISIONS

Hazardous Waste Siting in the UK, The Netherlands and Canada. Institutions and Discourses

ENVIRONMENT & POLICY

VOLUME 34

The titles published in this series are listed at the end of this volume.

Hazardous Decisions

Hazardous Waste Siting in the UK, The Netherlands and Canada. Institutions and Discourses

by

Dave Huitema

Institute for Environmental Studies,
Free University, Amsterdam, The Netherlands

KLUWER ACADEMIC PUBLISHERS
DORDRECHT / BOSTON / LONDON

A C.I.P. Catalogue record for this book is available from the Library of Congress.

ISBN 1-4020-0969-0

Published by Kluwer Academic Publishers,
P.O. Box 17, 3300 AA Dordrecht, The Netherlands.

Sold and distributed in North, Central and South America
by Kluwer Academic Publishers,
101 Philip Drive, Norwell, MA 02061, U.S.A.

In all other countries, sold and distributed
by Kluwer Academic Publishers,
P.O. Box 322, 3300 AH Dordrecht, The Netherlands.

Printed on acid-free paper

Table of contents

Foreword: of bullets and churches

The title of this book, 'Hazardous decisions', was chosen to reflect the character of the decision-making processes that are studied in this book. Far less hazardous, but still somewhat adventurous and uncertain, was the decision to start the research that is the basis for this book.

Having unsuccessfully attempted to avoid staying in university for some time, I was offered a position as a researcher at the Center for Clean Technology and Environmental Policy (CSTM) at the University of Twente in 1993. After working on various externally-funded research projects and even teaching a bit, the question was put by Hans Bressers, the Director at CSTM, of whether I'd be interested in writing a Ph.D. thesis on a topic that I liked. This question was accompanied by the statement that in order to have a career at a university, one should have a Ph.D. degree. Never having regarded myself as someone desiring such a career, the statement gave rise to serious reflection. Fortunately, I had the help of a T-shirt that I once brought from the United States, specifically from Auburn University's Graduate Student Organization. This T-shirt provided a 'wee test' for deciding whether one should become a graduate student or not. It went like this:

	True	*False*
I enjoy being a professor's slave	0	0
My idea of a good time is using jargon and citing authorities	0	0
I feel a deep need to continue the process of avoiding life	0	0

Having wholeheartedly filled out 'True' for all three questions I decided that I should perhaps have a go at it. A difficult decision was taken, or as we say in Dutch: 'the bullet went through the church'. In the years that followed it turned out that many bullets still had to follow and that I was not always such a good shooter.

We tried to get funding for my - still developing - plans from the Dutch Science Foundation (NWO). Although nobody apparently really disliked our plans, they weren't too keen on investing the sum of money we asked for, even though I had never asked for anything in my life so politely. But then, perhaps politeness should indeed not be the decisive criterion for funding research (I would not mind an experiment with this criterion though). Fortunately for me, Hans Bressers still liked my ideas for doing research, which all revolved around citizen participation in environmental decisions, especially decisions on Locally Unwanted Land Uses (LULUs). Where my fascination with this topic came from I don't know exactly, but I do know that I definitely considered a large degree of citizen participation necessary.

My original plan was to study the issue by using the concept of 'policy style' and I went to the University of Florida for a literature study on this topic, and perhaps some empirical work as well because I was considering an international comparison between the Netherlands (my own country), the United States and the United Kingdom. With the help of my hosts at the Department of Political Science, Albert Matheny and Walter Rosenbaum, I narrowed my focus down to decision-making on hazardous waste facilities and decided that Canada should replace the United States in my research, mainly because the Canadians were apparently experimenting with highly participatory decision processes (more so than in the US).

It might be rather ironic that my hosts at Florida, Albert Matheny and Walter Rosenbaum, were giving me much relevant literature, but that little of this seemed of interest to me, because it was not on policy styles. Only later did I realize that the concept of 'policy styles' is not that suitable for studying decision-making on hazardous waste facilities because it is a concept for use at the 'macro-level' and not at the 'micro-level' of local decision processes. The literature I picked up in Florida, largely about institutional and discourse analysis, was in fact much more useful. So I had been firing through the wrong church, which sounds like a violation of a rather basic principle.

Having been converted to a new 'church' (or less dramatically: having changed theoretical lens), I then went to the University of Essex in the UK, where Jeremy Richardson, Albert Weale and Hugh Ward were excellent hosts at the Department of Government and Thomas Christiansen a very good roommate. Having included the UK as a country where decision processes were far less participatory (and thus 'worse' in my own view) than those in the Netherlands, I started doing my first interviews there, which were mainly intended to identify suitable case studies for research. But then I read a highly critical review of a book that had a similar topic as my study. The critique was that cases of hazardous waste siting cannot adequately be studied without understanding their national context. This made me decide to devote some attention to the legal context of hazardous waste siting in the three countries of interest (which is of course only a part of the national context) and its development through the years.

The study of the UK system of environmental regulation and land use planning was not a simple issue, and I was warned various times (for instance by Andrew Blowers at the Open University) that the legislation was highly complex and easily misinterpreted. I felt personally touched by such warnings and decided that I should perhaps approach the UK system a bit less as an evil empire and maybe be a bit more 'objective' in my appraisals. I am not sure whether the continuously friendly talks with planning officers, environmental regulators, and academics were influential in this respect, but I certainly did start to develop more sympathy for some of the (indeed more hostile towards participation) thoughts underlying the UK system of environmental regulation. At Essex, this interest was shared with Tony Barker, who has written what must be the longest journal article ever on the topic (Barker and Couper, 1984: 113 pages), and was willing to spend hours sorting through this article with me.

One of the people that I also spoke to during my first visit to the UK was Judith Petts, who was then at Loughborough and later moved to the University of Birmingham. She is one of the best-informed persons in the UK hazardous waste scene and had much extremely useful information for deciding which case studies to do, and for actually doing them. Based at the Centre for Environmental Research and Training at Birmingham, I finally did these cases studies during a second visit to the UK. By then I had given up the hope that I could 'do' five case studies in each of the three countries involved, and changed that number to three, which was already a demanding number anyway. One indication of the involving nature of the work that needed to be done was the amount of paper that I mailed back to the Netherlands. I sent at least three boxes of 25 kilos of documentation on the case studies back home by mail and carried many more kilos myself.

Before my last visit to the UK, I had already visited Canada, land of what I expected to be the pro-participation orientated government. Looking for connections there, Hans Bressers gave me a paper from Audrey Armour of York University in Toronto. It turned out that she had been personally involved in actually designing one of the decision processes and that she had extensively studied the other processes as well. She was so kind to arrange for an office at the Faculty of Environmental Studies and despite the fact that she was on sabbatical leave, she helped me tremendously with my work and continued to give me wise advice. She stressed that the normative evaluation of what I was to observe would be the hardest bit of the work, which turned out to be true as I struggled quite a part with the Canadian case studies, which are unique in many ways, not the least of which is the fact that they greatly deviated from legally prescribed processes. To mention everyone who helped me along the way is almost impossible, but some people stood out in a special way as they took not only a professional, but also a personal interest. Dennis Foidart and Richard Sherbaniuk stand out in this respect.

Just to come to grips with all the information I gathered in those few months in Canada and later the UK and the Netherlands (thank you librarians and archivists!), I started writing several extremely lengthy descriptions of the nine decision processes that I eventually studied. Many pages were eliminated from these descriptions at a later stage. One famous sculptor once said that his work was easy: in sculpting a horse, he just cut away all the stone that wasn't horse. The final stages of writing this book felt a bit like that. I had a lot of stone, but still had to cut away all that was not book. The case study descriptions became more and more dense and analytical as time progressed, but I am quite sure that some of the readers of this work might have difficulty in believing this as they might find the current descriptions already quite lengthy. I advise them to quickly turn to the more analytical parts of the case study chapters, which were written in such a fashion that they can be read separately.

Having started to formulate my research ideas already in 1995, I had high hopes of finishing up my work in the year 2000. This became 2001 in the end (December 2001 is the last month in which parts of this book were written), partly because some of the three Dutch case studies that I included were rather complicated to master. Certainly in one case study, I felt like a rather unwelcome guest at the table because my interviewees either questioned my case selection, wanted to control the information I was gathering, or even refused to talk. Of course this made me all the more interested in finding out what went on, but this then led to issues that were not really that interesting from the perspective of this book (however salient they were). Another reason why this book was not completed in 2000 but only in 2001 is that, and this is perhaps inevitable, the further one progresses with a research such as this, the better one discovers what the real topic was that one was 'always' interested in. I am not sure that I am entirely there in this respect ('Life is a building site' says the title of a very funny German movie), but writing down an even mildly coherent book did require a second (or a third) look at many of the texts that I had produced before. Needless to say that I (eternally optimistic in this respect) had not planned for second or third looks.

If I were to say something about the message of this book, then it would be that it is still intended rather as a demonstration of certain dilemmas that are present in environmental decision-making and of the perspectives one can have towards them, than as a compelling case for a certain solution to hazardous waste siting problems or a certain perspective. It does not seem to me that (social) scientists have, or can have, nor even should have an important role in setting up real world institutional arrangements for decision-making. What we can do as social scientists is rather limited and would seem to be located principally in the sphere of critical reflection (e.g. by checking whether what people say and do is consistent) and perhaps agenda setting. In hazardous waste siting, or more generally decision-making on LULUs, I hope that this book can to some degree contribute to the setting of a climate, where participatory approaches are carefully reflected upon, which means they are not necessarily regarded as intrinsically bad, but neither automatically good.

All of the activities that I have spoken of so far have been performed from a very supportive environment at the CSTM. I have already mentioned Hans Bressers a few times, and this does not even do justice to the large role he played in making this work possible. Apart from him, the contacts with other colleagues were always good, especially with Rob van de Peppel, with whom I shared an active interest in pinball, cars, and other important things in life, and who became a close friend. Furthermore, I was fortunate to share an interest in citizen participation with the likes of Johan Woltjer, Larry O'Toole and Frans Coenen, which resulted in various common publications, the number of which hopefully continues to grow.

The group at CSTM that was there for the greater part of my stay was very close. This can already be gleaned from the fact that, as colleagues, Stefan Kuks, Josee Ligteringen, Ton van Snellenberg, Xandra Timmers and I even went on holiday together and had good times on such occasions. For some reason this group got smaller by one person every holiday. Nowadays

Stefan and Josee are the only ones left to go on holiday together. I wish them and others at CSTM all the best and thank them for the good times. If my time at the Institute of Environmental Studies at the Free University of Amsterdam will be equally nice, I can certainly be content. I thank Matthijs Hisschemöller and Harmen Verbruggen for given me a shot at it in Amsterdam and my new colleagues for their warm welcome so far.

The book that is in front of you has been prepared in cooperation with various people. At Kluwer, Peter Binfield, Henny Hoogervorst and Gloria Verhey were the ones to decide that they dared publish it. Ada Krooshoop at CSTM did excellent work on the layout of the text and incorporating comments. Su Carleton, in a very pleasant cooperation, corrected some of the English that 'had gone Dutch' in my drafts.

Auburn's Graduate Student Organization described the work on a thesis as 'the process of avoiding life' This is partly correct, as a university still is in many ways a very special working environment, with an unusual amount of freedom and emphasis on analytical thought, which separates it from 'normal' life. However, my own experience is that there is plenty of life going on as one writes a thesis, or more generally, works at a university. I remember very fondly how, on a rainy day, I was walking towards the university restaurant, and suddenly there was somebody walking along with me under the umbrella in the pouring rain. From day one you have been a pleasant surprise, Geertruide, and I hope that we can continue to walk under the same umbrella or rather in the sun and enjoy life together. Hopefully, my parents and brother will be around to enjoy life with us as for many more years as well. I thank Gabe and Sonja for their support and for the fact that they were always there, even (or especially) when life was tough.

Amsterdam, 24 June 2002

Chapter 1

Hercules, Leviathan or Promotheus?

This book is about taking good environmental decisions. Environmental problems exhibit many characteristics, among them complexity and uncertainty. These characteristics make it difficult to reach agreement on the way these problems should be solved; that is, if they are being acknowledged at all. Some say environmental problems are Herculean and must therefore be solved by a strong government, Leviathan. Others say environmental problems will surface in the markets of natural resources and lead to intelligent responses from man, using the fire that Promotheus stole from the gods to enlighten their choices.

In this chapter I will argue why I think attention must be paid to institutions that are used for arriving at environmental decisions. Coming from a representative democracy myself, I ponder on the qualities of decision-making by representative bodies in the environmental sphere and indicate what might be alternatives. I will introduce the issue of hazardous waste facility siting and use it to demonstrate the relevance of studying institutions for environmental decision-making.

I shall also make use of examples to illustrate how difficult it is to determine what 'good' decisions are; and that different - human - conceptions of good decisions exist and that these have a relationship to particular institutions. To choose a certain institution as a collective decision-making tool therefore has consequences for the outcomes of the decision process. The motivation for such a choice and its consequences are the main themes of this book.

1.1 The Age of Ecology

Introduction

We are living in what has been called the 'age of ecology', and environmental problems seem omnipresent. But what are the specific characteristics of these problems and how should these be addressed at a very general level of abstraction? Ecological systems have certain characteristics that may make it hard to predict and assess environmental problems. Added to that is the fact that there are various competing human perspectives on what ecological problems mean and how they must be addressed. This implies that some people maintain there is no environmental problem, whereas others consider environmental problems the most pressing ones on our societies' lists. Those who think that environmental problems warrant human action do not agree on what the action should be: an increase in awareness is one option, the design of better institutions another. My interest is in institutions themselves.

The Age of Ecology

Climate change, soil polluted by illegally disposed hazardous wastes, a widening 'hole' in the ozone layer, international tensions over the division of water rights - it is virtually impossible to miss the regular drumbeat of news proclaiming that the planet is confronted with severe problems concerning the natural environment and its capacity to sustain life and an acceptable social order. Ever since the end of the 1960s, environmental problems have been high on the agenda of most industrialized countries and much effort from citizens, governments and international organizations has gone into attempting to tackle them. In the early days of modern environmental concern, most attention was paid to the issue of pollution, but population growth and the depletion of natural resources were also seen as a problem. Since that early period of environmental concern, new problems have come to the forefront, partly because of the increasing availability of scientific data, partly because of changing societal perceptions. To mention a few 'new' problems: we now live with phenomena such as the hole in the ozone layer, acidification, and climate change. It seems fair to say that our relationship with our natural environment and the perception of this relationship has drastically changed in the past decades. Environmental problems have attracted much media attention and have, with some ups and downs, been at the top of the societal agenda in most of the industrialized world. This situation has led some to use a catchphrase label for our current era: 'the age of ecology'.

The idea that there is something as an 'environmental problem' is relatively new. As a matter of fact, many of the issues we now tend to connect to the environmental problematique, such as floodings and tornadoes, were once considered 'acts of god'. We have become aware that some of these phenomena are connected to human behavior and conscious human decisions. The German sociologist Beck (1992) argues that we now live in a risk society, where the main issue is not so much the distinction between labor and capital and the consequent inequities in welfare. But, instead, argues Beck, the spreading of risks caused by human inventions and technology such as nuclear facilities is the main issue now.

Characteristics of environmental problems

Environmental problems are those problems that are related to the relationship between human society and the ecosystems surrounding it. On the one hand, human beings undertake certain interventions in the ecosystems, for various purposes including agriculture, housing, and so forth, hereby arguably changing the way these processes run. On the other, they attach certain meanings to and envision certain functions for the environment. These meanings or functions can be either anthropocentric or also ecocentric, depending on the degree to which ecosystems are considered as something with an inherent value. I note that in either vision, human observation and problem interpretation play a large role. Environmental problems can thus be defined as situations where ecosystems are not able to fulfil the meanings or functions humans attach to them. It is said that environmental problems pose a special kind of social problem, that they are intractable and pose unheard of uncertainties over how to resolve them (see Press, 1994).

Why should such problems be so intractable or pose such great uncertainties? Part of the explanation for this is to be found in the characteristics of ecosystems. These characteristics are said to be interpenetration, emergence and homeostasis (see Dryzek, 1987). Interpenetration means that ecosystems are always embedded in other ecosystems and can be separated for analytical reasons, but not completely so in reality. Emergence relates to the fact that ecosystem properties can never be circumscribed by knowledge of the components of the system alone. New properties and characteristics emerge from unpredictable interactions within the subsystem. Finally, homeostasis implies that ecosystems are self-regulating, constantly adapting to both internal and external forces, but without necessarily having an identifiable purpose.

Because of the characteristics of ecosystems, environmental problems exhibit complexity, non-reductibility, variability, uncertainty and collectiveness (Dryzek, 1987). Complexity implies that ecosystems are extremely difficult to systematize and predict, and human activity itself is a contributing factor to this given its unpredictable character. Non-reductibility refers to the fact that it is not possible to 'provide compensatory solutions to ecosystem disruption' by manipulating the components of the system, because the results of partial interventions will be unpredictable. Variability means ecosystems may appear similar, but they vary in significant ways across time and space. Uncertainty means that there are problems of severe unpredictability, so that the future state of ecosystems is difficult to predict. Collectiveness relates to the fact that the interaction between humans and ecosystems involves and affects many actors, which raises the issue of how to behave 'collectively rational'. To demonstrate this aspect, the metaphor of the 'Tragedy of the Commons' is often used (Hardin, 1968). The metaphor relates to the idea that individually rational behavior leads to collective failure to protect shared and freely accessible resources. Finally, there is the problem of spontaneity that implies that ecosystems have a tendency to right their own imbalances, which may be obscured or undermined by human intervention.

The subjective nature of problems: environmental discourses

Many have observed that environmental problems are not just 'objective' presences, but a reflection of various factors, including the state of scientific knowledge, the economic situation, a society's culture and even language. Like any problem, environmental problems are thus 'subjective' and it is logical that different problem definitions exist. An example may clarify this. The presence of a 'hole' in the ozone layer was unknown until human-made observation equipment made it visible. But even when this equipment (science) provided data that suggested reduced ozone levels at certain locations in the atmosphere, humans initially considered these data incorrect and ignored them (also science). In the process that ensued, the idea that a problem might in fact exist was greatly helped by the easily comprehensible term 'hole in the ozone layer' (language). Efforts to then reduce emissions of ozone-depleting substances were influenced by economic dependencies (which countries produced ozone-depleting substances, which countries could produce replacements?) and by culture (some countries are less interested in environmental degradation than others).

Of the factors just mentioned as being of importance for environmental problems, language is most easily overlooked because a recognition of its importance requires a meta-evaluation of the way one speaks. However, language does have profound effects on our perception, recognition and ability to solve environmental problems. A 'hole in the ozone layer' sounds 'better' than a mere reduction of ozone concentrations in the atmosphere. And what of 'acid rain', a concept that had scientists studying the acidity of rain, whereas a lot of acid deposition is in fact dry deposition? In that case, language really functioned as a searchlight, pointing in the wrong direction.

Further highlighting the subjective perception of environmental problems, Dryzek (1997) identifies several 'discourses' on environmental problems. A discourse refers to a 'shared way of apprehending the world' (1997: 8), expressed by regularities in the language that people use to make sense of environmental issues. In any discourse, a certain vision on environmental problems is given, which includes the actual nature of the problem, reasons why the problem has arisen, and suggestions for improvement. Dryzek identifies nine different discourses, all very different ways of looking at environmental problems. To give an example: one discourse is 'survivalism', which implies that our planet is about to be destroyed because of environmental problems. Another one is called 'growth forever' and contends that human intelligence will prevent the destruction of our planet.

Important differences exist between the nine discourses. One of these differences, strangely enough, is whether environmental problems, with the emphasis on problems, are said to exist or not. Although every 'environmental discourse' is about environmental problems, some discourses deny that the situation of our environment is very problematic. 'Growth forever' for instance, does acknowledge the possibility that 'resources' (the environment is seen as a resource) will become depleted but trusts that human ingenuity will provide a solution, e.g. by turning to other resources. Dryzek called this a 'promothean' response, after Promotheus, who '*In Greek mythology (. . .) stole fire from Zeus, and so vastly increased the human capacity to manipulate the world for human ends*' (1997: 45). Those who feel there is little proof of environmental problems will happily embrace this position, just as those who think that any attempt to do something about environmental problems is a asking for new problems. Except for 'growth forever', the other discourses that Dryzek highlights do accept the existence of environmental problems, but are then very different in other respects.

One other difference is related to the role of 'rationality' in the discourse. Some discourses, such as 'Green romanticism', reject an approach to environmental issues that is based on 'reasoning'. Instead, they emphasize a new consciousness, which hinges on agency in nature itself and getting in touch with nature. Modernity (reason) is rejected and replaced with romanticism (feeling). Mathews (1995b) provides an example of people that think along these lines. She argues that in the modern vision of the world ('modern' in the philosophical sense of the word), individuals play the most important role[1]. As the world is seen as being made up of discreet, self-contained units (individuals), this vision allows for classification of these units along certain parameters. Mathews posits that modern thinkers treat 'mind' and 'matter' as the metaphysical categories that are most relevant for

[1] Mathews calls this the liberal vision.

classification[2]. Human beings are set apart from other beings because of their capacity to reason. Other classes of beings lack this capacity and are therefore inferior: '*human beings (. . .) transcend nature*' (ibid.: 73). So in her opinion, the idea that man should dominate nature is embedded in modern thought. She rejects this 'objectification' of nature. What is needed instead is that humans develop '*face-to-face relationships with a variety of non-human beings on a day-to-day basis, relationships which enable us to come to know those beings in all their variousness and individual uniqueness.*' (Mathews (1995b: 79). As a correlate of the wholesale rejection of rationality, Green Romanticists are often hesitant to devise institutions that could be used to address environmental problems. Instead they emphasize 'green values' and not 'green structures' (Goodin, 1992).

Most other discourses consider these structures very important, however. 'Structures' for environmental decision-making include 'markets', 'parliament', 'communities' and 'expert bodies'. Eight of the discourses described by Dryzek do contend that certain structures are better suited for taking decisions than others, although they differ in preference for a certain type of structure. Dryzek (1987) himself for instance thought that small communities might be best suited for environmental decision-making. Dryzek contended that such communities could best employ '*practical reason*', that is, engage in a 'rational debate' about both means and ends. Dryzek rejects most other institutions for environmental decision-making because the goals are not debated.

My position and interest here

This book is essentially about taking 'good' decisions in the environmental sphere. Because I believe institutions are important in bringing about 'good decisions', I am interested in the performance of different institutions in taking decisions, and as a correlate of that, in the decision process in which institutions are identified as the right ones for a problem at hand. Later on, I will devote some space to the question of the definition of good decisions.

Why focus on institutions? The importance of green values for achieving sustainability seems obvious. Still, I think that the relevance of institutions is also easily demonstrated and is an overriding factor. This is because institutions force humans in certain directions, even if they would wish

[2] It would seem to me that any line of thinking starts with certain basic classifications. To say that a classification between mind and matter (also called the subjective and the objective, see Pirsig, 1991) is the basis for environmental degradation seems somewhat far-fetched. It seems to me that any classification, if it is accompanied by feeling of superiority (for instance: red animals are better than brown animals) of one of the classes, can lead to environmental degradation. So it is not as much classification that is the culprit of environmental degradation, but feelings of superiority.

differently. One can point to the fact that the imperative of profitability considerably reduces the importance of values for people working under a competitive market regime. You may wish to produce in an environmentally-sound fashion, if others do not and produce cheaper than you do, you will be quickly out of business (that is, if the market functions 'properly'). In this light, some have described the market as a 'prison' (Lindblom, 1982). Needless to say, the market or other institutions can give incentives both towards a more ecologically-sound behavior as well. My interest in institutions is not dissimilar to Dryzek's (1987) interest in 'social choice mechanisms'. A social choice mechanism is '*a means through which a society (. . .) determines collective outcomes (i.e., outcomes which can apply to all its members) in a given domain*' (1987: 7). Such mechanisms are specifically of interest when preferences among the members of the society are not similar, which is quite likely for environmental issues. I acknowledge that many view a distinction between structure and values as being somewhat superficial, because the relationship between structures and values is one of mutual enforcement. Certain values, if held by a substantial number of people, can support the operation of a certain structure, while in turn the presence of certain institutional arrangements would seem to support certain values. Being aware of this, my principal interest still lies with structures, which I will later come to call institutional arrangements, and how these can help to achieve a better environment.

Why is it necessary to focus on 'good' decisions? For some, it is quite easy to see what constitute good decisions in an environmental sphere. Dryzek often appears to be among them. In his book 'Rational Ecology', he provides examples of unsustainable behavior with great ease. Among these examples: the cutting of rainforest, the generation of radioactive waste, etc. In the simplest terms, these decisions are 'bad' because they destroy or threaten ecosystems. This seems self-evident until one considers the alternatives. Take for instance the cutting of ancient forest in Northwestern Europe. Dryzek defends this as 'good' on the basis that a sustainable form of agriculture has returned to that land. If this were possible, why then would the cutting of rainforest be 'bad'? And what if the forests of Europe were still there, would Dryzek still approve of them being felled?

The examples and the text above suggest that environmental problems will often be 'different problems for different people'. In this sense, a definition of what the problem at stake is exactly, is a subjective definition, and environmental problems, in this sense, are therefore always subjective problems. The idea of subjectivity holds true even though 'objectively' speaking a forest may disappear, certain species might become extinct and even though characteristics of ecosystems may really change. Whether or not extinction is considered a problem is not clear a priori. Note for instance that some consider the eradication of species by human involvement insignificant in light of the destruction of species by natural causes themselves (see CPB, 1996). However, even the 'objective' part of environmental problems is not easy. To counter the 'circle the wagons' suggestion that I may have evoked until now, I do want to point out the importance of critical reflection. In many situations, the scientific evidence of what is actually happening in ecosystems is still developing and it is entirely reasonable that the question of whether there is a certain trend in ecosystems or not, must seriously be debated. In this sense, the ideal of 'good decisions' sometimes might imply extremely critical evaluation and rejection of the obvious and of the supposed problems and solutions.

Of course there are many types of social choice mechanisms for deciding on environmental issues, and most societies apply a range of them at the same time, often in co-existence. One can think of markets, expert bodies, courts, parliament, etc. In any case, an argument could be made that decisions need to be left to that specific type of institution. Dryzek has done extensive study on the degree to which several types of institutions can be expected to meet some conditions that are needed for ecologically rational decisions (1987)[3]. Dryzek considers, for instance, that institutions must allow for 'negative feedback', implying that information about environmental deterioration must not be neglected. On the basis of such criteria, Dryzek concludes that none of the existing institutions in the entire world can be expected to reach ecologically rational decisions. Instead, he pleads for communities based on practical reason. I do not find Dryzek's conclusions very convincing, for various reasons. One is the already mentioned tendency to link 'obviously bad' decisions to certain institutions, whereas such decisions may not be so bad at all. Secondly, environmental problems, as Dryzek later admitted (1997), are perhaps not the monolithic block he initially took them to be. There are some environmental problems that exhibit all of the characteristics of environmental problems that he attributes to them, e.g. climate change, hole in the ozone layer (these are global problems). But there are lots of 'smaller' problems that seem less interconnected and also less complex. Thirdly, the alternative of practical reason in communities seems attractive, but has never been tested on a large scale in reality (or has it?) and has easily identifiable flaws. Therefore, this alternative provides little incentive for jettisoning all existing institutions. Finally, I do not see why 'practical reason' should be located exclusively in local communities. If my impression is correct, the debate about proper

3 Dryzek actually examines 'social choice mechanisms', and for instance regards 'war' as such a mechanism. I will limit myself to more 'ordinary' institutions in this book.

institutions for environmental problems has thus not ended and is still worthwhile.

Coming from a representative democracy myself, my starting point in this venture is look at *democratic institutions*. Early environmentalists ('survivalists') were quick to point out that solving the environmental problematique required drastic measures that would never be feasible in a democratic society. According to them, democratic institutions would have to be replaced by technocratic rule. Since then, this call for the abolishment of democracy has waned somewhat, but there still is a strong current of thought that considers democracy and environment at odds with each other (i.e. considers that democracies do not take 'good' decisions).

1.2 Democratic institutions and their alternatives

Introduction

Somewhat in contradiction to Dryzek and many others, Press (1994) has argued that environmental problems are complex and require much human knowledge, but are not overwhelming in the technological sense because the know-how required is available. This may in fact be too bold a statement, but Press does point to an important aspect of the decision-making task: many environmental problems pose democratic dilemmas. This is partly because their solution requires the redistribution of resources over actors in society. In this section I will review some of the literature on the relationship between environment and democracy. I have not attempted to be complete in this review, but present the findings of others that have studied the topic. Specifically Lafferty (1996), Mathews (1995a+b) and Press (1994) were helpful here. Their overviews demonstrate that many social scientists are pessimistic about the prospects of taking 'good' environmental decisions under democratic systems, be it to different degrees and for different reasons. Some of the authors offer suggestions to improve democratic decision-making or offer advice for the introduction of different institutions. These institutions include the already mentioned communities, but also the market and experts.

Democratic decision-making

Democracy means rule by and on behalf of the people. It is an old way of decision-making in groups that emphasizes human adaptation-communication and social learning (Dietz, 1995: xvii). Citizen participation, that is, the taking part in discussions on public affairs by ordinary citizens, has been defined as the normative core of democracy (Fischer, 2000). Democratic decision-making presupposes the ability of every ordinary member of society to have the moral and instrumental competence to decide

(Dahl, 1989) and in this sense assumes equity between members of a society. Democracy is increasingly becoming a normative standard model for decision-making. Almost universally, other systems of decision-making need to be defended against the democratic ideal.

Before the twentieth century, three historical models of governance were associated with the democratic idea: the Athenian city state, the Roman republic and the Italian and Swiss city-states. Although a democratic ideal held sway until the age of Rousseau, democratic governance was an exception for the greatest part of the last two millennia. Especially the fact that democracy was thought to involve the participation of all (qualified) citizens in public affairs, ergo direct democracy, made the ideal less practicable. An important breakthrough was reached when the idea of representative democracy was invented. The beginning of modern (representative) democratic thought is not exactly traceable. However it does seem related to two major streams: reformation protestantism and the egalitarianism inherent in it on the one hand, and feudalism on the other, with the assemblies of privileged estates which were intended to be representative of social constituencies (Arblaster, 1994: 23-26). Under reformation protestantism, a doctrine of popular sovereignty developed, which posited that rulers were leaders, but at the same time servants of the people. Assemblies became the focus of the struggle for democracy. An example of an assembly of the privileged estates is the English House of Commons that gradually grew into its modern role, whereas the idea of direct democracy waned.

Absolutism was dominant in 17th and 18th century Europe and democratic ideas were considered quite radical. At the philosophical level, theorists struggled with implications of the democratic ideal, since most of them were afraid of 'mob rule' (Arblaster, 1994: 25-31). Arblaster suggests that the French revolution in 1789 placed democracy on the agenda of 'real politics'. It inspired the first successful slave revolt in Haiti and the political independence movements of South-America. The debates surrounding the writing of the American constitution in the same era, and after that the constitution itself, also inspired many democrats (ibid.). Although several traditions of democratic thinking have developed, it may be said that the democratic ideal has since then gained widespread acceptance, and democracy has become the dominant ideal form of decision-making in the industrialized world. Paraphrasing Arnstein (1969): democracy is a little bit like eating spinach, nobody would argue against it (at least not too explicitly). Leaving aside the different traditions of democratic thinking and thus for the time being working with a relatively vague notion of the concept of democracy, I first want to focus on the relationship between democratic decision-making and environmental concerns. Is democracy actually the proper institutional arrangement when it comes to decision-making on environmental issues?

Environmental problems: challenges to democratic decision-making?

The discussion about the implications of environmental problematique for democratic government is at least as old as the modern notion of environmental problems itself. It seems that there are many reasons why democracy and the goal of solving environmental problems are at odds. Lafferty (1996), Mathews (1995a+b) and Press (1994) identify most of the potential problems.

Lafferty (1996), in a paper about the concept of an 'ecological state', indicates that the coming about of such an 'ecological state' would imply a shift of constitutional structures and governing principles of existing states, and thus an adaptation of existing democratic norms and procedures. Lafferty perceives a very stark contrast between what is needed and what is currently practice in most liberal democracies. He stresses the complexity of environmental issues and thus the necessity for expert knowledge, the fact that environmental problems are not bound by administrative borders and that there is a need for prompt action. All these requirements are not things that democracies are known for incorporating. In addition, there is no-one representing future generations or ecological interests in democratic decision-making processes.

Although he disagrees with most of the pessimistic conclusions in the debate about democracy and environmental problems, and finds some empirical evidence for optimism, Press (1994: 9-10) gives a good overview of the reasons why democracy and environment are considered incompatible, and therefore pose challenges to the democrat. These challenges bear some resemblance to the problems brought to our attention by Lafferty. One challenge, the 'social justice challenge' infers that environmental problems are so urgent that the 'luxury' of democracy should be done away with for their resolution. Deliberation, reflection and compromise may all be good things, but not for environmental problems. Choices have to be made and action is needed in the short run. The second challenge questions the ability of lay people and their representatives to understand the environmental problematique. As explained above, environmental problems tend to be considered as complex and therefore difficult to grasp. An understanding may require training in the natural sciences and even then it is likely that a solution to environmental problems requires integration of and cooperation by people from several disciplines. An understanding of environmental problems does seem to require some knowledge of scientific terminology. The democratic emphasis on lay people and their representatives as the main principles of policy may not be compatible with this aspect of environmental issues. 'Uninformed' citizens may not understand environmental problems, and thus not make the 'right' decisions; they should perhaps leave it to the experts. A third challenge is an economic one. Economic realities may be such that many policy options are predoomed because market forces would punish choosing them. The implication of this challenge may not be so much that democratic decision-making should be abolished, but rather that this has no, or indeed *can not* have, real substance, as market imperatives essentially determine environmental policies.

Mathews (1995a+b), finally, also gives an interesting and in depth account of the relationship between democracy and the environment. She notes that government action in a democracy relies on a popular mandate and that this reliance creates a possible tendency between democracy and ecology. Apparently considering environmental problems as being in a serious state, she envisions that large changes in lifestyle at enormous expense might be needed to achieve sustainability. If these changes are too drastic and the costs too high in the eyes of the public then a democratic government would be in difficulties, given the fact that agreement with the will of the people is the proper justification for policies (1995a: 4; compare this argument to Lafferty's).

Mathews (1995b) goes on to analyze the core principles of democracy. She finds that the core values are freedom from higher authority and equality in the sense that each individual's opinion counts as much as the opinion of another (compare Lafferty again). Deeply embedded in this is a strife for autonomy from which stems the desire for self-rule. Self-rule is a goal in itself for democrats. Even if a benign paternalistic authority would be able to meet all their desires, this would not satisfy them. According to Mathews, democracy in this sense values individualism. Although real world democracies may show a somewhat different emphasis, Mathews argues that the ideological base of democracy is '*its over-riding commitment to individual freedom and autonomy*'. In turn, this implies that the function of society is to facilitate the self-realisation of individuals and the purpose of collective effort is rather limited: to secure conditions for self-rule. According to her, this implies that the basis for democratic societies is contractarian rather than moral (1995b: 68). This individualistic basis may be a problem when liberal democracies try to address collective problems, such as those relating to the environment. That is because everyone is considered to be free to interfere with the environment in any way they like.

Mathews admits democratic theory acknowledges that the liberty of one person can harm the interest of others and that this would not be just under the liberal principle of equality. Liberal philosophers have extended this 'harm principle' to the issue of environmental degradation, which then receives the meaning that each individual's use of the environment must be self-controlled so that it does not unduly limit the choices of others or the choices of future individuals. But Mathews argues that this still does not constitute a moral principle, because this line of reasoning makes concern for environmental issues part of an individual's interest. Of course, every individual may be asked to take into account the interests of others and to consider the prevention of environmental damage as part of their own interest. An individual is certainly free to defend his interests so defined, but they will count as nothing more than an individual interest in computations of the collective good. Another difficulty would then also be that ecocentric interests of individuals are considered lightweight compared with the direct material interests of other individuals, which are felt as much more urgent (compare Press's 'economic challenge'). As a consequence, modifying democratic thought by the 'harm principle' does not go very far in achieving protection of non-human life (Mathews 1995b: 70-71).

The table below briefly summarizes the arguments presented by Lafferty, Press and Mathews.

Table 1. Reasons why democracy and sustainable development are incompatible

1.	Democracy is geographically concentrated whereas environmental problems transcend borders (Lafferty).
2.	Democracy is about the self-interest of the current generations, but future generations are not represented and there is no means to make people take these into account (Lafferty, Mathews).
3.	Even if citizens care about the environment, their opinion will be treated just as any other opinion, whereas the environment is an overriding concern. Besides, democracy is slow, environmental problems are pressing (Lafferty, Press, Mathews).
4.	In a democracy there is equality and everyone's opinion counts. However ordinary citizens have no understanding of the problems at stake (Lafferty, Press).
5.	Even though decisions may officially be taken democratically, economic realities may dictate the outcome of decision processes (Press). Besides, most citizens care more about economics than about the environment (Mathews a+b).

This critique is quite strong and one may have doubts about the efficacy of democratic decision-making on environmental issues. Only Press finds that the negative conclusion about democratic decision-making may be ameliorated somewhat by the fact that not all environmental problems are threatening to end life on the planet and that democracy also has a potential to mobilize problem-solving capacity. He studied the value of arguments against citizen participation in the sphere of hazardous waste and forestry policies and finds that '*trees and toxics controversies are not "loaded" with the kind of urgency that characterizes a famine, flood, or the ozone hole.*' (Press, 1994: 130). That being said, Press does find that the economic challenge to participatory democracy is probably extremely strong. '*Consensus and compromise is so difficult to reach in environmental conflicts like the two presented here because people perceive real power not in terms of participation and citizenship but in terms of their economic feature.*' (ibid.: 121). Press does not really give institutional prescriptions on the basis of his work. Mathews, on the contrary, wants to replace democracies with coercion-free communities. Lafferty, instead, places great emphasis on the role of experts in decision-making. Such experts are sometimes nicknamed 'ecological mandarins' (Press, 1994).

It is interesting how the three authors just mentioned make little distinction between direct and representative democracy. Mathews explicitly denies that the distinction is relevant and suggests that her critique applies to both. Lafferty and Press take only the functioning of representative democracy as the basis for their critique and do not discuss the idea of direct democracy so much. It should be noted, however, that many authors do distinguish between representative and direct democracy and suggest that direct democracy is 'better'. Critics allege that representative democracy does not enhance debate and dialogue as does direct democracy. Burnheim (1995), for instance, faults representative democracy for 'interest aggregation', or 'power trading' and lack of dialogue. Interest aggregation is different from dialogue in the sense that there is little serious discussion of preferences or interests. Rather, paralleling markets, interests are taken as a starting point, and a compromise is sought that provides something of a middle ground between what various parties want. Other authors who see direct democracy as a force for better environmental decisions include Ward (1998) and Thompson (1995). Often, these authors link the idea of direct democracy to the idea that decisions should be made within local communities, which I regard as a structure in itself. Note that there are many authors (growth forever types?) too who see markets as the best solution to environmental problems (see Ostrom, 1990).

Some alternative institutions for environmental decision-making

Briefly summarized, it would seem that expert bodies, the market, and communities could replace representative institutions. I will briefly present some proponents of these institutions (a critique will follow later). There is one other alternative institution, and that is the court. Few are advocating the use of this institution, but I will still pay some attention to it.

The strongest expression of a desire to give greater control to expert bodies comes from 'survivalists', as Dryzek calls them. These argue that '*if ruin is to be avoided in a crowded world, people must be responsive to coercive force outside their individual psyches, a 'Leviathan', to use Hobbes's term*' (Hardin, in: Ostrom, 1990: 9) and '*even if we avoid the tragedy of the commons, it will only be by recourse to the tragic necessity of Leviathan*' (Ophuls, in: Ostrom, 1990: 9). The recommendation that usually follows this analysis is that strong centralized control and hierarchy are needed to regulate the use of natural resources (see Dryzek, 1997: 74). Nature is subordinated to human problem solving because it is humans who have the ability to understand and analyze ecosystems. People in general, however, would be subordinated to the state which guards the collective interest on their behalf. The proper basis for this control would be founded on expertise about ecosystems that can be found at the central level. A small group of enlightened experts, 'ecological mandarins', would take control and make wise decisions. So within the state, it would be the managers or the experts who take the decisions. Lafferty does not go as far, but does plead for a more important role for experts.

Another line of thinking highlights the importance of market forces and as a requirement for this, the establishment of private property rights. Many theorists think that common pool resources, which are collectively shared and freely accessible (natural) resources, can only be protected when individuals own parts of the common property. Only if ownership is assigned, will people have an interest in protecting it and not let others abuse their property. Thus some say it is '*by treating a resource as a common property that we become locked in inexorable destruction*' (R.J. Smith, quoted in: Ostrom, 1990: 12) and, relating to common grazing grounds, '*the establishment of full property rights is necessary to avoid the inefficiency of overgrazing*' (Welch, ibid.). The core of the argument can be traced back to Coase (1960) who shows that, in theory, negotiations between polluters and those affected can lead to a socially optimal solution, as long as property rights are legally clear. This is because those affected can offer a certain amount of money to the polluter to induce him to cease his activities[4]. If the amount is not large enough for the polluter to stop, then this must mean that he is deriving greater use from continuing his activities than from stopping and accepting the monies offered to him. Obviously, collective usefulness,

4 Throughout this book I am using the male sex ('he') for describing actions by principally sex-neutral actors, if legibility of the text demands it. I do so because I reject the use of 'he or she' for such situations because of style considerations. I subscribe to the view that an author should use his (or her - one exception to the just stated rule) own sex for situations wherein a sex must be defined because of linguistic reasons.

defined as the sum of all utilities of the members of the particular community at hand, is higher with the pollution than without it.

The third alternative type of institution I want to introduce here is the community. The point of view that the community should play a large role in political processes is advocated by so-called communitarians. Although communitarian ideology is more widely applied elsewhere than in the context of environmental issues (see for instance Etzioni, 1996; 1993), I will limit myself to treatment of communitarianist ideas regarding the environment. Two interrelated threads of the communitarian argument seem to be relevant for the present discussion. First, communitarians would stress dialogue between members of the community. Second, they would emphasize the local level. Both points of the agenda are summarized succinctly by Thompson: '*The need is for decentralised human-scale political units based on communities and districts with extensive autonomy and rights of self government, citizens participating directly in decision-making*' (1995: 46). To start with the issue of dialogue, communitarians argue that for ecocentric policies, it would be necessary to have people be able to empathize with others. Such empathy is only possible in small, face-to face, communities, where 'genuine interconnectedness' can be achieved through sustained experiences of mutuality and reciprocity (Mathews, 1995b). The relationships between the members of such a community are expected to be of a special kind, as they should be based on mutual recognition, understanding and communication between people (Mathews, 1995b: 77). The argument goes further. Apart from the fact that 'real' democracy is only possible at the local level, decisions must be made locally for various other reasons (see Ward, 1998). These reasons include room for experimentation with solutions on a small scale, the possibility of effective social sanctioning, and encouragement of persons to be less egoistic and materialistic. Along these lines, communitarians do not regard the liberal state as a possible solution to environmental problems, but as a major cause of them, because in the process of interest aggregation, the direct connection between actions and consequences is lost. The assumption is that those who have to live with the consequences of decisions and pay for them are most likely to take decisions that are beneficial for humans and the environment.

Although it is difficult to find advocates of decision-making through the courts, it must be noted that this institution does play a role in environmental decisions. Courts often become involved because they have powers to judge actions of actors (be they market parties or government agencies) from the perspective of certain legal principles or policies. Countries differ in the extent to which courts are allowed to test certain decisions (many countries do not allow constitutional testing of parliamentary decisions, for instance), but generally speaking court cases are attempts at hearing various sides of an issue. It is often felt that court cases therefore provide an incentive for parties involved to present only evidence that supports their view. In combination with the fact that courts have mainly a veto-power and little positive power and are often slower than the parties involved would desire, this makes the courts not a very popular avenue for decision-making with theorists.

Conclusions on the basis of this section

It is very easy to find criticism of environmental decision-making under representative democratic institutions and to find ideas for alternative institutions that might do better. Because of these 'competing' institutions, it can be expected that the question of 'allocating' the task of 'producing' environmental decisions to certain institutions is an important matter and has consequences for the 'quality' of those decisions. In such an allocation decision the (perceived) strengths and weaknesses of various institutions will play a large role. At this point, I wish to express my interest in this allocation process and the ideas that guide it. I will argue that an important component of these ideas are a vision of the qualities that the decision must have and consequently the 'language' in which these proposals need to be discussed. Before turning to this allocation process however, I will introduce a field where I think many of the relatively abstract issues discussed in this section can be made more concrete. This field is hazardous waste siting.

1.3 The siting of hazardous waste facilities

Introduction

One need not look far for places where the discussion on how to make environmental decisions takes place. Many decisions have an environmental dimension these days, and often there will be a desire from environmental groups or private citizens to have a say in what goes on. Examples include road construction, forestry management, genetically modified food, dam construction, etc. etc. In this book I will concern myself with just one of these topics. Although hardly an emblematic environmental issue (that is, a defining issue for environmental discourse at a certain time), hazardous waste siting has attracted a fair share of interest, exactly because the issue requires institutional choice. In this section I will introduce the concept of hazardous waste siting, sketch some background of the phenomenon, and demonstrate that various different institutions can be used to arrive at decisions. Finally, I will take a first step towards an idea of 'good' decisions.

Hazardous decisions

The issue of hazardous waste siting has received its share of attention in the public policy literature in the past twenty years (without being complete: see e.g. Hunold and Young, 1998; Inhaber, 1998; Williams and Matheny, 1995; Petts, 1994; Rabe, 1994; Fischer, 1993; Hisschemöller, 1993; Portney, 1991; Armour, 1990; Andrews, 1988; Morell and Magorian, 1982). Although much of this debate took place on the North-American continent, it is definitely not unique to that part of the world, as many European countries have also met problems in this sphere (the aforementioned literature attests to that).

Siting is the process in which decisions are made regarding the location (site) of, or licences for, certain facilities. Finding a site for or granting licences to hazardous waste facilities has become increasingly difficult in practice. In a nutshell, the reason seems to be that the democratically desirable practice of involving the local public in siting decisions results mainly in resistance because the local public opposes such facilities and is not interested in 'reasonable discussions'. Opponents tend to use their rights to block facilities, including the possibility of starting court cases. It is often considered that local opposition to hazardous waste facilities is inspired by selfish motives. In this context, many authors point out that most people agree to the need for hazardous waste treatment from an environmental perspective, yet are not willing to accept a facility in their area. In their resistance, local communities often try and influence their local and central governments to block any proposals for waste facilities in their area, whereas they do little to present alternatives. This phenomenon has been captured by the famous acronym NIMBY (not in my backyard) and related abbreviations such as LULU (locally unwanted land uses) and NIMTOO (not in my term of office). The latter term, of course, suggests that decisions on hazardous waste facilities are hazardous decisions, especially for elected politicians. If they approve hazardous waste facilities in an atmosphere of resistance, they may well loose their job.

What are hazardous waste facilities and why are people afraid of them?

Hazardous waste facilities 'treat' hazardous wastes. But what are these wastes? One (US) definition said that hazardous waste is: 'a solid waste, or combination of solid wastes, which because of its quantity, concentration, or physical, chemical or infectious characteristics, may cause, or significantly contribute to an increase in mortality or an increase in serious irreversible, or incapacitating reversible, illness; or pose a substantial present or potential hazard to human health or the environment when improperly treated, stored, transported, or disposed of, or otherwise managed' (Rabe, 1994: 10-11). Related terms are 'dangerous wastes', 'chemical wastes', and 'special wastes'. Unlike the US definition suggests, hazardous wastes may also be liquid. The meaning will be conveyed however. What makes hazardous wastes so hazardous? The hazards related to these wastes may stem from various factors, including their ignitability, corrosivity, reactivity or toxicity (ibid.). Whereas this description may sound somewhat frightening, it should be observed that normal chemistry and physics are at work for hazardous waste. This implies amongst others that the potentially dangerous properties of these wastes are dependent on concentration levels, on encapsulation, etc. Hazardous waste treatment facilities are intended to manipulate the waste in such a fashion that they become less dangerous.

Needless to say that hazardous waste facilities come in all sorts and sizes. An often used and relatively simple distinction, however, is between incineration, landfill and treatment facilities. In short, these facilities differ in what they do to hazardous waste: burn and destroy it (incineration), encapsulate and isolate it (landfill), or change its chemical or physical properties (treatment). In any of these cases a certain level of risk to the environment and to public health is involved. Incineration has for instance been linked to increased levels of toxins such as PCBs and dioxins in areas near them. Such toxins are related to the fact that destruction in an incinerator is never complete and to the fact that certain chemical reactions take place in incinerators. Landfill presents a certain risk to the ground beneath facilities and the groundwater reserves in them. For permanent encapsulation and isolation are never eternal and leakage is to be expected. Whether or not such leakage is problematic depends on how closely it is supervised, whether there are any groundwater reserves below the facility and the possibility of addressing the situation if a problem arises. In treatment facilities, no matter how small, there is always a chance of inadvertent mixing of wastes that creates potential hazards. For each type of facility, risks are also related to associated activities such as the storage and transport of wastes, where human and technical error may lead to certain problems, such as fires that create toxic fumes or even explosions. Figure 1 shows an image of an actual hazardous waste incinerator.

The examples of the aforementioned risks materializing are numerous, and as these things go, the failures are better remembered (and receive greater attention) than the successes. This has had important effects. Portney (1991: 156-157), for instance, presents data to suggest that hazardous waste facilities are very unwelcome neighbors for the US public. Among potential unwanted neighbors (varying from an airport to a nuclear power plant), only nuclear power plants are less popular than are hazardous waste treatment facilities. This data may differ for other countries, though I suspect a similar pattern though. Opposition in the American public is most strongly related to a perception of a health threat (ibid.: 89-92). Again, I have little reason to assume that this is different elsewhere.

Figure 1. Aerial view of the Swan Hills (Canada) hazardous waste incinerator. The siting of this facility is discussed in chapter 6.

The irrational public

It has often been noted that the fear of the public is 'irrational' in one sense. In the course of my research I have heard many experts in hazardous waste declare that the risk related to the operations of hazardous waste facilities is smaller than the risk of many other activities, which apparently have no effect on people. Among these activities, I have heard, is the *transport* of hazardous wastes and even more so, the transport of yet unused chemicals such as petrol (hazardous wastes are often spent chemicals and therefore exhibit less dangerous characteristics than unused chemicals). Associated with such remarks was a certain level of amazement about the choices of the public in topics to worry about. I have read in depth accounts of the distinction between 'real' and 'perceived' risks (see e.g. Lash et al.,

1996). The thrust of such accounts is often that the distinction between real and perceived risks is artificial, essentially because all risks are perceived risks (be it by different looking-glasses). Such a point of view has potentially important implications for environmental decisions because if 'real' risks do not exist outside human perception, then such real risks can less likely serve as a justification for imposing certain 'acceptable' risks on people (which admittedly goes further than saying that 'real' risks exist). Research in risk perception does find that popular risk perceptions are often affected by trust in institutions that control and or regulate risks (e.g. see Wynne, 1987). It has, for instance, been pointed out that the traditionally relatively deferential attitude of the British public in environmental affairs might well be related to the high levels of trust in companies and public regulators (Vogel, 1986). Such findings corroborate the idea that an interest in institutions is relevant. They also point to a need to be sensitive to the different rationalities involved in the decision process, a topic to which I shall return when I discuss the meaning of the good decision. First I want to demonstrate that various institutions are used to arrive at hazardous waste decisions.

Institutions used to reach decisions

One item that is quite well observable in much of the literature that I just quoted is that there is a certain concern that hazardous waste facilities must be constructed. Many of the prescriptions that are given in such literature are related to the introduction of institutional arrangements that make siting feasible. In line with this interest there is also much interest in evaluating the effectiveness of various institutional solutions in that perspective. This approach is perhaps epitomized by Inhaber's grotesquely titled book '*Slaying the NIMBY dragon*' (1998). For my current purposes, this literature has the advantage of providing an indication of the types of institutions that are used to create siting 'success'. The terminology to describe institutions used for decision-making varies widely across these studies. Despite that being the case, it was possible to sort out three main approaches. Note that each approach is presented as an alternative to current practice, which involves decision-making by representative institutions and the courts. The alternative approaches are *for now* referred to as the regulatory, the market and the voluntary approach, language consistent with the terminology used by Rabe (1994).

The regulatory approach has also been referred to as the pre-emptive (Portney, 1991; Andrews, 1988; Morell and Magorian, 1982), the conventional (Armour, 1990), or the comprehensive approach to decision-making (Williams and Matheny, 1995). The essence is decision-making by experts, on the basis of scientific information, at a highly centralized level, and affording a limited role for participation at the local level. The working of this approach usually involves identifying indicators for site suitability and attaching weights to these indicators. Then various possible locations are analyzed, seeing how they 'score' for each indicator. A total score is reached by multiplying scores for individual indicators by their weight. The number of indicators may be as high as 150 (Rabe, 1994:47) and the process is to lead to a site that is optimal from the perspective of these indicators.

The market approach is also referred to as negotiation (Portney, 1991) or compensation (Morell and Magorian, 1982). Under this approach, 'the' or 'a' market is the essential institution. With the market approach, the government stresses its neutrality towards sites and waste treatment technologies and relies on negotiations between a waste management firm and the local community to reach a siting agreement. Private firms must thus develop proposals for waste facilities and are required to devote resources to economic compensation for those affected by the facility. Where negotiations do not lead to agreement, parties have in some cases been required to submit their disputes to binding arbitration (Rabe, 1994: 36). Note that where such a market approach has been introduced, this was always a decision of the government. Such a decision may be reached passively (the government chooses not to use its power of eminent domain) or active (the government relinquishes its power of eminent domain). This insight is consistent with the contention (see amongst others Diesing, 1962) that markets tend to need government to define rights and obligations.

The voluntary approach is sometimes called local veto (Portney, 1991; Andrews, 1988; Morell and Magorian, 1982), or the co-operative approach (Armour, 1990). Here, the emphasis is on the community. Government picks up the role of actor responsible for hazardous waste siting, but in doing so only approaches volunteer communities and no coercion takes place. Extensive public information sessions are provided, during which citizens can raise questions and determine their support for or opposition to the proposal. The approach tends to be supplemented by institutional arrangements to ensure safe facility construction and management, compensation packages, long-term measures providing community oversight of facility operation, and reassurances against exploitative practices.

The definition of the good decision: connected to institutions

So far, I have conspicuously avoided any specification of the concept of a good decision. The discussion on the 'rationality' of citizens indicates, however, that attention to the definition of a 'good' decision is very much desired. If citizens are irrational, who then *is* rational, and how would we ensure that these people take the decisions and not citizens? Often, the answer is that experts, who know more about hazardous wastes and are impartial, should decide, instead of citizens. So apparently experts have a different view of good decisions than citizens do. What about communities then, and what about markets? If these institutions are allocated the task of deciding hazardous waste siting, can we expect different results? The answer is likely to be yes. That is because every type of institution is associated with a different kind of rationality (see Diesing, 1962). That different kind of rationalities exist, may sound strange at first, but is easy to see when one realizes that the various groups involved are not likely to have the same motivation. Being rational implies striving for reaching one's goals, but if these goals are not the same, then rationalities differ. Earlier I said that citizens who oppose a hazardous waste facility are often considered irrational *in one sense*. It was pointed out that hazardous waste facilities probably do not present citizens with a greater amount of risk than other facilities. If they were to devote energy to a reduction of risks, then they should perhaps concentrate on other issues. Equally often however, it is pointed out that such citizens are very rational in *another sense*. This is because people can be expected to reject developments that present them with negative net benefits, as hazardous waste facilities tend to do.

What now is the relation between these two senses of rationality just described and institutions that take hazardous waste decisions? As I already noted, remarks about the irrationality of the public in the first sense often come from experts. So allocating them with the task to take a decision they are likely to ignore the objections from the public. However, if we allocate the decision to a market instead, the outcome will be different. Under a market, assuming that certain citizens have control over certain resources required for hazardous waste decisions (e.g. land), it is unlikely that these citizens will themselves accept net negative benefits from the decision. So the decisions produced by that market would almost certainly be decisions that do justice to the economic losses that these citizens might suffer from the hazardous waste facility. Note that this implies that citizens without a 'vested interest', that is property, even though being affected, would not see their interests acknowledged. And what if a local community takes a hazardous waste decision? Or elected politicians? Without elaborating on these examples, it will be clear that selecting a certain institution to take a decision, also likely implies that a certain idea of the good decision is present. More generally speaking, I am assuming that a choice for a certain institution also implies a choice for a certain language. By that I mean that my expectation is that, embedded in the allocation of decision-making powers to a certain institution, will be a preference for a certain type of argumentation. So whereas in a market, it will be perfectly acceptable to reject outcomes that are not in one's own interest, this is likely to be different in a community, where one needs to motivate a position in the debate by referring to the interests of the community. This insight implies that institutional choices have two aspects that can perhaps be referred to as 'hardware' and 'software'. The 'hardware' is the material part of the institution, the aspects that can be touched. Parliament, for instance, resides in a building ('The Parliament'), has a chair, ordinary members, expert committees and is likely to meet on a regular basis. The 'software' is the non-material part of the institution, the language that can be used by its members to produce outcomes. Certain argumentations will be acceptable, others will not. So far, I have mainly used the word institution with the 'hardware' in mind. I will continue to use the word in this sense. However, both the hardware and the software are needed to make an institution function. From now on, I will refer to the software as 'language' or 'discourse'. The combination of institution and discourse will be called 'institutional arrangement' from here on.

Not ecological rationality but sustainable development

Given the fact that my interest is in good decisions in the environmental field, I feel it is good to devote some space to the way the different types of quality might relate to each other and to the concept of 'ecological rationality'. To sum up, I am assuming that different institutions produce different types of qualities, that these qualities are incomparable with each other and that the concept of ecological rationality should be replaced with different notions of decision quality. I have already pointed to Dryzek's attempts to define ecological rationality and to assess whether various institutions are capable of producing such rationality. Dryzek's approach leads him to condemn certain institutions (e.g. the market and the bureaucracy) because they did not lead to ecologically rational decisions, e.g. by allowing nuclear waste to arise (1987: 111). Personally, I think there are various reasons why nuclear waste has arisen, but to say that their presence is intrinsically linked to a certain institution would be wrong. My explanation would be that human insights in nuclear facilities and nuclear wastes have changed. To say that the production is ecologically irrational is itself a reflection of ideas that currently live in many circles (but not in all). The irony of the situation is that Dryzek condemns instrumental rationality, emphasizes dialogue and practical reason, but at the same time refuses to accept that the very implication of this stance is the rejection of standards outside human decision-making. If ecological rationality were such a clear concept as apparently transpires from a discussion on nuclear waste, then surely we would not need practical reason, we would need *instrumental* reason! If discussions in the environmental sphere, however, should involve both attention to goals and means to achieve these goals (i.e. practical reason), then ecological rationality can not be placed outside the sphere of human decision-making and should become subject to the discussion itself. I note that 'sustainability', a much more popular concept than 'ecological rationality', tends to be defined in exactly such a way: human discussion leads to a definition of what *is* sustainable and what not. Most of the time, sustainable development is defined in such a way that it is a composite concept, containing an economic, technical, and social dimension. If this is true, then looking at hazardous waste decisions from the perspective of such 'partial' rationalities is warranted. Furthermore, if human discussion is to lead to 'good' decisions, then this brings us immediately back to different institutions that should take such decisions. I am personally not at all convinced that local communities, even if they were so bent on dialogue and in touch with nature, take better decisions than, say, experts. Note that adherents to the local model (communitarians) base themselves on decision-making in the scientific community as a normative ideal. But why then would experts not be able to engage in dialogue and arrive at good decisions? The close association of 'ecological rationality', in itself a concept that I do not reject, with local decision-making, puzzles me. And, whilst a choice for a certain institution implies a certain idea of what the good decision is, this does not imply that the quality of the decision that this institution will produce can not be judged by yardsticks that have greater affinity with other institutions. A question one then could ask is whether a decision arrived at through, for instance, a market, is also a good decision from the perspective of the community or from the perspective of experts.

The choice of institutional arrangements and their actual functioning

The issue of 'institutional choice' has been called the domain of politics (Diesing, 1962), and choosing the 'right' institutional arrangements is the task of politicians. Diesing (ibid.) argues that political rationality demands that institutional choices lead to the solution of particular problems (i.e. responsible hazardous waste treatment) and leave the legitimacy of the political system in tact (i.c. does not lead to alienation between politicians and the electorate or arouse a desire for systematic political change). At a somewhat lower abstraction level, I would like to add, political rationality also demands that the individual representative tries to avoid taking decision that can cost him his job. In Western democracies, it is easy to see that, in the sphere of hazardous waste decisions, elected politicians do face the task of devising institutional arrangements for hazardous waste decisions. This is because Parliament has the right to regulate siting processes and has often chosen to exercize this right. Hazardous waste decisions tend to be guided by land-use planning and environmental regulations. These regulations specify which institutions are relevant and what language must be used, i.e. they specify the relevant institutional arrangements for concrete decision problems. These institutional arrangements must then be made to function at the concrete level, when actual siting decisions are made. It is easy to see that institutional arrangements are under stress in actual siting debates, which take place at what I shall call the 'concrete choice level'. Both the process of 'electing' institutional arrangements and actual siting decisions are of interest to me.

So it can be expected that ideas on 'proper' institutional arrangements for hazardous waste decisions are debated at two levels: the level of political choice regarding regulations (collective choice level) and the level of the individual decision (concrete choice level). At both levels, institutions and discourses are present, which are therefore 'nested' or 'layered' (Ostrom (1986) observes this fact for institutional arrangements; Healey (1999) for discourses). The difference between the two levels of debate is that the institution for taking binding decisions at the collective choice level - at least in Western democracies - is clear beforehand (Parliament). By contrast, Parliament can freely choose a certain institution and a language to be used at the concrete choice level from the whole pallet of options. The two levels of decision-making interact: the actual practice of taking decisions provides, if not the actual model for, then certainly feedback to the political process. In either case, I expect that any position on the proper institutional arrangements for hazardous decisions needs to be defended on the basis of a certain language. In other words: I expect elected representatives, acting at the collective choice level, whilst not needing to defend their own position as an institution, to choose a certain language that they think accurately describes the problem of hazardous waste decision-making. This chosen language then influences their selection of certain institutional arrangements, thus both a certain institution and a certain language, for making decisions at the concrete choice level.

How these choices could take place is the topic of the next chapter, but for now I wish to note that I subscribe to a particular view of decision-making. I side with those who argue that 'power' in modern societies is (a) a product of constant human interaction where power can be acknowledged *or not*, and (b) argumentation plays an important role in this sense that decisions need to be motivated no matter how powerful the decision-maker is. In concrete terms: because members of Parliament believe that hazardous waste decision-making is best understood as problem-solving with scientific means, they may identify a combination of expert bodies and scientific reasoning as the proper institutional arrangement for taking hazardous waste decisions. This, however, does not automatically result in acceptation from the public. A choice to let a certain decision be made by experts, even from the highest authorities, needs argumentation. This argumentation is not automatically accepted by the public, as are, by the way, the actual choices that such experts make. So suppose that decisions on hazardous decisions are indeed awarded to expert bodies, then this decision needs a motivation, i.e. that experts are (politically) neutral. This assertion will need to be defended in Parliament where there are likely to be other opinions, but also in concrete siting discussions, where people will want to test this neutrality.

Review of this section

In this section I have done two things. The first is that I have made some of the relatively abstract issues from previous sections somewhat clearer by using a concrete example. This exercise has resulted in a confirmation of the idea that institutions are very relevant in environmental decision-making and are a highly relevant topic. In achieving this confirmation, I also demonstrated that hazardous waste decisions are a good field in which to explore some of the topics which interest me, specifically the relation between institutions and good decisions. A second result of this section is that I have moved closer to an understanding of what good decisions are. I have rejected the idea that there is a norm for 'good decisions' outside the capabilities of human beings to gather, interpret and evaluate information. I have also demonstrated that different institutions will produce different qualities. The selection of institutions for collective decisions is a political choice process, which will need a certain motivation. I have posited that such a motivation will not only be needed at the level of legislation that guides hazardous waste decisions, but also at the level of concrete decisions.

1.4 Outline of this book; research questions and structure

Introduction

In this chapter I have sketched the background of environmental decision-making by paying attention to the nature of these problems. I have also indicated that my interest in 'good' environmental decisions will be narrowed down to an interest in the quality of decisions rendered by various institutional arrangements. My starting point was representative democracy and I have discussed some of the suspected failings of this institution and presented some alternatives. The efficacy of these alternatives has not been discussed at great length yet, but this will happen in the remaining chapters. The section on hazardous waste decisions demonstrated that in this field, various institutions to arrive at decisions have been tried. This makes the hazardous waste field fertile ground of interest and I will study empirical cases of hazardous waste decisions later in this book. The discussion of hazardous waste decisions also revealed that it will be difficult to view institutions and decision quality as an independent and dependent variable respectively. The reason is that there is likely to be a connection between the choice of a certain institution and a certain type of quality as the institution will be operated on the basis of a certain language that stresses a certain conception of quality. This implies that I need to move one rung up on the abstraction ladder and look at the way institutional arrangements and different concepts of good decisions are linked up. Given the fact that different institutional arrangements produce different types of quality, it is likely that there will be debate about the 'right' arrangement for a hazardous waste decision. This debate can be expected when there is debate about the rules that guide hazardous waste decisions (likely to be found in land-use planning and environmental regulations) but also when there is an actual attempt to site a facility. At the collective choice level, Parliament takes binding decisions about such regulations. These choices will be inspired by a certain discourse that identifies the relevant institutions for making decisions at the concrete choice level and the discourse that needs to be used there. The next chapter will be devoted to these discourses and the way they influence the choice of institutional arrangements.

International comparison

Not only do I want to look at the debate about the allocation of hazardous waste decisions to certain institutional arrangements, I also want to look at the way different institutions performed in practice (in their own terms, but also in terms of other institutional arrangements). Because my initial - and wrong - assumption was that the use of institutional arrangements would be nationally colored, I embarked on an international comparative work, focusing on countries that use strongly differing arrangements for hazardous waste decisions. Coming from the Netherlands, the inclusion of that country was my starting point. In considering countries that might contrast well to the Dutch approach, the rich literature on hazardous waste facilities provided many ideas for countries that would be good to compare the Netherlands with. In order to get more contrast in terms of institutional arrangements involved in hazardous decision-making, I decided to also look at Canada and the UK. The view that I had of these countries was that Canada placed much more emphasis on communities and their interests in hazardous waste siting than the Netherlands (Armour, 1996; Rabe, 1994; Fischer, 1993). I expected the UK to have greater emphasis on the market and on experts and their argumentations than the Netherlands (see Vogel, 1986; Allen, 1992). When I later found that there was variation in approach to hazardous waste siting within these countries (especially within Canada), I decided to focus on those cases that exhibited use of institutional arrangements that I wanted to study in both countries[5]. The reader is thus warned that the cases that will be described later in this book do not necessarily represent the way decisions

[5] Case selection is explained in appendix 2.

are made within one country as a whole. Instead, they are (hopefully) representative for the various institutions that were used to take decisions.

Outline of the rest of this book

Chapter 2 of this book is a continuation of this chapter, and mainly seeks to develop the discourses that guide the development of institutional arrangements. Based mainly on the work of Williams and Matheny (1995) I have identified three discourses on social regulation. Each of these discourses identifies relevant institutional arrangements that should be used in the field of hazardous waste. Working with Dryzek's (1997) and Hajer's (1995) methodology for analyzing discourses and their effects, I will describe the major aspects of every discourse and the way in which discourses influence the allocation of decision-making responsibility over the various institutions. The resultant institutions can be described within the framework for institutional analysis that was developed by Elinor Ostrom (1990) and I will derive expectations about the way decision rules will look like, when certain discourses are dominant at the collective choice level. I will also indicate what type of quality is stressed by the three discourses and develop a set of conditions for good decision-making that need to be met from the perspective of each discourse.

Chapter 3 is devoted to a search for the existence of the various discourses and their affect on institutional arrangements in the Netherlands, Canada and the UK. The analysis in this chapter is partly historical as, in each country, institutional arrangements have a long tradition that goes back to the first social regulations near the end of the 19th century. Already at that time some basic decisions were taken about market and governmental responsibilities. The analysis shows that in either country, experts mainly implemented government regulations. These experts were often initially quite reformist in intention, but soon developed a close relationship with industry. In this relationship a shared understanding of reasons for government interference in the economy developed, which largely revolved around the idea of the government assisting industry. In this chapter I further look at important turning points in this relation. Certainly government planning in all three countries received a great boost after World War II. And each country has had its share of protest against paternalism from experts and the political system in the 1960s. It is at that time that both a sense of environmental problems and a desire for greater citizen participation in planning and environmental decision-making developed. It is at this time that the existence of a hazardous waste problem was discovered in the three countries. Just as a historic background, I will chronicle some of the developments in hazardous waste policy since then. The analysis will show that hazardous waste siting in all three countries - at least partly - was to take place through the land use and environmental licensing system. These systems were allowing increasing amounts of citizen participation, but never to such an extent that elected representatives or experts were assigned just a minor role in the decision-making process. In each country, environmental impact assessment was later added as a requirement for hazardous waste facilities. Here too, a fair share of citizen participation was regulated, but never to such an extent that, for instance, the local community would become the dominant actor in the process. In fact, the duty of preparing an environmental assessment report has been placed firmly on the shoulders of private developers in all three countries. Differences between the three countries do definitely exist, with the Netherlands leaving relatively little space for the market to decide about facilities and the UK and Canada considerably more. the Netherlands further stand out because of their relatively easy access to the courts. Canada is remarkable in that developers, as a requirement of government regulations, sometimes fund citizen participation. Remarkably enough, little traces of a more community-oriented decision process can be found in Canadian regulations. The UK is remarkable because of its inquiry system, where many environmental conflicts are to be fought and because of the bias in its regulations that demonstrate a hesitation to let government interfere with private development. The language that needs to be used in all three countries, in debating hazardous waste proposals, tends to stress technical and economic arguments. At the collective choice level, a mixture of what I will call managerialism and conservative pluralism dominates. This implies that the interest of the community is not deemed a relevant argument for siting debates by the Parliaments in all three countries.

Chapters 4 to 6 are devoted to three case studies from every country (for a description of case study selection see appendices). The question in those chapters are whether the prescribed institutional arrangements for hazardous waste decision-making were used in actual decision processes and what were the effects. I am particularly interested in the question whether the use of certain arrangements was challenged from the perspective of rivaling discourses and to what effect. It is difficult to do justice to every case study in the short space here, so I will refer the reader to these chapters and make only a few brief points here. The Canadian practice of community-based decision-making originates in Alberta and is the consequence of a failure of existing institutions to solve a hazardous waste problem. An innovative approach was used, which stressed the community as an institution and enhanced discussion on the basis of ideas about the community's interest. The approach led to community consensus and agreement in three cases. Interestingly, the fact that community consensus obtained so much priority did imply that other types of rationality were more or less ignored, implying that in either case the outcome was not an unequivocal success from all perspectives. Also, the degree to which economic considerations motivate people to accept hazardous waste facilities and the degree to which communities are susceptible to manipulation warrants attention if one applies 'voluntary siting' once more. In the UK, the role of the inquiry - a quasi-judicial institution - is extremely large. Although discredited quite often for being biased against opponents and not offering a good opportunity for communities to express themselves, the various inquiries that I have studied do discuss many important issues and come the closest to achieving the ideal of 'dialogue' of any of the institutions I have reviewed - be it that they focus almost exclusively on technical arguments. In the Netherlands, finally, decision-making processes frequently end up in court, despite the fact that the courts do little to bring closure to siting disputes.

In chapter 7, I summarize and analyze the case studies from a comparative perspective. Finally, in chapter 8 I draw my conclusions and give some recommendations even though I am very aware of the strong limitations that my research design places on the making of such recommendations.

Chapter 2

Knowledge, competition or dialogue?

Institutional variations and their relation to discourses on facility siting

In this chapter I will introduce discourse analysis as a method that could help study hazardous waste decision processes. In discourse analysis, attention is paid to the language that people use and the effects it has on their interactions. Discourse analysis focuses on regularities in what is said or written, and these regularities are discourses. I will demonstrate that discourses imply a preference for certain institutional arrangements and therefore contain a concept of 'good' decisions. This is because discourses identify relevant actors, prescribe their relation and acceptable arguments, and imply expectations of what the motives of each actor are. At the collective and operational choice level, it is expected that several 'discourse coalitions' exist that use 'story lines' to convince others that their perspective is right. I will argue that actors involved in hazardous waste decisions have three discourses at their disposal. These discourses are the managerial, the pluralist and the communitarian discourse. I will describe these discourses in this chapter and describe the institutional arrangements associated with them. I will also provide a critique of these discourses, which together with the description of the discourses, leads to expectations surrounding the actors involved in siting decisions. These expectations will be empirically examined in later chapters. Associated with each of the three discourses are different story lines that summarize the understanding of hazardous waste decision-making embedded in the discourse. These will be described as well. At the collective choice level, choices to assign hazardous waste decisions are likely to be influenced by one or more discourses. These choices are then embedded in legislation regulating hazardous waste decisions, i.e. they are 'institutionalized'. At the concrete choice level, these institutional arrangements still need to be upheld however. The concept of 'structuration' is relevant here, and again, story lines and discourse coalitions are likely to play a role in structuration.

2.1 Discourses on environmental decision-making

Introduction

In the previous chapter I have explained my interest in institutional arrangements and specifically, the institutional arrangements used to arrive at hazardous waste decisions. I noted that a choice for a certain institution also implies a choice for a certain language and a conception of quality. I argued that choices for institutional arrangements will need to be motivated by arguments. In the case of hazardous waste siting, institutional choices are likely to be expressed through formal legislation, which in a sense, provides the conclusions of a debate on different institutional options at the collective level. The institutional arrangements finally assigned with the task of producing hazardous waste decisions are however likely to be challenged again at the operational level. The 'practice' of actually deciding about hazardous waste proposals therefore is yet another place where arguments are made about who should make the decision and how it needs to be justified. If institutional arrangements devised for decision-making are challenged, an interesting tension between various rationalities can be expected and one of my interests is how such debates can be successfully closed. In this section I take a closer look at the concept of discourse.

Discourse analysis

Discourse analysis is a relatively new approach to studying decision processes, at least in the policy sciences. I will base much of my account on the work of Hajer (1995) who did much to explain the bases of discourse analysis and make it more applicable to policy processes. In my opinion, discourse analysis is based on various different assumptions. One such assumption is that power in modern societies is not necessarily to be found in 'traditional' places such as the monarch, the parliament, the army, etc. Power has transformed from being based largely on the ability to exert force to being based in language. This language allows people to make sense of what they experience and their roles in it. An example may clarify this somewhat. In hospitals, medical professionals work together and develop a language of describing their experiences. In this language, the word 'patient' is used to describe a particular subset of the visitors to the hospital. When someone is being defined as a patient, there will be certain expectations regarding the way this patient behaves, whether or not the patient himself likes this. One of these behavioral norms will be that the patient must listen carefully to nurses and more than doctors, and likely that he must not smoke. Doctors and nurses on the other hand, will be obliged to behave in a certain way as well. These norms could imply, for instance, a refraining from too personal relationships with patients and perhaps a degree of secrecy on the condition of the patient towards others. The web of norms and expectations associated with the words 'patient', 'doctor', and 'nurse' is very powerful. If somebody does not know or chooses to go against the norms, he can be given a hard time and is likely to start behaving 'as a patient should' after a while because of the conditioning by others. The patient is then 'disciplined' and his behavior is likely to be trouble free. It is easy to see that power is not exercised here in the classical sense, as there is no one who carries a big stick to punish people who 'won't listen'. The power of the doctor over the patient is very much based on agreement of the patient with his presence in the hospital and with his role of patient and the behavioral norms embedded in that agreement. Communication will often be aimed at getting information about these norms. To make the patient to behave in a certain way with force would clearly not be very efficient in a hospital and that is why the terms 'patient' and 'doctor' and the associated norms are so handy. The point here is not that force in modern society is not at all a basis of power. The point is rather that certain shared ways of understanding have developed and that these make the exercise of force much less necessary. In this sense, discourse is disciplining.

Attention to discourses implies attention to language, the effect it has on the understanding of the world and the various ways in which people can normatively evaluate reality by use of language. Hajer (1995: 44-45) indicates that discourse is normally taken to mean 'discussion' or 'a mode of talking'. When social scientists perform 'discourse analysis' however, they refer to the analysis of regularities and variations in what is being said or written, and to an attempt to interpret the social backgrounds and effects of specific modes of talking. Reflecting both aspects, Hajer defines a discourse as '*a specific ensemble of ideas, concepts, and categorizations that are produced, reproduced, and transformed in a particular set of practices and through which meaning is given to physical and social realities*' (ibid.: 44). The term 'practice' in this definition is not so much defined, but more or less illustrated by examples: academic teaching is a practice, but laboratory experiments, protest meetings and peer-reviewed journals are also practices. Perhaps a practice is thus to be seen as a social activity in the broadest sense, all activity where people live together and interact with each other.

Especially when interactions are repeated, people are likely to develop a common understanding of what they are doing, why they are doing it, how it should be done, and who should do what. This common understanding is 'carried' by discourse. According to Dryzek, discourses always contain four elements. Discourses recognize basic entities, contain assumptions about the relationship between the entities, carry assumptions about agents and their motives, and finally contain metaphors and rhetorical devices that convey the ideas embedded in the discourse. To illustrate what is meant here I briefly apply these concepts to hazardous waste facility siting. Basic entities that a discourse on hazardous waste facility could recognize (or not) include the individual, the community, market parties, communities, the government and society. Related to this are the notions of individual, common and collective interest. Regarding the relations between basic entities, the assumption in a discourse may be that the individual is more important than the community or vice versa, etc. Regarding the motives of agents, some discourses will assume that individuals and communities are self-interested, while others may assume that they are willing to act in some collective interest. An example of a rhetorical device in hazardous waste siting is the already mentioned catchphrase 'NIMBY-ism'. The NIMBY-label is a rhetorical device since it paints opponents as selfish people and denies the possibility that their opposition stems from more fundamental concerns. Note that the label could have disciplining force. Those seeking to site a hazardous waste facility can try and use the term as a means of silencing the opposition or gaining the upper hand against them in an attempt to win the support of onlookers.

Foucault

The most important scholar associated with discourse analysis is Foucault. Foucault was mainly concerned with the societal discourses on social discipline, punishment and sexuality. The overall societal discourse on these topics, to his mind, consists of various 'component discourses' which together somehow translate into a larger system. These component discourses arise, according to Foucault, in 'smaller' practices (such as perhaps the hospital I mentioned) where the attention of the analyst should consequently be focused upon. Foucault criticized political theory for paying too much attention to 'institutions' and too little to these smaller practices where much power resides. The usage of the word institutions here suggests that institutions are 'big practices', the highly visible places to make policy, such as parliament. Foucault's argument is that societal change is not a matter of univocal change (instigated by the big practices), but rather a result of the plurality of micro-practices. The process whereby micro-practices result in transformations of societal discourse is regulated by definable rules, which is called a discursive order by Foucault. Whereas this is often taken to be the case, with these statements Foucault does not deny the importance of institutions per se, but observes that they should be studied in tandem with the language that drives them. Williams and Matheny (1995: 9-10) posit: '*Foucault (. . .) was always interested in the ways in which various discourses reflect and influence both individuals and institutions. For him, "all social institutions and practices, including the law, the political system, the church, the family, the media (. . .) are located in and structured by a particular discursive field." This concept of a discursive field stresses the competing "ways of giving meaning to the world and organizing social institutions and processes." In this way it emphasizes the relations between language, social institutions, and power.*'

Hajer (1995: 48-52) points to the fact that the role of an actor in Foucault's theory is somewhat ambiguous. On the one hand, Foucault has rejected the idea that an individual is a 'sovereign subject' who manipulates 'passive' discourses for his own use. Discourse is not to be seen as a medium through which an individual can manipulate the world, it itself is part of reality and 'constitutes' the discoursing subject. This is because discourses contain internal rules that make them function as a structure to behavior. Discourses '*make it impossible to raise certain questions or argue certain cases; they imply exclusionary systems because they only authorize certain people to participate in a discourse; they come with discursive forms of internal discipline through which a discursive order is maintained; and finally there are also certain rules regarding the conditions under which a discourse can be drawn upon*' (ibid.: 49). On the other hand, Foucault seems to have seen room for individual strategic action to create a new discursive space within which discussion can take place. The example given by Hajer is the one of homosexuality, where a negative discourse with various elements has been countered by homosexuals to define their own identity and legitimize homosexuality.

Whatever the position of Foucault himself, many political scientists who refer to Foucault's work are very keen to point out that discourses are being strategically used to benefit certain interests. In fact, some political scientists see strategic use as the only function for discourse as they define discourse as '*a language or system of representation that has developed socially in order to make and circulate a coherent set of meanings about an important topic area. These meanings serve the interests of that section of society within which the discourse originates and which works ideologically to naturalize those meanings into common sense*' (Fiske, quoted by Williams and Matheny, 1995: 10). I have two reactions to this definition. One is that it gives the sensation of recognition, because it has often been observed that various scientific disciplines 'make' different problems out of unsatisfactory situations. The economist for instance will look differently at environmental problems than a psychologist. The question what 'the' problem is, will receive entirely different responses from these two. In their problem analyses, they will focus either on the economic or on the psychological aspect, which by coincidence, they themselves know a lot about. So to solve the problem, the economist will ask for economic solutions, to be advised by him, whereas the psychologist will ask for psychological solutions, also to be advised by him. In this sense, discourse favors the employment of those who use it. A second reaction I have relates to the motivation of people. Fiske emphasizes the strategic component too strongly in my opinion, in the sense that the motive of individuals is not always based on their own self-interest, but might very well be 'societal improvement' and not so much to fill their own purse. Whether or not people try and influence discourse for their own benefit might also depend on the topic at stake. In the environmental sphere, Dryzek (1997: 11) notes that powerful actors, if they see their interests threatened by a particular discourse, can attempt to override developments 'at the level of discourse', which then implies that a different 'account' of environmental problems and their causes starts being propagated. According to Hajer, the way in which this happens can not be by mere reference to vested interests. This is because interests are 'constituted' through discourse, which means that terms need to be used to define and express them, and thus derive their power from a discourse. The position of an interest needs constant reproduction, for instance in controversies, and if the process of reproduction 'fails', this position might well change. Hence the importance of 'the level of discourse' to use Dryzek's terminology. I consider the idea of strategic use of discourses in this sense credible.

How is societal discourse made?

The above logically leads to the question of how people then try and influence societal discourse on a certain topic. Hajer (1995: 52-58) has done much work to further this discussion. He argues that there is a gap between Foucault's abstract work and the study of concrete political events. Specifically the influence of various discourses upon each other and the role of the actor in that process deserves further detail according to him. He finds a basis for doing so in 'social-interactive discourse theory', which stems from social psychology and which focuses on the level of interpersonal interaction, and the role discourse plays therein. Authors from this discipline do not argue so much for a 'linguistic turn' (examining discursive systems), but for an 'argumentative turn' in social science. This implies that political scientists should examine the way actors seek to persuade each other. '*The argumentative interaction is a key moment in discourse formation that needs to be studied to be able to explain the prevalence of certain discursive constructions*' (ibid.: 54). Those engaged in discussion try to make others see the problem at hand from their perspective. But they also seek to define the position of other actors in a certain way, try to 'frame' the discussion so that the problem becomes understandable and in doing so they make certain elements appear fixed or appropriate, while other elements appear problematic.

Hajer (ibid.: 55-56) states that social-interactive discourse theory contains a 'immanentist' view of language, which holds that discourse needs to be reproduced in interaction. Therefore concepts only have a meaning to the extent that they are being used and accepted in interactions. If they are not reproduced in such situations they no longer hold prescriptive voice. The consequence of this is that the 'power structures of society' can and even must be studied through discourse. '*It becomes imperative to examine the specific idea of reality or of the status quo as something that is upheld by key actors through discourse. Likewise it becomes essential to look at the specific way in which appositional forces seek to challenge these constructs*' (ibid.: 55). Social-interactive discourse theory contends that actors hold routinized understandings of reality, embedded in particular images, metaphors, concepts and 'story lines'. Story lines are narratives that allow actors to draw upon various discursive categories to give meaning to specific physical or social phenomena and that help to make sense out of complex discursive systems (Hajer, 1995: 56). A story line implies that people do not have to draw on comprehensive discursive systems in order to understand phenomena, but that they have a compressed and simplified idea of 'what is going on'. This then implies (automatically?) the other elements of the discourse.

Like the hospital example showed, routinized forms of discourse express power relationships in an effective way because they help avoid confrontation by setting a certain mode of expression as the standard. This is because the implicit assumption during interaction usually is that the next speaker will answer within the discursive frame set by the first speaker. Hajer takes the argument further by indicating that routinized understandings can become embedded in institutional arrangements. This is because institutions function on the basis of such routinized understandings. To argue against such understandings is to argue against these institutions. Therefore, says Hajer, discursive interaction is not simply a debate between clever, creative individuals, but may be frustrated by role-playing on the part of those who hold a particular routinized understanding of a phenomenon. We already begin to see the contours of the relationship between discourse and institutional arrangements. I will return to this topic after the concept of discourse coalition has been introduced.

Discourse coalitions

Hajer's interest is in the constitutive role of discourse in political processes, with a central role for the discoursing subjects within structures that both enable and constrain their actions (ibid.: 58). He conceives of politics as a struggle for 'discursive hegemony' in which actors try to secure support for their definition of reality. Discursive hegemony is achieved when two conditions have been met: 'discourse structuration' and 'discourse institutionalization'. The first means that the credibility of actors in a given domain requires them to draw on the ideas, concepts, and categories of a given discourse. In the example of NIMBY-ism for instance, opponents to a facility may be forced to argue the position that they are not 'NIMBIES'. Structuration thus refers to the 'software' of institutional arrangements. Institutionalization means that the discourse is translated into institutional arrangements, which Hajer describes with examples such as policy programs or departmental organizations, i.e. what I have called the 'hardware' of institutional arrangements. Although not stated in so many words, it seems that Hajer assumes two things at the same time: there can only be one dominant discourse at a given moment and in a particular domain, and secondly that there always is a hegemonic discourse.

Hajer continues to argue that the story line, in combination with another 'middle-range' concept, discourse coalition, can explain why certain discourses obtain a hegemonic position and why this may perhaps change again over time. If a phenomenon is relatively complex, understanding it may require knowledge from various fields, and nobody is able to master all this knowledge completely. Therefore, knowledge that is developed from several fields needs to come together in a story line that explains what is happening. Story lines are the metaphors in politics, to which (for instance) scientific findings are to be translated. Story lines simplify and thus help actors to develop an understanding of the problem and giving permanence to a debate. They play an essential role in the clustering of information and they create information networks or coalitions between actors of a given domain. The working of this can be illustrated by Hajer's topic: acid rain (ibid.: 64 and further). Acid rain itself is a story line that transferred the understanding of industrial emissions and has connected emissions to, for instance, the death of fish and the corrosion of buildings. Where in the past smoking stacks were a sign of wealth, smoking stacks have now become something negative. The story line will change the perception of actors of phenomena, for instance, because they can now relate things they saw happening to 'acid rain'.

By constructing a problem, but also by establishing 'blame', 'responsibility', and other morally-laden concepts, the story line empowers actors, who may for instance start to think of themselves as victims of acid rain, whereas the position of others may be threatened. Note that because of the story line, actors may re-interpret their own interests, which is thus not defined a priori. This is a recurring theme with Hajer, who argues that this sets his form of discourse analysis apart from more common theoretical approaches in the policy sciences. Discourse coalitions, argues Hajer (ibid.: 65), are formed when various '*previously independent practices are being actively related to one another, if a common discourse is created in which several practices together get a meaning in a common political project*'.

Various actors will start to rally behind the idea that there is something like acid rain, even though they do not have complete (disciplinary) understandings of the phenomenon. According to Hajer what happens in such a process is that the appeal of the story line is particularly relevant. If 'it sounds right' other actors will develop an affinity with a certain story line and associate with it. This relatively simple mechanism is called 'discourse affinity'.

Some limits to discourse analysis

I think that discourse analysis has a lot to offer when it comes to analyzing hazardous waste facility siting. Williams and Matheny (1995), on which I will build here, have amply demonstrated the point. However, discourse analysis does have some limits, as Dryzek (1997: 10-12) points out. His first warning is that a focus on language and discourse is sometimes taken so far that reality is all but denied. In the case of hazardous waste decisions this warning seems hardly necessary. It is easy to see that, though hazardous waste decisions may mean different things to different people, this does not imply that certain acts do really take place, e.g. that a license is issued.

Dryzek's second warning is that interests and power are important and provide very real barriers to people, even if they are 'captured' by a certain discourse. This relates to Hajer's point that interests are 'constituted' through discourse. This seems to imply that these interests do not exist outside discourse. Like Dryzek, I have my doubts about this, because I think that interests partly derive from fundamental human needs and are thus 'real', even outside a discourse. Whether or not such interests are accepted as a legitimate concern in a policy decision is an entirely different matter. If the 'constitution of an interest' is to signify this process of defending an interest in certain terms and attempting to gain acceptance for it, then I would concur with Hajer. In this sense, I would subscribe to a 'mild' version of discourse analysis.

My impression is that the issue (discourse versus interest) becomes most salient at the point where Hajer theoretisizes how exactly a discourse becomes hegemonic. Story lines and discourse coalitions may well be very important, but to say that such coalitions arise only because of 'affinity' seems somewhat unrealistic to me. I would assume that what 'sounds right' to an actor, depends partly on the way he was brought up, educated and has worked, but it also depends on what the consequences for him personally are. Although it may be correct to say that his interests need constant confirmation in a discourse, to deny that such interests are already present at the moment when we start a debate seems silly. Similar remarks apply to the topic of power. To give an example, the power of a monarch is not only dependent on discourse, it also depends on the ability to exert physical force. Although the subjects of a particular monarch may not like him at all and actually want to overthrow him, they are still likely to obey (at least for some time) if he has a large army and is able to suppress them. Again, the power of the monarch will be explained and motivated in a certain discourse, but this discourse *can* have little to do with reality. The ability to challenge this discourse is limited however, because of sheer physical force.

A third warning, if you may call it that, is that Dryzek goes against Hajer's idea of the presence of a hegemonic discourse. Hajer argues that discourse hegemony can be established by looking at structuration and institutionalization. Structuration is achieved as the terms of the discourse are needed to express one's self credibly, while institutionalization occurs as a discourse leading to organizational and policy implications. In the case of hazardous waste siting, I have already noted that the choice of a certain institution as the decision-making device and the language that it will use will be embedded in formal regulation. I have already demonstrated in chapter one that many different institutions can be used for hazardous waste siting. Hegemony would imply that only one discourse informs the institutional design process. The idea that the debate and the choices at the collective choice level (e.g. in Parliament) about hazardous waste siting are informed by only one discourse seems unlikely to me. And even if all lawmakers are clearly captured by one single discourse, then this does not imply that everyone at the concrete choice level agrees. I will therefore work on the assumption that hegemony, if present at all, will only be temporary and partial at most.

The connection between discourse and institutional arrangements

Hajer seems to imply that a discourse *itself* contains rules specifying the situations wherein the discourse can be drawn upon and in which situations this can not be done. If that is correct, then it is entirely imaginable that several discourses claim relevance to a certain situation. This implies that there may very well be 'competition' at the 'inter-discursive' level and the two (or more) discourses that claim relevance to the topic need to somehow fight it out. At some point there will be a decision on the issue of discourse relevance. So in a way it may be perfectly true that a discourse contains rules regarding the situations in which it is appropriately referred to, it is also true that such rules are not always legitimate in the eyes of others. Turning this discussion in the direction of facility siting, there are likely to be various competing discourses on how facility siting should be done, and it is likely that there is a decision on which discourse is relevant.

Given his topic, acid rain, Hajer did not need to distinguish between a collective choice and a concrete choice level. Indeed, prima facie it would seem that his theoretical framework, which emphasizes the role of story lines in explaining why hegemony occurs, could describe discussions at both levels. However, one needs to realize that hazardous waste siting is different from acid rain (Hajer's topic) because 'institutionalization' of discourse is a conscious process which results in formal regulations (land use planning law, pollution control regulations). But what then is the relationship between discourse and institutional arrangements? I think this relationship is to be found in the sphere of rules. I find one - previously referred to - quote by Hajer particularly interesting. Hajer notes that discourses are restraining because '*they make it impossible to raise certain questions or argue certain cases; they imply exclusionary systems because they only authorize certain people to participate in a discourse; they come with discursive forms of internal discipline through which a discursive order is maintained; and finally there are also certain rules regarding the conditions under which a discourse can be drawn upon*' (Hajer, 1995: 49). This means that, if a certain discourse is considered relevant, then this discourse will identify those who are relevant to a certain issue, what their relationship should be to others that are relevant, and how actions should be undertaken and decisions be made. So what distinguishes two discourses from each other is that they emphasize different actors, different roles, different relations and different methods of interest aggregation. Discourses can in this sense be seen as preferences or prescriptions for certain types of configurations of rules. Institutionalization then implies that the rules that a certain discourse would prefer actually become law. Interestingly, institutional arrangements themselves are often understood to be rule configurations (see e.g. Ostrom, Schroeder and Wynne, 1993; Tang, 1991; Ostrom, 1990 and 1986; Kiser and Ostrom, 1982). If this is correct, then discourses can be seen as preferences for institutional arrangements. Institutionalization of a discourse can then be studied by analyzing rule configurations, with some rules pertaining to the language that must be used in a decision process, some pertaining to the institutions that may be involved.

In the rest of this section, I will develop this point further. Various steps need to be taken in order to fully develop a theoretical framework that can be used to analyze hazardous waste decision processes at both the collective and concrete choice levels. One necessary step is the development of a rule typology that can be used to analyze institutional arrangements. This step can be based on the work of the just quoted authors. A second step is the development of a typology of 'hazardous waste siting discourses', that is, discourses that specifically pertain to the field of hazardous waste siting and that express preferences for clearly distinct institutional arrangements. Finally, I will need to connect both typologies and demonstrate what different institutional implications the various discourses would have. This last step requires that the rule typology is further developed by indicating the possible 'values' that various rule types can obtain. To clarify this somewhat: some rules pertain to the flow of information in a decision process ('information rules'). Knowledge of the existence of such a type of rule is not extremely helpful however, unless one can distinguish between various possible 'values' of this rule type. For instance, information rules could emphasize 'ordinary knowledge' or 'scientific knowledge' in the decision process. By distinguishing such possible values, one can sharpen the image of possible contrasting institutional arrangements[1].

Which types of rules exist?

One of the advantages of the work of Ostrom et al. is that it gives ample attention to the development of a rule typology. In turn, their typology has been used and evaluated by many others, who provide helpful comments. I will refrain from engaging in or even just summarizing the debate about the types of rules that really matter, but simply present my position. Ostrom et al. have introduced a rule typology that exists of seven types of rules. Comments by others (see e.g. Heilman, 1992: 18; Sabatier 1992: 249; Hupe, 1990: 292) suggest that several of these types are overlapping and that certain types of rules alone can carry the analysis. Taking their experience to heart, I posit five types of rules that are particularly relevant for analyzing institutional arrangements. These are:

1. Aggregation rules. These prescribe formulae for weighing individual choices and calculating collective choices.
2. Authority rules. Prescribe which positions can take which actions, how actions are ordered, processed, and terminated (i.e. competencies). These rules imply the creation of positions (e.g. licensing agency) and identify the scope of the authority (e.g. licenses are only about safety aspects).
3. Boundary rules. These set the entry and exit conditions for actors. They determine who is in and who is out, and by which conditions.

[1] This is one of the points of the just quoted authors, who claim that the different expectations scientists have about the performance of various institutional arrangements are often based on slightly differing assumptions about the 'value' of certain rule types. To give an example: when people speak of introducing a market in a certain domain, they often implicitly assume that those engaging in this market have complete information on prices. When this assumption fails, the predicted effect of the market may be different than expected. Ostrom c.s. plead for precision in defining institutions. I will heed this warning as much as possible here, but continue to work with admittedly emblematic descriptions of various institutional arrangements such as 'the' market. I address the - warranted - comments on the imprecision of many institutional analyses somewhat by working with Ostrom's rule typology and giving precise indications of their values.

4. Information rules. They establish information channels, state the conditions under which they are open or closed, create an official language, and prescribe how evidence is to be processed.
5. Pay-off rules. Prescribe how benefits and costs are to be distributed to participants in positions given their actions and those of others.

The definition of these rule types is relatively abstract, but I think that their meaning can be explained by examples from the hazardous waste siting sphere. Aggregation rules relate to tensions between individual interests and ways of determining collective choices when these interests are different. At a low level of abstraction, aggregation rules can stress majority or consensus in certain fora (e.g. in a municipal council) as an appropriate condition for taking a decision. At a somewhat higher level of abstraction, aggregation rules are about the philosophical motivation for the decision. One possible aggregation rule in hazardous waste siting is 'the greatest good for the greatest number of people'. In other words, the good decision is the decision that maximizes utility for all. Even though some may lose a bit because of a certain decision, this may be acceptable if the rest of society greatly benefits. Another aggregation rule could stress compensation. In this sense, aggregation rules point to the 'type of rationality' that needs to be achieved in the decision process. Following Diesing (1962), five types of rationalities can be distinguished, including social, political and economic rationality. They will be explained from section 7.3 onwards[2].

Authority rules are important in siting debates. It is my impression that a quite fundamental distinction is one between a situation wherein market parties can operate rather freely and one wherein government strongly intervenes. In some situations, market parties, such as hazardous waste firms, can freely choose locations for their operations and can also determine which precautions will be taken at the facility. At others, it will be governments that control locations and prescribe safety measures. Note that in such a case, the position of 'waste regulation authority' may be created automatically. The authority of other people to have a say on siting decisions (e.g. the community) seems often closely associated with the level of control that governments have. Governments tend to have greater desire for democracy than companies and therefore allow for more participation, which sometimes leads to 'degrees of citizen power' (Arnstein, 1969). These authorities will have a certain scope. This implies that some 'states of the world' can be affected, while others can not (Ostrom, 1986). Building further on the example of citizen participation for instance, their authority to participate is not unlimited but restricted by the scope of other authorities. One problem that is often observed in the facility siting debate is that

[2] For the reader who desires a quick overview: table 9 in chapter 7 summarizes them.

citizens are rather disinterested in talking about waste policy at first, but become passionate when the concrete implications of a policy become clearer and affect their lives. When a specific site is proposed, they tend to have a desire to restart a debate on more general policy issues, which is then outside the scope of their authority.

Boundary rules are also likely to be relevant. Again building on the example of citizen participation in hazardous waste siting, legal rights to participate are sometimes assigned to 'anyone', sometimes to 'all aggrieved parties', or sometimes to the 'license applicant' only. What this means is that the number of actors participating in the process is likely to differ with variations in boundary rules, and therefore different interaction types may occur. Information rules will also have a clear effect on the decision-making processes that will take place in hazardous waste siting. This is because even if someone has authority to take part in the decision-making process, this person must be notified of the various decisions that are being made and when they are being made. Hazardous waste siting may require various 'types' of information such as scientific information, but also local information. Rules can identify which of these types is considered more relevant and where they are to be obtained.

Pay-off rules might exist in a hazardous waste decision situation in the sense that the process itself, especially in a situation where permission is needed from various authorities, will involve time and money. Pay-off rules prescribe who needs to pay for these costs. In the case of citizen participation for instance, the rules may specify that these citizens need to pay for themselves or that they may be supported. In the case of environmental assessment procedures, the rules prescribe who pays for the assessment document. I restrict the meaning of pay-off rules here to the issue of the costs of the process. The costs and benefits of the outcomes of the decision process are adequately covered by aggregation rules in my opinion.

Earlier I said that in order to be useful for my purposes, a rule typology should reflect that institutional arrangements have two dimensions: the institutions involved and the language that can be used. Although this is not an exact science, I do feel that certain types of rules distinguished by Ostrom et al. are about the institutional side of institutional arrangements, whereas others are more about the language side. Especially boundary, authority, and pay-off rules seem to be concerned with the roles of various institutions in the decision process. Authorities will be allocated to certain institutions or not, and the entry of certain institutions (or the actors that represent them) will be stimulated or not. Information and aggregation, in my understanding, are much more about the language part of institutional arrangements as they prescribe the type of information that is to be exchanged and the type of arguments that are relevant in the discussion. As stated before, I suspect that these two types of rules are coordinated in the sense that rules which emphasize the participation of expert bodies in a decision also stress the type of information that these bodies can deliver (scientific information).

Rule configurations

It is important to note that the five rule types work configurationally. This implies that if one type of rule changes, the total set of rules is affected, and the rules will have a different effect (Ostrom, 1986: 14). To make things a little bit more concrete: suppose that one type of rules relates to the exchange of information between participants in a decision-making process. Suppose additionally that the decision-making process is aiming to be democratic, which for now can be assumed to mean that ordinary citizens have a position in the decision-making process and that they share full decision-making authority. It is easy to see that even in a situation where ordinary citizens have full decision-making power, the importance of this rule is heavily dependent on other factors. Among these, the degree to which they are informed about the decision-making process and the degree to which they have access to information that is relevant to the substance of the matters that are being decided on. No access to information would render the democratic character of the decision-making process an empty idea.

Ostrom et al. are eager to point out that a great number of institutional arrangements are possible because each type of rule can take every kind of shape or 'value' and because institutional arrangements are often 'nested'. This nesting of institutional arrangements is related to the fact that various arrangements may exist at the same time and be interwoven with each other and interdependent. That this should be the case is easy to see for facility siting as well. Some issues will be left to the market, others to the community or experts. This argument reinforces the notion that discursive hegemony is not very realistic. Despite this realization I will continue to use 'standard' terms, such as 'the market' or 'the community' to describe institutional arrangements in this book. For the moment, I shall be dealing with them as if they were hegemonic and as if entire institutional arrangements have been shaped in line with them, with only here and there some side remarks. In a sense I am therefore developing 'ideal models' that will not be found in reality, but yet can be found to analyze reality.

Which discourses on facility siting exist?

In analyzing which discourses might influence the crafting of institutional arrangements in hazardous waste siting, I will build on the aforementioned work by Williams and Matheny (1995). In their analysis of hazardous waste debates, they introduce three discourses that can be used to describe both the regulation of hazardous waste problems and hazardous waste siting. These three discourses are: a managerial discourse, a pluralist discourse and communitarian discourse[3]. Based on an analysis of the development of regulation in the US, Williams and Matheny define each discourse and place it in a certain historical context.

Managerialism (in the US) is typical for the early twentieth century. At that time, large industrial firms started developing, which monopolized the emerging national markets. There was much resistance against the excesses of modern capitalism, which had created 'poor and restless masses'. These masses themselves sought access to the political system to improve their circumstances (ibid.: 12). The middle class found itself caught in the middle between the poor and the industrialists. It was suspicious of the 'barons of big business' and their motives, but also fearful of the intentions of the masses. It is within this concerned middle-class that the 'Progressive movement' arose. The Progressive movement sought to address emerging societal tensions by largely de-politicizing political issues. The Progressive movement claimed that political quibbling, which was not in the public interest, should be replaced with scientific answers. Inspired by the emergence of scientific management, the Progressives considered it necessary that the scientific method should start playing a greater role in politics. Using the scientific method, public interest could be encouraged. Progressives equated the public interest with an increase in the production and consumption of goods and services. Because experts were better at understanding these processes, they should largely replace politicians as policy makers in the field of regulatory policy. These experts would then start to 'manage' production and consumption processes 'in the public interest'.

Williams and Matheny locate the emergence of the pluralist discourse (again in the US) in the 1930s. Adherents to this discourse are inspired by academic publications on the relation between markets and democracy (e.g. by Schumpeter). They resist the notion of an identifiable public interest, but rather see 'right' decisions in the political process as a result of competition between different interests, just like on a market. The concern of this discourse is not so much delegation of decision-making to experts or managers, but much more with '*a set of fair and open allocational decision-making processes*' (ibid.: 21). Williams and Matheny note that the emergence of this discourse is partly in response to the changed character of regulatory problems. Whereas in the beginning of the twentieth century, the focus was on economic regulation, this slowly changed in the twentieth century towards an emphasis on social regulation. Social regulation is about the negative consequences of economic developments and often involves issues that are difficult to evaluate in monetary terms, involve uncertainty

[3] I have noted a similar distinction in Korsten (1979: 111).

and have redistributive effects. Because of these characteristics, the division between the technical and political parts of policy problems is hard to maintain and the special claim to knowledge on behalf of the public interest by experts becomes suspect.

Williams and Matheny place the roots of the communitarian language in the US at the very start of the American republic. They indicate that this republic was largely founded on an adverse attitude towards centralism, especially towards the central state. Added to that is a rejection of individualism. Instead of centralism and individualism, local self rule within communities was emphasized. The idea of an objectively definable public interest is resisted and replaced by development of a common understanding of the interest of a community. To achieve such an understanding, civic participation was needed and the importance of religious and family ties stressed. Williams and Matheny argue that the communitarian language has been present in debates on social regulation in the 1970s, specifically on environmental policies. At that time, the 'public interest lobby', i.c. environmental groups, used communitarian language to challenge the then existing decision-making arrangements in the US. They argued that powerful economic interests had invaded the political process at the expense of local communities. At the same time, they argued against the special position of experts, as they were not expected to be able to understand all values involved. '*Events like the first Earth Day in 1970 marked the emergence of environmentalism as a powerful social movement reflecting a more general questioning of the worth of technological progress and economic growth, central values underlying both the managerial and pluralist discourses. Environmental groups rejected the ability of experts to capture the values at stake in vital environmental issues or the possibility of balancing such issues against other competing economic considerations. Instead, articulating the democratic wish, they called for participatory democracy, arguing that only by including the people directly in the regulatory process could environmental values be preserved*' (ibid.: 28).

The essential difference between the three discourses are their philosophical basis for justifying decisions (Hunold and Young, 1998) and - related to that - their ways viewing decision-making, either as solving objectively definable problems, as a battle of interests, or as a problem for a community. The reader will have already noted that each discourse implies a preference for certain institutional arrangements *and* thus a certain understanding of what qualitatively good decisions are. In this sense, a discourse connects the dependent and independent variables I was interested in: institutions and decision quality. That this connection existed was also my conclusion in chapter one. In the next sections, I will elaborate on the institutional prescriptions and conception of decision quality that is embedded in each of these three discourses.

Summarizing

In this section I have presented the idea of discourse analysis. A discourse has been defined as 'a shared way of apprehending the world', and as 'a specific ensemble of ideas, concepts, and categorizations that are produced, reproduced, and transformed in a particular set of practices and through which meaning is given to physical and social realities'. In this book I am concerned with the practices of law making on hazardous waste facility siting (which takes place at the collective choice level) and the practice of actual decision-making. At both levels, these discourses might be present and compete for hegemony (which they are unlikely to achieve). The concepts of story line and discourse coalitions are particularly important in this process. It is expected that a certain institutional arrangement will result from the collective choice level debate that must consequently be used at the concrete level (institutionalization). The question is then whether the required way of expressing oneself credibly under these institutions actually holds up in concrete decision-making. Certainly, it seems likely that the official procedures prescribe which entities are relevant to the decision, which arguments they may use and who in the end decides bindingly, but given the context of hazardous waste regulations, this will be challenged. It then becomes interesting to see whether individual decision-making processes feed back into the discourse at the collective level or more generally how challenges of the discourse implicit in the decision rules are handled.

In the next section I will elaborate on the three discourses on facility siting. I will analyze them, using Dryzek's framework for discourse analysis, which involves amongst others a look at actors, their relations, and the rhetorical devices that are being used. This implies that I explore which basic entities are recognized, which relationships are assumed, and which story lines are available for these discourses. I will describe the prescriptions that the discourses contain. Finally, I will also develop a critique of the discourse. In doing so, I develop certain testable assumptions about the relationship between institutional arrangements and quality, which will be explained in the final section of this chapter.

2.2 The managerial discourse on decision-making

Introduction

Dahl (1991) argues that the position of values and moral ideas in modern political analysis is one of the most stubborn problems to be confronted. He describes the problem in connection to the historical and scientific developments taking place from the 19th century onwards. Until that time he argues, moral ideas were considered as objective as empirical knowledge because both were in the end based on '*divine revelation, nature, or self-evident intuitions*' (1991: 119). Since the 19th century, what he calls a more 'subjectivist' view has become dominant, implying that values can not be rationally justified and therefore maintain a problematic relationship with science. This development is connected to the decline in religious faith, which resulted in the loss of an 'Archimedean point' of foundation for values. It is also connected to the triumphant progress of science, which, ironically, given the loss of an Archimedean point of foundation, resulted in the idea that only scientific knowledge could lay claim to any objectivity and that the position of science should be the basis for politics and ethics. The idea was that as science progresses, more and more would become known of the natural and social processes, and more and more will we be able to control these processes, thereby eliminating inequities or failures in the systems surrounding us. In its most extreme form, positivism (as this movement is called) leads to determinism, implying that human behavior and choice are determined by natural laws, which in the course of time will all be discovered. And once discovered, these laws will allow us to control, or manage, our surroundings. As we understand these natural laws, there will be no need for ideologies, as these are only particular and subjective interpretations of social phenomena that at some point will be *properly* and not tentatively understood. As an implication, administration and government should not be run by politicians or the people, but by those who understand science, who know by which natural laws to manage society. Hence the name managerialism.

Positivism in its purest form has since long ceased to exist, but the idea that expertise should be extremely important in decision-making is still very much alive. Fischer (2000) for instance, indicates that many 'political' choices are actually made in 'policy communities', which he equates with decision-making by experts. Today, most managerial thinkers acknowledge that values can not be scientifically discovered, but need to come from outside the scientific process. The standard view is then that elected politicians should 'provide' these values while managers do the rest. Williams and Matheny write that as early as the beginning of the twentieth century this was already the ideal of many managerial thinkers. Because adherents to managerialism are resistant to viewing issues as being political, one can imagine that the role of politicians was to be as limited as possible. Scientifically-versed administrators are better at taking decisions according to managerialists. This is because they are the ones who can understand the technically complex issues at hand. Also, they are the only ones that could be expected to consider the common good, whereas others are expected to pursue their own interest. Thus expert decision-making processes should be isolated, as much as possible, from interest-group pressure and citizen participation. Participation by 'lay people' is not considered necessary, or should at most be limited to the political process that defines the goals to be attained by the scientists.

Managerial thinkers do not consider the possibility that the achievement of the broad goals by administrators requires many 'political decisions'. In fact, they purport the neutral and scientific - and thus not political - criteria of efficiency and dependability as the yardsticks for policies. The philosophical basis of the discourse is located in *utilitarianism*. Utilitarian reasoning implies that to study the desirability of actions, one should focus the consequences of these actions, and that these consequences must be evaluated against standards of what is desirable. Classical utilitarian reasoning holds that such policies should be chosen which maximize some chosen value (for instance money); the 'good' policy is the one that brings the greatest net gain for the entire aggregate, or 'the greatest good for the greatest number' (Dahl, 1991: 125-126; Rawls, 1971: 22-27).

Briefly summarized, a study on the evolution of managerial thought in the US revealed the following patterns of thought among managerialists (see Williams and Matheny, 1995: 13-17):

1. It is believed information and education are the keys to successful regulation. The ideal is a society where class and ethnic differences have been transcended via the application of neutral, scientific principles.
2. Conflicts during the decision-making process are not considered legitimate. In order to be able to make the right decision, the decision-making process should be isolated from 'petty political squabbles of self-interested groups'.
3. The basis of state legitimacy is to be found in the production and consumption of goods and services, not in the nature of democratic government. Economic efficiency is another term for the public interest. Consistent with this, the original emphasis on engineering expertise later shifted, or was supplemented with an emphasis on economic expertise.
4. Democracy is viewed as a mode of decision-making designed to arrive at the public interest: it is a means to an end, rather than an end in itself. As economic issues are best handled by 'hardheaded technocrats', democratic decision-making is not the best way to arrive at the public interest.

How do experts themselves go about their decisions? The managerial discourse emphasizes that decisions should be 'rational' within the politically prescribed set of goals. 'Rational' decision-making in this sense has been described many times, for instance by Allison (1971: 29-38). The basic concepts of a model of rational action are goals and objectives, alternatives, consequences, and choice. The goals and objectives of a decision-maker provide the starting point for the analysis. Alternatives are the possible courses of action that the decision-maker has. Consequences relate to the fact that costs and benefits are attached to certain alternatives. Choice is the process of selecting an alternative. The choice is rational, if the alternative is selected with consequences that rank best in terms of the decision-maker's preferences.

Table 1. Elements of the managerial discourse

Basic entities recognized:	• Market economy. • Expert administrators. • Expertise. • Politicians. • Interest groups. • The public interest.
Assumptions about relationships:	• The public interest is served by the production of goods and services. • Politicians set goals for administrators. • Administrators obey politicians but should be isolated from pressure groups. • Administrators hold the relevant (=scientific) knowledge for decision-making.
Agents and motives:	• Politicians know about values only. • Administrators strive for the public interest. • Pressure groups are not interested in the public interest.
Rhetorical devices:	• The decision must achieve the greatest good for the greatest number. • Petty political squabbles of self-interested groups. • Neutrality and science. • The greatest good for the greatest number.

Note that adherents to this discourse see the scientific method as a good way to solve problems. This implies that to solve problems, use must be made of scientific knowledge and information, that is, knowledge or information that owes its origin, testing, degree of verification, truth status, and currency to distinctive professional techniques (compare Lindblom and Cohen, 1979: 12). Scientific knowledge and information tend to be organized and carried forward in written texts. To be accepted as scientific, knowledge and information must transcend the limits of a particular situation or culture and be generally applicable (see Fischer, 2000: 193-208). Partly because of the expected general applicability of scientific knowledge, managerialists do not foresee disagreement between experts about the predicted consequences of certain alternatives nor about the ranking of alternatives. This is because the scientific method discovers truth and to disagree about the truth would be silly.

Using Dryzek's method of discourse analysis, I can present the managerial discourse as depicted in table 1.

Institutional implications

The proper institutional arrangements to regulate decision-making on hazardous waste are easily discernable in the above. The *aggregation rules* under a managerial discourse will be heavily influenced by the notion that there is a discoverable public interest. This public interest is congruent with a situation where the greatest number experiences the greatest good, so the correct decision is the one that maximizes utility. Judgments of which decisions constitute the greatest good to the greatest number are to be made by experts, who are in a special position, as they are aware of the prerequisite knowledge. As described in the above this expertise should be both technical and economic in character because combined they allow for application of the scientific method and rational decision-making (calculation of the greatest good). Since there can only be one 'truly best' choice, consensus about this between experts is needed, since lack of such a consensus would imply that the facts have not been properly established yet.

Authority rules under a managerial discourse would be heavily influenced by the separation between goal setting (politics) and utility maximization (administrators) and by the idea that decision-making should be isolated from interest groups. Politicians would have to set the goals and objectives of a policy, but of course lack the necessary technical and economic knowledge to choose the alternative that brings the greatest utility. In hazardous waste decision-making, this could imply that selection of waste treatment techniques and the selection of particular sites should be left to the experts, whereas politicians formulate broad goals. Isolation from interest groups implies among others, that for instance, the waste industry should not be allowed to bring forward siting proposals, as their proposals would be based on self interest. Only bureaucrats can be expected to put forward proposals that are not influenced by subjective factors such as the desire to maximize profit.

The *boundary rules*, as a correlate of the authority rules, would emphasize the necessity that experts are isolated from pressure. Politicians would give them 'homework' in the form of goals and objectives to be reached and after that should remain silent. Citizens are not welcome in the decision-making arena as they would also tend to speak for their own interest. *The information rules*, finally, can again be expected to reflect an emphasis on objective, scientific knowledge. Within the community of experts, this information should flow freely, so that indeed the best decision can be taken. So *within* the expert community, rules that normally apply to scientific research must also apply (information must be freely accessible to all, methods used to derive the information must be explicated). *Pay-off rules* will be such that the state retains and pays for the experts because they need to be independent from special interests.

Table 2. Rule types and values they obtain under the managerial discourse

Rule Type	Prescription for value under managerial discourse
Aggregation	• The decision must achieve the greatest good for the greatest number. • Expert consensus required.
Authority	• Politicians set goals, expert administrators decide about the rest.
Boundary	• The goal-setting process is open to politicians. • Other decision processes are closed to everyone except expert administrators.
Information	• Scientific information is the basis for the decision. • Information is to be shared between experts.
Pay-off	• Experts are paid by the state.

Some limits and a critique of the managerial discourse

Before proceeding with a description of the institutional implications of the two other discourses, I will briefly summarize some aspects that are seen to cause particular problems to the managerial discourse. These aspects are first of all related to the utilitarian logic underlying the discourse, and secondly related to the emphasis on science embedded in it.

To start with the first aspect. Utilitarianism has been attacked by many, including John Rawls, who '*made a bigger splash in political philosophy than previous critics*' (Dahl, 1991: 126-127). The thrust of Rawls' criticism as he develops it in his 'theory of justice' is that people have inalienable, inviolable, natural rights that can not be overridden by the requirement that utility for an aggregate must be maximized. He connects his argument to the believe that some things are right, good or just, even though they might not produce the greatest happiness for the greatest number (see Dahl, 1991: 126). Obviously, this criticism could make sense in the sphere of hazardous waste facility siting, where attempts may be made to concentrate certain health risks upon a few for the benefit of a collective.

The idea that scientific knowledge should (largely) determine choice has been attacked too. Williams and Matheny (1995: 17-19) argue that social regulation (including hazardous waste policies) concerns itself with the limitation of economic activity which causes unwanted consequences. The managerial discourse owes its origin to a period when, in contrast, economic regulation was the major theme. The difference between social and economic regulation is explained as follows: '*The issues in airline or utility regulation, for example, involve questions concerning who will bear the monetary costs of monopoly, what are the economies of scale, and how they can be fairly distributed among consumers and producers*'. In contrast: '*Environmental, consumer, health, and safety regulation all entail slippery quality-of-life questions that make it difficult, even in theory, to capture the value of the resources at issue*' (ibid.: 19). This implies that the calculation of the 'greatest good for the aggregate' is really impossible. What it also means is that scientific knowledge can only go so far in determining outcomes. Dahl (1989: 69-70), in discussing the role of experts as policy makers, writes that both instrumental knowledge (possessed by the expert) and moral understanding (possessed by anyone) are necessary for policy judgements, '*neither alone can ever be sufficient*'.

But science does not only fail because valuation is impossible. It is also because effects of hazardous wastes per se are subject to scientific uncertainty, i.e. because there is a time lag between causes and possible health effects. Probability estimates of the relation between exposure to substances and effects can differ by several orders of magnitude, for instance depending on assumptions that are made about the relation between animal tests and effects on humans. Note that this often implies that scientists will disagree when they estimate effects. Fischer (2000) and Lindblom and Cohen (1979) have extensively attacked scientific information as a basis for policy decisions. These authors claim that 'scientific knowledge' is not as special as it is portrayed to be. The reasons for that are manifold. First, scientific knowledge owes its existence to common sense, which is certainly not only available to scientists. Second, scientific knowledge only covers a small subset of the issues to which common sense speaks. Third, the requirement of general applicability (needed to qualify as scientific) is not always functional as some things work in local situations but may not work elsewhere. The authors just mentioned speak passionately about the need to base decisions not on scientific knowledge, but on ordinary knowledge (more on that concept when I describe the communitarian discourse).

Precisely because scientific information only goes so far in deciding issues in social regulation, the distinction between 'political' and 'technical' issues is not always clear. For instance, risk assessment is often based on a host of assumptions, including assumptions of the behavior of human beings in a particular situation. Such assumptions are not neutral, but really signify a certain level of trust in these human beings. By saying that risk assessment has demonstrated that there are no significant risks, one is really asked to accept the assumptions in the assessment and to trust the human beings who operate a certain facility. Dahl (1989: 69): '*instrumental judgements depend on assumptions that are not strictly technical, scientific, or even very rigorous. Often these assumptions reflect a kind of ontological judgement: the world is like this, not that, it tends to work in this way, not that*'. Whether or not trust should be given to the experts who made these assumptions is a political question, not a technical one.

This connects well to the observation of many that 'science' is in a sense abused in environmental conflicts (see e.g. Burns and Ueberhorst, 1988). 'The' public and politicians tend to expect a high degree of certainty from scientists. If asked to advise on certain complex environmental issues, scientists are not expected to qualify their statements too much, as this undermines their own credibility in the eye of a critical public. Therefore, scientists are stimulated to downplay uncertainty, which they would probably not do in the 'safe' confines of the scientific community. In extreme, but also not that rare, cases, the tendency to downplay uncertainty can be used to advance the agenda of particular interests. Burns and Ueberhorst (1988) are especially concerned about the possibility that people commissioning research exert pressure on scientists and influence the outcomes. Even subtler - but perhaps more common - is the influence that those commissioning research have on the questions that are being addressed in studies. Very often, these questions presuppose the answers. Although under a managerial discourse, experts are organized into the state and will be paid for by the state, this does not exclude the possibility that interest groups influence their behavior. Furthermore, it is often pointed out that the state serves environmental degradation if need be (e.g. Brubaker, 1995).

2.3 The pluralist discourse on decision-making

Introduction

The pluralist discourse differs considerably from the managerial discourse. It does not assume that there is such a thing as a public interest that is discoverable a priori. Instead, '*good regulatory policy emerges (. . .) from policy-making procedures that allow open access to all affected interests. In this way, regulatory policy balances the vectors of political pressure brought to bear by organized groups*' (Williams and Matheny, 1995: 20). As there is no public interest, the 'right' policy must emerge from a clash between the parties involved. 'Good' decisions are a result of individual interests that are presented with a certain force. The greater force a particular interest exercizes, the greater its impact on decision-making will and must be.

The philosophical basis for pluralism is twofold. In the first place there is a strong emphasis on individualism. If there is something of a collective interest, it is the sum of individual interests and not something outside the individual members of society and their interests. Dahl (1989: 74): '*If we accept the human-centered view, then, to understand the general good of a collectivity requires knowledge of the interests of persons, and nothing more*'. Therefore, for pluralists the 'common' or 'mutual interest' are more proper terms to use than the 'public interest'. The position of individual interests in pluralism is remarkable. Certain interests on the one hand are assumed to be mutually exclusive (they are competing) but at the same time be comparable to each other, because otherwise no deal would be possible. This leads me to the second basis of pluralism: pluralists pay much attention to distributional issues. The risks a facility brings can perhaps not be redistributed, but the inequality of the risk distribution is acknowledged and an attempt must be made to compensate for the costs. Therefore, the basis of a good decision lies in *distributive equity*: the decision that is taken is a just decision when all interests are satisfied in the distribution of costs and benefits. It is considered unjust to force a facility on part of a collective for the benefit of others or for the greater collective (Hunold and Young, 1998: 84).

If decision-making is a matter of balancing interests, various actors could be engaged in this act. Balancing of interests could be done by the parties themselves, by representative institutions, or by the courts. Parties are expected to resolve the issues affecting them independently since they are considered to know their own interest best. When they interact, the emphasis is on the information these parties bring to the negotiating table. There is an incentive to use this information strategically, for instance, by remaining silent about intentions or by filtering out unwelcome information. Parties are largely considered responsible themselves for discovering strategic use of information and countering information with their own findings. Offering financial compensation may be a good way of coming to an agreement, but if the parties cannot agree they may present their case to a third party, for example a judge. The courts are regarded as good decision-making institutions because the parties present will try and build a case that is as strong as possible for themselves, whereas the judge can decide the 'right' balance between these competing views.

Many pluralist thinkers acknowledge that representative institutions, such as Parliament, must also play a role in society. In fact, in political science, pluralism is a distinct line of thinking on how democracy can and must function. I will call these thinkers 'political pluralists'. The pluralist line of thinking on democracy was developed by Schumpeter (see Hisschemöller, 1993) and taken further by others, including Dahl (1991). In describing the functioning of representative democracy, Schumpeter modeled representative institutions after markets. He placed much emphasis on a continuous competition between individuals who want to acquire political power through the popular vote. These individuals present the public with certain proposals and request their support by means of a vote. If they get it, they are allowed to rule for some time. The essence of a democracy, in this pluralist vision, is that politicians must be accountable to the public and therefore replaceable if they do not achieve their agenda or if the public considers another agenda better. In this model, ordinary citizens should remain inactive, except for issuing a vote every now and then (democratic consumerism). The reason why passivity of the public was considered desirable is that Schumpeter considered citizens incapable of reaching a well-founded position in political matters (Hisschemöller, 1993: 11-13). Closely associated with this argument is the idea that citizens would not want - and could not - get involved in political matters too much. Indeed, having all citizens involved in all problems at a certain time seems somewhat too demanding. Voting is then perhaps a good alternative, full-time representatives can be elected who actually take the decisions.

Do political pluralists believe that these representatives must take all decisions? Hisschemöller argues that the traditional pluralistic view of decision-making lends only very qualified support for representative democracy. This is related to the fact that Schumpeter considered democratic decision-making to be a method for decision-making, i.c. a means, not an end. He considered this means of decision-making only practicable for relative simple issues, where there is a certain value consensus. For other types of problems (were value consensus is lacking) decisions should not be made by the democratic method. Instead, a non-democratic agency should decide issues (ibid.). Modern pluralist writers such as Dahl tend to be more accepting of citizen participation than the old master, especially when people want to defend their own interests. However, here too fear exists over what people will demand: citizens are expected to overestimate the importance of their own interests and become too radical. So here too, issues that involve values are not to be solved by the democratic method. This does, however, not lead a 'modern' pluralist to the conclusion that the democratic process should be abandoned. Rather, it leads to the advice that people formulate their demands to the political system in moderate terms. Pluralists reject the notion of 'absolute values'[4]. Democratic debate about absolute values is considered impossible and is expected to lead to overly rigid positions. Pluralists point out that any policy goal can only be achieved at certain costs, for instance, because other policy programs must be reduced somewhat. These costs, and the expected benefits, must be considered. Reasonable discussion is only possible when people are willing to perform such comparisons and not consider their values 'absolute'. As a correlate to this position, modern pluralists are cautious of drastic changes in policies informed by the idea that some 'absolute value' must be protected. Rather, policies should change incrementally.

Newton (1976: 76-85) has summarized the assumptions underlying pluralist thinking. The reader will note that these assumptions go farther than just the view that decision-making should be based on a clash of interests. In order to produce the 'right' result, many preconditions must be met. Typical assumptions by (political) pluralists are:

1. Although political resources may not be distributed equally among the members of society, the inequalities are not cumulative. Because so many types of resources determine political influence, the lack of one resource (for instance, money) can not, in a pluralist society, lead to a lack of influence. All political groups have weight and leverage in the political system and therefore nobody is left out.
2. The modern industrial society contains such an enormous diversity of organized interests that any given political issue is likely to attract the attention of two or more groups whose interests are involved and fight out the issue between them. The proliferation of groups is considered a long chain of actions and reactions between interests.
3. The formation of allies and enemies rotates with the issue, the time and the place at stake. Because political resources are distributed with non-cumulative inequality, proponents (or opponents) of certain decisions need allies. The pluralist assumption is that parties do not form stable coalitions, but that parties that have cooperated on one issue may just as well fight over the next issue.
4. A connection is assumed between 'social pluralism' and 'political pluralism', meaning that many social groups (sport, hobby, ethnic, etc.

[4] I have not systematically studied how this relates to Rawls' point of view that some things are right, good or just, even if they do not provide the greatest happiness for the greatest number of people (see earlier). Perhaps these 'things' relate to 'basic rights' with an absolute value, whereas 'other' rights must not be formulated in absolute terms. This thought of course makes a proper distinction between the two classes of rights necessary, which would seem a rather difficult enterprise.

associations) will share a common interest that they are willing to defend in the political arena.

A discourse analysis of pluralism might look something like that presented in the table 3.

A first critique of pluralism

The assumptions that underlie pluralism (especially its variant as a political theory) can be summarized under the heading equal opportunity, political equality and freedom. The assumptions can be - and often are - attacked, even by some who believe that the essence of good decision-making is a clash of interests. The critique of pluralist thinking mostly centers around its assumption that social inequalities do not translate into political inequalities (see e.g. Newton, 1976).

Williams and Matheny indicate that there are two strands of pluralism, which have different views on this issue. What they call conservative pluralists tend to emphasize decision-making by people themselves, especially on the market. Conservatives see the market economy as an extremely efficient social allocation mechanism for resources and this mechanism must therefore be left as undisturbed as possible. According to conservatives, the existence of markets is not a threat, but a precondition for democratic systems because markets create the individual freedom needed in a liberal democracy. The role of the state should be minimized and may only step in when the markets fail. There should be effective limits on government interference; the proposed criterion is the market criterion, which implies that costs of regulation should not exceed the benefits. Williams and Matheny note that the application of this criterion often implies a resort to experts schooled in economics (1995: 21-22). Therefore, there is a certain affinity between conservative pluralism and managerialism.

Table 3. Elements of the pluralist discourse

Basic entities recognized:	• Market economy. • Private interests. • Individuals. • Interest groups. • Democracy. • Politicians. • Courts. • Due process, openness. • Distributive equity.
Assumptions about relationships:	• The general good is a function of private interests. • Individuals want to achieve their private interest and know this interest best. • Decision-making is a clash of private interests, in which actors use information strategically. • Private interests are equal and comparable to each other and can be weighted against each other. • Individuals can do weighting of interests themselves, or leave it to the courts or representative institutions. • Individuals with similar private interests combine to form interest groups. • Interest groups influence politicians. • Politicians compete for support from individuals and interest groups. • If politicians lose support from individuals and interest groups they must be removed. • Politicians will make decisions that maximize their support and therefore they will give weight to the interests that are presented most ferociously. • The democratic system can not decide issues that involve absolute values. • Social equality translates into political equality. • Democracy can only resolve conflicts between values that are not absolute.
Agents and motives:	• Politicians want to stay in power. • Individuals can not and do not want to get too involved in politics, but will vote. • Individuals want to defend their private interests. If need be, they found interest groups that influence politicians or go to court to defend their interest.
Rhetorical devices:	• People are the best judge of their own interest. • Due process. • Fair fight. • Social diversity guarantees the right outcomes. • Freedom and equality.

Liberal pluralists do see tensions between democracy and capitalism and are sensitive to the idea that economic inequality may lead to political inequality. They therefore envision a much stronger role for the state in decision-making, partly by erasing the limits to process.

Pluralist arrangements

It seems to me that the distinction between a conservative and a liberal (or progressive) variant of pluralist discourse is helpful, also because adherents to the two variants would emphasize different institutional arrangements. Although both variants deny the existence of an objectively definable public interest, they disagree over the question of what is the proper mechanism for interest aggregation. Conservatives would stress the outcomes of the market economy, or create an artificial market for waste facilities (for instance in the form of an auction) - but then only when this solution is more efficient than the solution the market provides. Liberal pluralists would rather emphasize the need for government interference to guarantee due process. Liberal pluralism then implies the introduction of a supervisory layer over market decisions. In line with pluralist logic, such supervision would be exercised by representative institutions or by courts (not by managers of course). In the end, a third party should decide, based on a weighting of interests that are presented to it. The following two tables and the explanations following them present my idea of how both advocates of conservative pluralism and liberal pluralism would view adequate siting institutions.

Table 4. Institutional implications of conservative pluralism

Rule Type	Value under conservative pluralism
Aggregation	• Market parties must agree with each other on an exchange of resources between each other.
Authority	• The authority to decide rests with the parties involved (affected) themselves.
Boundary	• Only parties that have property rights, which they can exchange, are allowed to the market.
Information	• The decision should be based on information brought forward by the parties. • Information may be used as the parties see fit.
Pay-off	• Parties pay for their own activities.

The market reliance of conservative pluralism is the issue most strikingly obvious in this set of rules. First of all, *aggregation* rules provide that deals between market parties who exchange resources are the prime mechanism for good decisions. Parties themselves are considered the best judge of their own interest, and it is up to them to decide which amount of compensation they demand for certain resources. If they find the compensation acceptable, the decision is considered ethically acceptable. At the societal level, the functioning of markets is supposed to lead to economic rationality, which is about the 'purification' and 'extension' of the economy itself (Diesing, 1962: 23). Purification implies that activities must be performed as efficiently as possible, extension implies economic growth. Diesing (ibid.: 9-13) observes that there is a close relationship between economic and technical rationality (which is perhaps more in the managers' ballpark). According to Diesing, technical rationality relates to efficient ways of achieving acceptable ends. Technical rationality is thus purely about making the achievement of a certain goal (e.g. production of a good) more efficient. Such improvement may continue endlessly, until a society excels in the production of that good. This can only be translated into economic efficiency if the purpose of producing that good can be compared to the purpose of producing another good, and resources can possibly be re-allocated. If this actually occurs and more can be produced of another valued good, then technical progress translates into economic progress and the economy is extended and purified (ibid.: 23). Decisions that enhance economic growth are thus economically rational.

Authority rules forbid the government to get involved in decision processes, as the government is no better judge of the interests of individuals than these individuals themselves. Note however that even the strongest anti-government thinkers do acknowledge a legitimate role for the state in the definition and enforcement of property rights. These property rights are needed for any market to function. *Boundary rules* stress ownership as the logical prerequisite for participation in the decision-making process. As land-ownership and the property rights following from it are easy to identify as interests being affected, this is likely to be an important boundary rule in the decision-making process.

Information rules under a conservative pluralist discourse are in line with the above and geared towards information that parties hold or obtain themselves. Without government interference in the siting process, there is a reliance on the ability of individuals themselves to detect gaps, holes or insincerity in the information presented to them. This is somewhat parallel to a court-case (but then without a judge) where the parties have an incentive to present 'their case' in the most favorable way (for themselves), with 'the market' being the mechanism in, finally, determining (reputations of the) reliability of the parties.

As expressed before, liberal pluralism is a discourse that is rather more sensitive to possible tensions between the market and democracy. This implies that some institutional measures will be proposed which are intended to redress the balance somewhat. These measures are likely to be focused on leveling the playing field that was made unequal by differences in economic power. Attempts to redress the inequality are to be achieved by a degree of government interference in the process and/or the enhancement of the participation of disadvantaged groups. Government interference implies that a certain degree of authority is taken away from the private sector and placed in the hands of the state. If this happens in a fashion that is consistent with pluralism, then this could infer various institutional implications. The decision (on the government's part of the total decision) could either be placed in the hand of elected representatives or in the hand of the courts. Attempts to enhance the participation of disadvantaged groups can then take the form of rules regarding free access to the decision-making process in court or the representative institution. Alternatively, there can be attempts to increase the access to information and resources of the disadvantaged groups. Note that these attempts can also 'stand alone' in the sense that the government does not take away authority from the private sector, but does assist certain groups. Three types of resources are most likely to be distributed unequally and may require redistribution: information, time and finances. For all three resources liberal pluralists are likely to incorporate a range of compensatory measures such as free access to information, provision of experts to act on behalf of citizens, financial support for citizens, the right to take leave from work to participate, etc.

Quite likely, the decision-making process will, under the liberal pluralist discourse, exhibit a layered character, where each additional layer checks whether all interests have been accurately weighed in the previous decision process. The intensity of review may differ greatly, partly depending on how certain balances between the private and the public sector have been struck, partly based on the balance of powers between the three branches of government.

Table 5 provides an attempt to represent the possible institutional implications of liberal pluralism.

Table 5. Institutional implications of liberal pluralism

Rule Type	Values under liberal pluralism
Aggregation	• The decision-making process is divided in phases: • For the initial decision, market parties must agree with each other on the appropriate amount of compensation. • After that, review of the initial decision takes place judging whether the various interests involved have received due process.
Authority	• Market parties make the first decisions. • Review of these decisions takes place in representative institutions or in the court. • All interests affected by the decision may present their case to the representative institution or in the courts.
Boundary	• All affected interests may participate. • Attempts are made to include interests that had been left out initially.
Information	• The decision should be based on any information brought forward by the parties. • Measures are taken to guarantee freedom of information and/or to see that participants have sufficient expertise to judge information brought forward by the proposer.
Pay-off	• Some parties may receive financial assistance in order to make their participation possible.

Another critique of pluralism

Much criticism is mounted against pluralism. It is often argued that the individualistic foundation of the discourse tends to lead to an ignorance or under valuation of common pool resources. In chapter one I already discussed Mathews' (1995b) comments on this. Her critique was that in a system based on individual preferences anybody can declare environmental protection in his/her interest, but that this does not count for much in the end,because it would then only be 'in the interest' of a single person. Representation of interests in this scheme depends on individuals bringing the interest forward and it seems questionable that all interests are always brought forward. Indeed: why would anyone bring forward 'the' environmental interest and why would they organize to protect this interest? Doesn't a society need a certain sense of 'citizenship' to preserve the integrity of the society as a whole? If self-interest is to be the basis of decision-making and is self-justifying, then it is unlikely that citizens will think together in solving many problems.

More critique is possible. As a consequence of their preference for incremental decision-making, pluralists assume a great deal of ratio in the situation that exists at a certain time (see Hisschemöller, 1993: 11).

Preferably then, policies must be made in a relatively stable arena, where the leaders of established interest groups produce incremental changes to policy (Williams and Matheny, 1995: 43). But, say Williams and Matheny, social regulation is about altering the status quo, if the status quo was not working. And limiting policy change to incremental change only may be dangerous if fundamental changes are necessary. They also point out that hazardous waste siting, mobilization of citizens in groups is likely to be shifting and unstable. These groups are not likely to accept stable interest groups as their representatives nor the acceptability of the status quo.

From a moral point of view, the idea that people should be their own judge of their own interest is sometimes considered problematic. Hazardous wastes for instance may cause effects that are technically and scientifically difficult to understand and predict and therefore there may be barriers to accurately understanding one own's interest (are citizens competent enough?). Secondly, people in an economically disadvantaged position may be more willing to accept certain outcomes than those who are wealthy. It would then become possible that differences in wealth translated into differences in (health) risks, which tends to meet moral objections (standards of fairness). This fundamental point applies to both liberal and conservative pluralism.

Some attention to one 'pluralist institution', the court, will highlight a critique that may have wider ranging meaning for the pluralist discourse. Especially the fact that in court, parties presenting only their own view on a certain case is sometimes expected to lead to problems. For certain, if party A and B are before the court, and party A withholds information from the court that may be beneficial to party B, then one could say this is not very efficient and information gets lost. Taking this example further, it may well be that many aspects of the matter are not brought before the court, whereas these aspects may be highly relevant. Diesing has equated 'legal rationality' with the ability to prevent conflict or to provide solutions for acute conflict when other methods of conflict resolution fail (1962: 146). Prevention of conflict must occur by rules, and these rules provide the framework for social and economic relations by defining rights and duties that belong to certain relationships between members of society. The rules must be clear to those that are supposed to follow them and must be internally consistent. When conflict arises, legal rationality requires that the courts clarify the rules and end disagreement. In so far as a legal system achieves these purposes, it is legally rational. These requirements are hard to achieve in practice however. It is often noted that the 'adversarial' nature of legal proceedings does not really generate a sense of a common interest in the parties. Because of this, closure of decision-making processes through the courts might be hard to achieve. Often there is a second round of judicial review, and persons are stimulated to use these options fully, i.e. by lawyers. The fact that courts often indeed restrict themselves to the clarification of rules (and therefore not really decide the issues between parties) increases this tendency. In addition, Dahl (1989: 188-191) presents a normative argument against decision-making through the courts. He thinks that judicial review implies 'quasi-guardianship' as one person (the judge) is placed above the parties. Judges are not democratically elected, and therefore, the greater their authority, the more they reduce the scope of the democratic process.

Earlier I presented Newton's summary of the four assumptions undergirding pluralism. I do not wish to present a full elaboration of all available empirical evidence for or against these assumptions, but find that Newton's study of politics in Birmingham at least presents interesting indications. Newton reports that the resources, which are most closely associated with political activity, are distributed with cumulative inequity in Birmingham. In simpler terms: interest groups tend to have either high or low amounts of all relevant resources at the same time (large membership is correlated to a large income, the existence of a paid staff and being well established, see Newton, 1976: 77). Newton also attacks the idea that there is competition between various groups. He finds that interest groups themselves have difficulty identifying other groups that are acting in opposition to them (ibid.: 79-90). Therefore, the idea that on every issue there will be force and a counter force creating a balance, may not be empirically justified: '*the pluralist system of action and reaction, organization and counter-organization, proposal and counter-proposal is a very diffuse and underdeveloped one which fits where it touches, fails to touch in a number of places, and which may well have large holes in it*' (ibid.:80). Newton does not find much coalition building as predicted by pluralist theory. However, even if there had been so much coalition building as predicted by pluralists, the pluralist belief that this should lead to some sort of balance between interests is still doubted (recall the argument of 'power-trading' discussed in chapter 1). Finally, Newton finds that social pluralism does not translate into political pluralism, as membership and activity in voluntary organizations is pre-eminently a middle-class activity.

Of course the differences between conservative and liberal pluralism have implications for the criticism the two versions meet. Conservative pluralism is more vulnerable to the traditional critique of pluralism than its liberal variant. Conservative pluralism seems almost blind to the effects economic power has on the liberty of taking decisions and its reliance on the market as a correcting force is naive in the sphere of hazardous wastes, where open competition is questionable and the quality of treatment often seems to require supervision. As will become transparent in the next chapter, there are severe problems in establishing responsibility for negligent behavior in the waste market, for instance because many firms do not exist anymore once the consequences of their behavior become apparent. Also, as has been argued many times already, private firms are becoming increasingly mobile and less and less rooted in local communities. One of the main purposes of local governments is likely to be maintaining employment and although hazardous waste facilities do not generate much employment, the amount may still be considerable on a local scale and therefore, the market may indeed function as a prison.

In concluding this section I briefly want to point out that not all 'polyarchies' are exactly similar to those in the United States, the country that Williams and Matheny write about. It has been observed for instance that the Netherlands have a (past) tradition as a 'consociational democracy', which implies that deep cultural divides in the country were solved by political handiwork from those at the top. Under such a system, each cultural group has veto power in order to maintain integrity of the society and outright competition is restricted. Also it has often been pointed out that 'interest group liberalism' in the United States is quite an extreme case in the sense that it is highly adversarial. There are many countries with a more or less 'corporatist' tradition, which implies that the state hands out a certain part of its authority to affected interests. The interests then effectively start making policy, or advising government on policy. For now, I will assume that these other types of polyarchies still work under pluralist logic and decision-making there is not inspired by a distinct discourse. I feel this is justified because even in such different systems, the 'weight' of certain interests in decisions is very much dependent on the force with which they are expressed. This works directly (in the decision process itself) or indirectly (in the design of decision-making arrangements, where actors without power will be withheld access). If not completely inspired by the pluralist discourse, then I expect certain elements of the two other discourses distinguished here to be embedded in institutional arrangements. Especially corporatist thought is sometimes considered to be inspired by communitarianism, the discourse I will consider.

2.4 The communitarian discourse on decision-making

Introduction

The communitarian discourse is in certain ways similar to the pluralist discourse, but different in others. Like pluralists, communitarians do not believe in objective truth. Unlike pluralists however, they do not believe that the determination of the good decision should be the result of clashes between individual interests. Instead, they would emphasize the role of the community, which is a public entity, small enough to allow (at least potentially) repeated, face-to-face, interaction over issues involving both individual and collective interests. A central idea to communitarian thought is that the definition of self-interest by individuals is forged through these repeated face-to-face interactions and political activities. By participating in deliberations, citizens become aware of the relation between their own self-interest, the self-interest of others and the collective interest (Williams and Matheny, 1995: 47-49; see also Kelman, 1991). '*Through public discussions citizens often transform their understanding of the problem and proposed solutions, because public communication forces them to take account of the needs and interests of others and may also give them information that changes their perceptions of the problem and alternatives for solving it*' (Hunold and Young, 1998: 87). The changes of perception taking place are expected to lead to consensus in the local community on what the problem is and what are the proper alternatives for solving it. This consensus, if achieved in a coercion-free way (Mathews, 1995b), is the criterion for a good decision according to adherents of this discourse. The basis for a good decision is thus deliberative or communicative democracy.

Communitarian thought and hazardous waste decisions

The communitarian discourse may be particularly relevant to social regulation, because the effects of such policies fall on communities. Williams and Matheny (1995) describe the effects of communitarian thought in the context of the development of social regulation since the Second World War. Referring to Lowi (1979) and Wilson (1978) they argue that the coming about of social regulation in the US is best described as interest group liberalism. Interest group liberalism infers that legislative bodies have delegated increasing amounts of policy-making authority to administrative agencies as the tasks of the modern state expanded (the state became more liberal in this sense). The consequence of this was that policy-making became dominated by interest groups that have the political resources to gain access to administrative agencies (Williams and Matheny, 1995: 30-31). The term 'captured agencies' must be seen in this light. The passage of legislation under interest group liberalism is a puzzle as those who have to bear the costs have a strong incentive to organize and those who may benefit have little incentives to do so. As a consequence, the best organized interests (businesses?) with a clear stake in the outcomes of the policy-making process gain most influence over policy. This may lead to policies that are favorable to them, but much less so to specific communities that have to bear the costs of a weak environmental policy.

Williams and Matheny find that mainly new groups (social movements) that mounted an attack against the perceived negative implications of interest group liberalism in the 1970s have used the communitarian discourse. The communitarian discourse offered them a powerful rhetorical instrument as it appeals to the interest of the community, which was widely felt to be under threat. Communitarianism is appealing to many as it is characterized by a suspicion of central institutions (big government) and a yearning for the 'lost values' of attachment to a social and political community. The appeal to direct democracy and local self-governance was certainly appealing to communities resisting some developments such as roads. Under communitarian thought, a community can only be expected to accept certain risks associated with such developments if there is informed consent that is achieved without coercion. Communitarians suggest that conflicts of interest within communities are illusory. This is because an enlightened citizenry that governs directly on its own behalf can discover their common interest: '*conflicts of interest among the people will disappear in the process*' (ibid.: 27-29). This is not to say that individuals should submerge their autonomy in the collectivity a priori, but that because of their active participation, they will come to see the shared interest of the community.

The above makes clear that communitarianism is closely connected to the ideal of local direct democracy. The table below presents an analysis of the discourse.

Table 6. Elements of the communitarian discourse

Basic entities recognized:	• The community. • Self-governance. • Dialogue, deliberative democracy. • Direct democracy. • Common interest.
Assumptions about relationships:	• The community must be self-governing. • Members of the community are engaged in coercion-free relationships and dialogue. • Members of the community meet face-to-face on a daily basis. • Conflicts in the community are superficial.
Agents and motives:	• Members of the community care about each other (and about nature).
Rhetorical devices:	• Harmony in the community. • Self-governance. • Direct democracy. • Dialogue.

The reader will note that the motives of members of the community are sometimes expected to relate to nature. It is here that authors such as Mathews and Dryzek fit in, with their emphasis on local communities that have a special relationship with nature. However, not all communitarian thinkers argue that the community itself should maintain relations with nature in order to take good decisions. Sometimes the desire to maintain the integrity of the community is considered to automatically imply a desire for a balance with the natural environment surrounding it. Diesing's definition of 'social rationality', for instance, primarily relates to the integration of a certain society, i.e. is, the web of relations it enspans. Decisions are 'socially rational' if they enhance integration and avoid conflict. However, another element of social rationality in his conception, is adaptation to the environment of a community, which includes the natural environment (ibid.: 88-90). The natural environment provides the absolute borders within which society can develop, in Diesing's view.

Communitarian institutions

Table 7 presents the institutional implications that a pure communitarian discourse would have.

Table 7. Institutional implications of communitarianism

Rule Type	Values under communitarianism
Aggregation	• There must be consensus resulting from dialogue in the community. • Decisions must be in the common interest of the community.
Authority	• The community takes all decisions affecting it. • There may be no relations of domination within the community.
Boundary	• All members of the local community may participate.
Information	• The decision should be based on ordinary knowledge. • In discussions both values and facts must be expressed.
Pay-off	• Participants bear their own costs.

Aggregation rules under a communitarian discourse would reflect the idea that dialogue or deliberation, where the common interest of the community is discovered, is the accurate basis for the decision. This rule also reflects the idea that deliberation will lead to consensus. The dialogue wherein this happens is infused not only with facts, but also with values.

Values and facts are discussed at the same time and participants in the debate must also explain from which values they infer conclusions.

Obviously, the *authority rules* will emphasize local self-governance. This is because the capacity to maintain face-to-face relations in which people can develop an appreciation and understanding of each other's positions and are present long enough to see the consequences of their choices only exists at this level. There is a hostility towards the central level, which is not to be involved in the decision-making process. The relations within the community are to be coercion-free because coercion would undermine the free dialogue.

The 'ticket to participation' is membership of the local community. As bonds between members of the local community and the wider community are not as strong as the bonds within the community, *boundary rules* will exclude the wider community from the process. Members of the wider community are excluded because they will not experience the consequences of the local decision to the same extent and will therefore have an incentive to make the community accept a facility. *Information rules*, finally, are likely to emphasize ordinary knowledge. Ordinary knowledge has been defined as the opposite of scientific knowledge: it is '*knowledge that does not owe its origin, testing, degree of verification, truth, status, or currency to distinctive (. . .) professional techniques but rather to common sense, casual empiricism, or thoughtful speculation and analysis*' (Lindblom and Cohen, 1979: 12).

Why would ordinary knowledge be superior and what kind of local knowledge is supposed to be relevant? To start with the latter issue: only local people are able to assess the consequences of specific proposals for their own community, as they are the ones who live in the terrain. The ordinary knowledge that is considered necessary is thus local knowledge. The reason why this type of knowledge should be superior is exactly because it is local. Even though not scientific and generally applicable, local knowledge is often developed over generations and is based on close observation of, for instance, the effects of human intervention in the landscape. This knowledge is highly specific for a certain locality but can be expected to cover almost anything that goes on there. Many of these events may well be outside the remit of the (limited) scope of our scientific knowledge.

What's more, Fischer (2000) argues that cultural values and meanings are often tied in to local knowledge. The landscape or the particular form of agriculture present in a locality may thus have a high degree of meaning to the local people. Many anthropologists have observed that with the introduction of scientific knowledge, e.g. in agriculture, these meanings get lost, exactly because scientific knowledge is supposed to work in any culture. Communitarians would thus resist the use of the scientific method, in part because they want to protect the uniqueness of their own culture. In a similar vein, Diesing's (1962) work suggests that communitarians are also likely to resist the application of the market mechanism. This is because they value relations with each other and even with certain non-human objects. These relations imply that persons and things have a particular quality that needs to be respected. This particular quality implies a certain incomparability, which is needed in an economic exchange. The approach to rational decisions is then not an exchange of interests, but the adaptation of demands. In a sense communitarianism is thus a discourse resisting what is normally called progress (technological and economic advancement).

A critique of communitarian thought

The response of the political system (in the US but perhaps also elsewhere) to the communitarian challenge has been such that opportunities for participation have been increased and representatives (for instance of the environmental movement) have been invited to discuss policy. In the process however, environmentalists have become absorbed into the same political system they first rejected, thereby supporting and not undermining interest group liberalism. Environmental groups have grown, become active at the national level and have started lobbying in Washington. While they were at it, a distance has been created between environmental groups (which needed to become more and more organized and professional) and newer local groups, who had difficulty in viewing the nationally operating environmentalists as their representatives (see also Fischer, 2000: 94). It is argued that this 'failure' must be attributed to the lack of specification in the communitarian discourse. It is said that the discourse does not specify how an enlightened citizenry is to be mobilized and maintained (Williams and Matheny, 1995: 51). It is argued that communitarianism is a strong banner to organize resistance behind, but does not do much to productively engage in decision-making and solving problems. In this sense, communities where people do have a proven incentive for interaction (e.g. the scientific community and representative institutions) might be equally effective in achieving dialogue and consensus. In other words, these critics wonder out loud why dialogue within a community should be anything special.

In addition, communitarianism has been attacked for its emphasis on the local level. In modern societies there is a broad national context within which social regulation takes place. Certainly, it seems possible that issues are debated at the local level, but how are all the outcomes of local debates to lead to a coherent national or international strategy? And how would local communities respond to the pressures stemming from an ever more internationally orientated business community that could 'hop' from community to community. Some adherents to the communitarian discourse have foreseen these problems and plead for 'assocationalism', which is the connecting of various independent communities (see e.g. Ward, 1998). This solution strikes some as quite unrealistic, however.

Communitarianists have also been attacked for their failure to consider scientific and technical knowledge in decision-making. Many have argued that scientific knowledge is crucial to environmental decision-making (e.g. Williams and Matheny themselves).

2.5 An agenda for the rest of this book

Introduction

In this final section I want to summarize some of the issues that I have discussed up till now. I also want to outline the purpose of this highly theoretical chapter for the remainder of this book. The essence is to be found in the starting point of this chapter, which was that the design of institutional arrangements for decision-making comes about in a human choice process, where competing discourses on the essence of hazardous waste, or environmental, decision-making are present. My conclusion in chapter 1 was that choices for certain institutional arrangements to address a problem also imply consequences for the outcomes of the decision-making process. In this chapter I have introduced discourse analysis as the proper method of studying the choice of institutional arrangements. Discourses are shared apprehensions of the world and as such identify institutional arrangements that are considered suitable for solving certain problems. In the process, they also identify criteria for what are good decisions. Table 8 recapitulates the three discourses that are expected to hold relevance for hazardous waste decision-making and their institutional implications.

Each of the three discourses implies a certain understanding of the hazardous waste problem and the ways to address them. In terms of the institutions I have identified in chapter 1, 'expert bodies' can clearly be placed in the remit of the managerial discourse. These expert bodies are likely to be made responsible for hazardous waste siting when those responsible for crafting institutional arrangements consider hazardous waste problems not value-laden and puzzle with an objectively definable right solution. Scientific knowledge will be stressed in the communications on hazardous waste issues. If those responsible for devising arrangements decide that 'representative bodies', 'the market', or 'the courts' are to be responsible for hazardous waste decisions, then they are influenced by pluralist thought. Within this line of thought, people make 'truth' together, be it in competition with each other. They can put their own interests first and act strategically in communication with others. Finally, 'the community' as a decision-making institution can be placed in the communitarian line of thinking (where else?). Here people also make truth, but now in cooperation and mainly using arguments and information that are rooted in the local community. Note that in each case, different notions of qualitatively good decisions are to be found, mainly in aggregation rules. Only in one case, managerialism, is it expected that this truth has an existence apart from human beings and needs to be discovered. I find that the concept of 'ecological rationality' in the sense that Dryzek used it (see chapter one) also seems to be a concept that exists independent of humans. In this sense, the notion is managerial in character, which is odd because Dryzek passionately pleads for ecological democracy.

Table 8. Discourses and institutional implications

Discourse➔ **Rule-type ➘**	**Managerial**	**Pluralist**	**Communitarian**
Aggregation	Greatest good for the greatest number; objectively determined by experts	Distributive equity; agreed upon by market parties, a judge or representative institution	Deliberative democracy; consensus in the community
Authority	With experts	With the market parties, judge or representative institution	With the community
Boundary	Only experts	Market parties, parties in a court case, members of a representative body	Only community members
Information	Scientific knowledge based on statistical evidence	All information that parties think is relevant	Ordinary and interactive knowledge based on experiences of community members
Pay-off	Experts paid by the state	Parties pay for themselves, or receive help from the state	People bear the costs for their own participation

One issue that was discussed above is the nested character of institutional arrangements. The implication is that various institutions and languages may be present at the same time in the sphere of hazardous waste decisions. It will be interesting to see - using discourse analysis - what motivated the structuring of these nested arrangements, what might have caused changes in the balance between certain institutions, and how the various institutions and their associated languages affect each other and interact. In the real world, traces of various different discourses in the set-up of arrangements can be found at the same time. Dryzek (1997), for example, notes that when environmental problems surfaced in the agenda in the early 1970s the response of most countries was to organize scientific expertise into the bureaucratic hierarchy so that 'rational management in the service of the public interest' could take place (1997: 73-74). After some time, there was a development towards greater involvement of the public by adding forums for citizen participation. Dryzek notes however that the effects of increased participation have been limited because participatory exercises have in reality often been used only to gain greater legitimacy for expert decisions (Dryzek, 1997: 86). Dryzek also notes that it was attempted to introduce state-founded markets. Where environmental goods and services did not exist yet, the government would step in and set parameters of quasi-markets (e.g. for tradable permits), which will then help to achieve public goals. Here too, progress has been limited.

The rest of this book

As already indicated in chapter one, I will be looking at three different countries and their responses to hazardous waste problems. These three countries, Canada, the UK and the Netherlands have been chosen because they seem to use different institutional arrangements for hazardous waste decisions. I write 'seem' because the image of how decision-making in the UK and Canada takes places was very much based on the 'siting' literature that does not always speak explicitly about the legislation that regulates siting processes. It is this element however that will receive much attention in the next chapter. There, I will analyze the rules that guide hazardous waste decisions in order to see which institutions are considered most important in the process and what language is supposed to drive the process. I will also analyze the discussions that underlie the choices of the three countries. In doing so, I will try to see whether the three discourses were indeed present in the debate and whether it is possible to detect the story lines associated with certain discourses. I will also look for discourse coalitions, as it may well be that certain discourses are used by different groups than other discourses. Part of my purpose here is to see whether discourse analysis as advocated by Hajer can work in this sphere as well. All of these issues will be the topic of chapter three. At this point already, I want to point out that analysis of formal legislation in Canada did not reveal a hint of the supposedly more community-based siting efforts in that country that have received much attention.

In each of the three countries I have subsequently studied three cases of hazardous waste siting on the basis of interviews with those who were closely involved, examination of official records relating to the decision process, and by collecting paper clippings[5]. The results can be found in chapters 4 to 7, with the UK cases being discussed in 4, the Netherlands in 5, Canada in 6, and a comparative analysis in 7. The three cases serve to study the way institutional arrangements are upheld in actual siting efforts. I am interested, also at this level, in discourses people use, the effects they achieve by using this discourse. Also, I will analyze the quality of the outcomes of the decision process from various perspectives embedded in the three discourses. Although it will be impossible to obtain 'absolute measures' for quality, I will look for opinions of those involved in the technical, economic, social, legal and political rationality of the outcomes. The table below presents an indication of what I understand these terms to mean.

Table 9. Conceptions of the good decision

Type of rationality	Relevant questions
Scientific rationality	Is the decision acceptable on the basis of proven scientific facts (e.g. environmentally safe) and does it deliver the greatest good for the greatest number (not too expensive given the benefits)?
Market rationality	Was agreement between free parties achieved and compensation given, does the decision enhance economic growth?
Political rationality	Did the decision leave the support for the political system (and for individual politicians) intact and was the problem solved?
Legal rationality	Were the rules clear and was conflict, once it occurred, resolved?
Social rationality	Was the decision acceptable to members of the local community, was it perceived to be in their interest, was there consensus on the decision?

Actors

The theoretical framework of Ostrom et al., although geared towards the analysis of institutional arrangements, is explicit about the fact that institutional arrangements are set in motion by actors. It is clear that in the empirical cases (and probably also in the collective choice debates) that I will study, this will also be the case. The actions of actors can therefore be expected to, at least partially, explain how a certain level of quality has come about. It is quite obvious that the three discourses differ greatly in which actors they consider relevant for hazardous waste decisions and in what they expect from these actors. The critique of each of these discourses is partly related to such assumptions. One often heard objection towards discourse

[5] Case selection is explained in appendix 2. Appendix 1 provides a list of interviewees.

analysis is that it tends to obscure reality and replace it with a wilderness of mirrors. It is in fact often placed in an increasingly suspect corner: that of postmodernism. The difficulty that some would have with postmodernism is that it studies perspectives and perceptions without actually pronouncing a verdict on their moral acceptability. In the postmodern world, many fear, everything is just an opinion or a point of view. If this view of postmodernism is correct, then conflicts between contrasting points of view, if they need to be resolved, can only be resolved by style: the best presented and most affluently spoken opinion deserves to win. This often gives an eerie feeling as rational discussion would seem difficult. I wish to note, however, that some of the expectations that are embedded in discourses are in fact empirically testable. I am particularly interested here in the degree to which some of the expectations about 'agents and their motives' and 'relationships between entities' have borne out. I observe that in various instances discourses contain positive expectations about the capabilities and motives of one set of actors and negative expectations on other actors. In this sense, the discourses provide a critique of each other and this critique can also be used to sharpen the image of expectations embedded within the various discourses. In addition, in this chapter I have described a more general critique of the expectations in each discourse, which also contains empirically testable assumptions. The difference between this more general critique and competing discourses is that the general critique does not provide a full-blown and independent discourse.

The table on the next page provides an overview on some of the expectations that could possibly be examined in the case studies, depending of course on what these actors do in a particular case.

The chapters that present the case studies will, where possible, reflect on these expectations. Finally, this book will conclude in chapter seven with an analysis across the cases and feedback into the theoretical framework that was outlined in this chapter.

Table 10. Some expectations embedded in discourses or their critique

Actors	Assumptions in discourse (from which discourse, or other source?)
Experts	*On the positive side:* • Are expected to be objective and neutral, will separate values from facts (managerial). *On the negative side:* • Will not be able to separate values from facts (pluralism, critique of managerialism). • Will deliver the results that the persons paying for them want (critique of managerialism).
Market parties	*On the positive side:* • Know their own interest best and will defend it if need be in court or in politics (pluralism). • Can be expected to define their own interest so that it includes the interests of others (pluralism - the harm principle). *On the negative side:* • Will have difficulty in defining their interest in complex issues (managerialism, critique of pluralism). • Will not think of anybody but themselves (critique of pluralism).
Politicians	*On the positive side:* • Will be accountable to the people because they need their support (pluralism). *On the negative side:* • Are accountable only to special interests (managerialism, critique of pluralism). • Will only use hazardous waste decisions for political gain, even if it is strictly a technical issue (managerialism).
Voters	*On the positive side:* • Will hold politicians accountable if they do not deliver the right decisions (pluralism). • Will form interest groups to influence decisions in the right direction (pluralism). *On the negative side:* • Do not have an interest in decision-making and are not sufficiently educated to form an opinion (pluralism and managerialism). • Will behave selfishly: favor economic development over the environment and will express 'absolute values' (critique of pluralism). • Will form interest groups to further their own particular interest and prevent achievement of the public interest (managerialism).
Parties in a court case	*On the positive side:* • Will make the strongest possible case for themselves and thereby support truth finding (pluralism). *On the negative side:* • Will make the strongest possible case for themselves and thereby hinder truth finding (critique of pluralism). • Will not rest until the last legal resort has been tried (critique of pluralism).
Members of a community	*On the positive side:* • Are engaged in coercion-free relationships and are in dialogue with each other (communitarianism). • Have local knowledge, which is crucial for the decision process (communitarianism). *On the negative side:* • Will be parochial in their outlook and not look beyond their own community (critique of communitarianism). • Do not have scientific knowledge, which is crucial for the decision process (managerialism). • Do not want to participate in the decision process (pluralism).

Chapter 3

Hazardous decisions

The problems of hazardous waste, institutional responses and the ideas underlying them

This chapter addresses the way in which the three countries of interest here have built institutional arrangements to deal with hazardous waste decisions. The first argument I shall make is that these arrangements were not so much created specifically for hazardous wastes, but rather already existed. This justifies a look at the origins of pollution control and land use planning arrangements and the ideas underlying them on the basis of an analysis of the existing literature. In the 19th century a reform movement arose in all three countries that wanted to counter the excessive consequences of industrialism. The managerial discourse was a powerful asset to these groups. Regulations were adopted in the public interest. Land-use regulations tended to be more controversial than pollution control. Intervention of the courts subsequently shaped the institutional context. In the UK, administrators were left with greater discretion than in the two other countries. To some extent, this pattern is also visible in the pollution control regimes in the three countries, where the use of guidelines is much greater in the Netherlands and Canada than in the UK. Despite differences in discretion, a mixture of conservative pluralist and managerial language guides the actions of administrators in all three countries. The dominant axiom is that regulations must increase utility. The position of the courts vis-à-vis the administration differs somewhat in the three countries, with the courts in the UK being most deferential and those in the Netherlands least deferential. The community finally receives hardly any attention in all three countries. Only certain limited rights of citizen participation have been granted. These are however largely an extension of existing rights for property owners and do not reflect communitarian discourse, which is largely absent from the collective choice debate in all three countries anyway.

3.1 The arrival of hazardous waste problems

3.1.1 Introduction

This section is about the way in which the UK, the Netherlands and Canada have dealt with hazardous waste in the past and how they do it now. The pattern that was followed seems broadly similar. The basic institution for making societal decisions in each of the three countries is the market, with Parliament having the powers to interfere if need be. There is a strong hesitation to introduce controls unless there is proof that such regulations would enhance the public interest. In this sense, conservative pluralism is the dominant discourse. Despite that being the case, the three countries are all three mixed economies, implying that the state and the market are both seen as important institutions for delivering goods. In very general terms, the state tends to step in only in cases of market failures in all three countries. The degree to which this occurs differs somewhat with the Netherlands having the most active state (total tax receipts are 41.9% of the Gross Domestic Product), followed by Canada (36.8%) and then the UK (35.4%)[1].

In each of the three countries, there are clearly identifiable periods in which the state influence on the economy grew, whereas in others it declined. Broadly speaking the end of the 19th century (start of environmental regulations) until the 1920s and the period after the Second World War (until the 1970s) were periods of expanding state influence and the 1920s-1930s and 1980s-1990s were periods of declining state interference. In terms of institutions, an increase in state interference implies that representative bodies and experts were given a greater role at the expense of the market. The role of the courts in the structuring of institutions has differed somewhat in the three countries, but has generally been important. Communities are largely ignored as an institution for decision-making, although there have been developments towards more participatory decision-making in all three countries.

3.1.2 Regulations concerning hazardous waste

Given the above it should be no surprise that waste disposal was largely a free activity in all three countries until far into the 20th century. Little distinction was made between different sorts of wastes until the 1970s. Dumping at sea, discharge to surface water and landfilling of all wastes were the most favored options. Landfilling (or landraising) was seen as a beneficial activity for instance, because it made the habitation of marshland and the restoration of the landscape after mining activities possible. Waste

[1] See OECD basic structural statistics, August 2001, at www.oecd.org.

disposal was also often a very local activity, with each municipal conurbation having relatively easily accessible dumping grounds that often lacked environmental precautions and supervision.

It has been argued that especially since the rise of the petrochemical industry, broadly placed around the Second World War, wastes have changed in nature. In the 1950s and 1960s the first awareness of the downside of the use of chemicals per se arose. Not much later, the way these chemicals were handled after their use came to the attention of the public as well. The carelessness in waste management came to the attention of the public through incidents. These included the discovery of drums of wastes at or near residential areas such as Nuneaton in the UK (early 1970s) and Lekkerkerk in the Netherlands (early 1980s), but also the discovery of a 'toxic blob', a mass of black oily substance floating in the St. Clair River in Canada in 1985 (Doern and Conway, 1994: 221). In many cases, wastes had been disposed of illegally to spare the costs of more responsible waste disposal, such as landfilling on municipal landfills.

In all three countries, the siting of landfills had largely been a matter of common sense - spontaneously grown landfills existed everywhere - guided by a desire to avoid nuisance to the public. Limited controls by land use plans and local regulations were added, if they existed at all. Because of the scandals just mentioned, the perception of wastes as relatively innocent material changed profoundly and there was pressure for more stringent control from the government. Especially environmental groups, both local and nationwide, started pleading for greater government involvement[2]. In the UK, the responsible government department, the Department of the Environment (DoE) seemed hesitant to introduce new regulations. DoE acknowledged the inadequacy of the existing regulatory system but their most optimistic forecast was that measures would be enacted only some time before 1975 - largely because industry would have to be consulted. Nothing could be done meanwhile since the disposal of toxic wastes was 'perfectly legal' in their opinion (Kimber et al., 1974: 203). There were attempts from several local governments to ban the disposal of certain hazardous substances at waste tips by using their authority under the Public Health Act 1936, but the DoE actively suppressed such activity. In Canada and the Netherlands regulations to improve landfill operations were introduced by the provinces, followed by national action in the Netherlands. Unlike the

[2] The pattern differs somewhat between the three countries, with The Netherlands and Canada already taking limited action on improving waste disposal before the issue became one of great public concern (but speeding up efforts after eruptions of public outcry) and the UK just responding to public pressure. This difference may be caused in part by the fact that public concern arose first in the UK and other countries could therefore learn from the UK's problems before the public became concerned.

UK, where the co-disposal of municipal and hazardous wastes continues until today, the Netherlands and many Canadian provinces sought the separation of hazardous wastes from other wastes. Reflecting of the dominant conservative pluralist discourse, the provision of separate hazardous waste disposal facilities was initially considered an issue for the market in all three countries. In the Netherlands and Canada, private parties picked up this responsibility, but with little success. Grudgingly, Canadian and Dutch authorities started accepting some responsibility for hazardous waste disposal in the 1980s, but mainly in the form of subsidizing private initiative. This move was never made in the UK.

The siting of hazardous waste facilities has remained a private sector responsibility in all three countries, but with close government oversight. This oversight is exercised through the land use planning system, which indicates that local authorities must approve of the site, and through the pollution control system, which implies that the proposed facility must meet safety norms of government and receive a license from the Environment Agency (UK), the province (Canada), or local authorities and the province (the Netherlands). The Netherlands stand out here as it is the only country that also embarked on capacity regulation. This implies that facilities need not only be safe from a pollution perspective and acceptable in planning terms, but must also 'improve' the Dutch hazardous waste treatment infrastructure. This implies that proposals from the private sector that do not fill a gap in the existing treatment market, or alternatively, do not apply innovative treatment technologies, will not be licensed by the Ministry of the Environment. The table on the next page provides an overview of the relevant legislation in all three countries. For Canada, I have limited myself to the federal regulations and regulations applied in the province of Alberta.

In all three countries, the controls on hazardous waste have led to improvement of practices, but have not solved all the problems and in some cases have introduced new ones. Space does not permit me to go into detail here. But all three countries have had problems with fly-tipping, unaccounted waste transports and the correct definition of waste (see e.g. ECA, 1980; Royal Commission on Environmental Pollution, 1985; Wynne, 1987; Wynne and Hortensius, 1987; House of Commons Environment Committee, 1989 (a+b) and 1997; Sherbaniuk, 1992; Petts and Eduljee, 1994; Davoudi, 2000). On top of that, there have been, at least to some degree problems with the issue of (hazardous) waste siting. In the UK, the waste industry itself likes to speak of a 'waste crisis' implying that getting government approval for waste facilities is practically impossible (see for a more general discussion Petts, 1994 and 1999). In Canada, several provinces have struggled immensely in this field (see Munton, 1996; Rabe 1994) and in the Netherlands concern over the siting of waste facilities was particularly great in the 1980s (see e.g.: Van Baren, 2001; Van der Heijden, 1998, Driessen et. al., 1990; Leroy, 1983). Part of the problem appears to be related to the diminishing amount of trust or deference in the regulatory system on the part of the general public. But there are more dimensions, including the relatively fragmented nature of the regulatory system, which has for instance led to a degree of concern about overlap between pollution control and land use planning and to concern about the proper degree of centralization needed.

Table 1. Overview of the relevant legislation

Country	Type and name of regulations
United Kingdom	*Land use regulations*: Town and Country Planning Act 1990; Town and Country Planning (Assessment of Environmental Effects) Regulations 1988(SI 1988/ 1199); Planning and Compensation Act 1991. *Pollution control*: Deposit of Poisonous Waste Act 1972; Control of pollution Act 1974; Environmental Protection Act 1990 and the Waste Management Licensing Regulation 1994 (SI 1994/1056); Environment Act 1995.
Canada (Federal and Alberta)	*Land use regulations*: Federal: none; Provincial: Alberta Municipal Government Act; Alberta Planning Act. *Pollution control*: Federal: Canadian Environmental Assessment Act; Environmental Assessment Regulations Order (1973 and 1984); Canadian Environmental Assessment Act (SC 1992, c. 37); Provincial: Hazardous Chemicals Act, Hazardous Waste Regulations (Alberta Regulations 505/87) Alberta Environmental Protection and Enhancement Act; Environmental Assessment Regulation (Alberta Regulations 112/93 and 243/93), Alberta Cabinet Guidelines for Environmental Assessment (1977, 1985).
The Netherlands	*Land use regulations*: Land use planning act of 1962 (revised 1985). *Pollution control*: Nuisance Act of 1896; Chemical Waste Act of 1976; Environmental management Act of 1993.

3.1.3 The rest of this chapter

Introduction

In the rest of this chapter, I will describe and analyze the place of various institutional arrangements in hazardous waste decisions and the underlying discourses. This description will be at a relatively general level and is based on literature review. When I say that the analysis will be at a general level, I mean that it will go beyond the topic of hazardous waste and beyond the period since the 1960s. The main reasons being the fact that hazardous waste regulations are affected by the broader and older pattern of regulation and the ideas lying underneath already existing regulations. To some extent, I was also forced to keep my research at a relatively general level because the amount of literature specifically about institutional arrangements and hazardous waste was relatively limited.

Working with existing literature has affected my work in yet another way. I have discovered that discourse analysis is rather demanding upon the analyst as both 'factual' developments such as the introduction of acts must be studied, and also the interaction process leading up to the act. There are relatively few authors in political science or legal studies who performed discourse analysis of the interaction processes that are interesting from the point of view of this study. The more common perspectives in political science are interest-based and in legal studies somewhat 'naturalistic', that is, mainly descriptive and explaining the course of events as if they were only logical. I have found that discourse analysis has some affinity with critical studies, which often tend to apply a Marxist perspective to policy. Indeed, authors such as McAuslan, who studied the UK planning system and who gives a critical account of the emergence of that system that seems far more thorough than many other accounts which appear very naturalistic as just described. But even such studies tend to have a large focus on 'interests' and explain much of the way interactions develop from such interests. Only here and there is attention paid to the language aspect of collective choice debates and are quotes given that can shed light on the use of language by those engaged in the debates. Because the alternative of studying the many choice processes in question personally was out of the question due to reasons of resource, I have had to rely therefore on studies that describe events pretty much from other perspectives and the account below must therefore be seen as an attempt to reinterpret some of the existing material. Here and there however, I found that there are clear indications that interests do motivate the actions of people, with the language they use being a result of this interest. I will return to this topic at the end of this chapter, it being an important an important issue of attention in my case studies, in which I have found similar indications.

Constitutional backgrounds

Perhaps at my own peril, I have focused my study on the collective choice level only, and not at the constitutional choice level. There are certainly distinct differences between the three countries in terms of constitutional rules and I freely admit that they are of importance. One need only realize that Canada is largely a federal state, with the provinces having certain constitutionally assigned authorities (see e.g. MacGregor-Dawson et al., 1997). Compare this with the UK, where the role of local government is almost completely dependent on central government, and the differences are clear. I do refer to such differences, but have not systematically assessed them because of time constraints. I feel that this can be justified in part by the fact that all three countries are considered polyarchies (that share a certain degree of corporatism, by the way) and thus share certain basic constitutional characteristics. To analyze these in greater detail would lead me too far from the topic of hazardous waste and therefore I limit myself further to the debate about pollution control and land use planning in the three countries. The focus will be on the various institutions I have identified in the previous chapters (the market, representative institutions, courts, experts and the community) and the language they use.

Mixed economies

In each of the three countries these institutions play a certain role, be it in an ever-changing relationship. Experts and elected representatives must work together, for instance within local government. Environmental regulation is defined largely as a technical matter, which implies that experts play a great role which is somewhat at odds with the formal supremacy of elected representatives.

The courts are used to determine certain conflicts. They have played an important role in identifying norms for government behavior in all three countries, especially in the early days of environmental regulation but have continued since then. Furthermore, during the last three decades access to the courts has widened. This has happened either as a consequence of court verdicts or as a consequence of explicit legislation. The motivation for this move is sometimes an extension of property rights in the environment to environmental groups and sometimes a recognition that more knowledge is available to the citizenry. The position of the courts differs somewhat between the Netherlands on the one hand and the UK and Canada on the other. Apart from the obvious existence of penal courts, the Netherlands have long made a distinction between courts assigned with administrative issues and courts assigned with private affairs. Administrative courts were seen as part of the administration, which affected the issues they could consider greatly. Currently there is a development in the Netherlands towards administrative courts that are outside the administration, which implies that similar limits apply as in the case of private courts. In the UK and Canada, countries which share a certain law history because of their past colonial relationship, the distinction between administrative and private courts is not relevant, although use is made of similar tribunals or quasi-judicial institutions there.

The community has not received much emphasis in the environmental regulation debate in the three countries. Yet in the 1960s and 1970s, citizen participation increasingly became an ingredient of the decision process. The degree to which that occurred is different for the three countries, with the UK being the most hesitant to grant participatory rights. In each of the three countries a certain tension is felt between increased citizen participation and decision-making by representative and expert bodies. Because of that, citizen participation processes are usually structured in such a fashion that the results are just one input in the decision process of representative institutions or experts. It should be noted that the desire to increase citizen participation in environmental decision-making did not lead to a changed perception of environmental regulation. The dominant form of communication is rather technical and citizens must conform to the standards if they want to be taken seriously. In that sense, the argument that a certain decision could potentially be against the interest of the community is given relatively little weight. Such claims must be substantiated with 'technical' evidence in order to be taken seriously. Environmental assessment in Canada and the Netherlands has added some rights in citizen participation, but this instrument too is very much about the technical aspects of the decision at hand. In the following section I will discuss the role all relevant institutions (and the language they operate in) in separate sections. Because regulation always involves markets, parties and state institutions in an interwoven way, I will discuss these institutions together, before turning to the role of the courts and communities. This occurs in sections 3.2. Section 3.3 discusses the role of the courts, 3.4 the role of the community and section 3.5 presents the conclusions of this chapter.

3.2 The role of the market, experts and representatives

3.2.1 Introduction

The historic roots of government regulation in the sphere of pollution control and land use are to be found in the late 19th and early 20th century. In all three countries of interest here, industrialization had taken place or was taking place in that era. Competition between private firms was unbridled, with governments largely remaining inactive except for some classical government tasks such as defense and infrastructure. The limited controls that were present stemmed mainly from the courts - where the conflicts over incompatible land uses were fought out. Judicial procedures could however only be afforded by a limited segment of the population, with the rules of evidence providing rather strong complications to actually win cases. Certainly in the UK, the courts were moreover unsympathetic to those bringing complaints about industrial pollution. Unbridled competition subsequently led to a host of problems in all three countries, mainly those related to the housing conditions of the poor and to labor conditions - including the handling of toxins in the factories. This situation resulted in a challenge to the dominating conservative pluralist line of thought being challenged by progressive pluralist and managerial thought. Section 3.2.2 highlights the interplay between the existing and emerging societal discourses in that period and describes the institutional arrangements that resulted from the debates up until 1940. Because the Second World War had quite an impact on the way the three countries looked at government interference in the economy and the potential for state planning, I will treat the developments after 1940 in a separate section, 3.2.3. Section 3.2.4 summarizes and reviews the entire section 3.2.

3.2.2 Shifting discourses and institutional arrangements

Middle-class reformers

In all three countries, the deplorable conditions of the poor led to concern from the developing middle class. Many of its more articulate members - united in the Netherlands for instance in the 'hygienist movement' - were committed to political and social reform to redress some of the sufferings and injustices produced by industrial development. Their concern led to or stimulated the publication of various reports that chronicled the condition of the working class. Examples include the 'Report on the Sanitary Conditions of the Laboring Poor' in the UK (see Rydin, 1998: 14). Among the reforms deemed necessary by this discourse coalition was the transformation of controls through the common law or private law (which were only useful to the upper-class), to public law that would be more accessible to less well to do. The concern for the working class seems to have been inspired by various motives including outright compassion (it is noted that there were many Evangelical Protestants among the reformers), but also personal interests. Progressive pluralist concerns with distributive equity definitely played an important role.

Changing story lines: from distributive equity to utility

It seems that the reformers' initial story lines, revolving around distributive equity, were replaced by utilitarian ideas in the course of a few decades. In a study of the planning profession in Canada, Grant (1989: 12) observes, for instance, that the early planners there considered planning a device for promoting human values and a means to stabilize society. '*The town planner believes that decent living conditions for the workers are not only possible but absolutely imperative if revolution and disaster are to be avoided*' writes an editor of a Canadian planning journal in 1921. Grant notes that such a commitment to distributive equity had been omnipresent among Canadian planners but was rapidly diminishing at that time. Concerning the Netherlands, Querido (1965) indicates the reformers there initially used progressive pluralist language as well, with a slight managerial tint. '*It had proven to be impossible for the state to restrain its own role to that of état gendarme. The dangers (. . .) that result from the completely free interplay of societal forces are becoming evident, the state must protect the weaker party in the production process, namely the non-owners, the workers, against the stronger party, the owners of the production resources, the employers, in order to safeguard the welfare creating productivity*' (236-237).

These story lines were not at all successful in convincing opponents. In the Netherlands, the ideas of the reformers were not accepted, certainly not by the factory owners: '*as long as industry blossomed, it could not be realized that a more sustainable welfare was equally dependent on labor force as on entrepreneurial capacity. Measures to reduce working hours were described as "sick philanthropy"*'(ibid.). The political majority in Parliament at the time (the Liberals) was very much anti-government interference. A prominent liberal at the time: '*I would not even want to accept just the appearance that this Parliament considered itself under the duty of fulfilling the needs of the capable poor. There is much too much helping these days, unfortunately. Parliament can only take away some general barriers, but individuals themselves must be able to fulfil their needs*' (De Ruijter, 1986: 86). Such staunch conservative pluralist language must be understood in part by the fact that political representation in the Netherlands and in the other two countries was intimately connected to property as only taxpayers could vote, and only property owners paid tax. In other words, the dominant political ideology of the time was permeated by the importance of property. As all Members of Parliament were property owners, they could easily understand the desire to exercise property rights as freely as possible. McAuslan notes that the ideational justification for this constellation in the UK stemmed from the very influential writings by Locke in the late 17th century, in which the protection of ownership and the rights connected to such ownership stood central. He also observes that because of the intimate connection between political power and property ownership, the protection of private property could be equaled with the protection of the constitutional order. The Dutch 19th century idea of the 'Rechtsstaat' (a State ruled by law) as described by Van Gunsteren (1976: 89-108) reflects the same thoughts. The pleas for an increased role of government, beyond the narrow task conception that the liberals of the time envisioned, were thus strongly rejected.

It is interesting to see how this resistance was broken. The way in which this happened differs somewhat between the three countries, but the common thread was a reinterpretation of the public interest and use of the story line of utility. As I have indicated in chapter two, state interference presents a dilemma for conservative pluralists and they are only willing to accept such interference if it increases general welfare. This is exactly what those who were propagating government controls, started arguing, thereby saying that regulation was in the public interest and not (only) in the free flow of market forces. Utilitarian arguments, stressing the idea that regulation was in the public interest, became quite forceful. Around 1870, the leader of the Dutch liberals and writer of the Constitution, Thorbecke, announced the introduction of rules regarding public health that would '*prohibit a municipal council to sacrifice general interests to willfulness or narrow-mindedness*' (Querido, 1965: 233). Legal historian Dicey, writing about this era in the UK noted that '*the principle of utility became an argument in favour, not of individual freedom, but of the absolutism of the state*' (see Vogel, 1986: 233). Grant indicates that in 1914, the Canadian government hired a British planner by the name of Thomas Adams to the Commission of Conservation. Apparently in contradistinction to the more utopian Canadian planners of the time, Adams 'sold' the idea of planning in Canada mainly on the basis that it was a means to promote social efficiency. '*Adams set to work to convert the nation to his utilitarian vision of planning*' (Grant, 1989: 12). Planning was to promote orderly development and growth, and it was in this sense that planners such as Adams tried to convince the business community and government leaders that planning would pay off economically. Fortunately for Adams, the utilitarian ideas about planning tied in well with the ideas of the Canadian 'reform' or 'progressive' movement of the day that sought to improve local government, mainly by professionalization and depoliticizing public issues (see Hodge, 1998: 128-129).

According to Querido (1965), the reasons why reformers in the Netherlands could make reference to the public interest are multiple. Certainly, they included the fact that 'material well being was no barrier to a contagious disease' such as cholera and it was therefore in everyone's interest to fight the disease. Querido adds that the workers had started to unite in cooperative and trade associations and were engaged in a battle for suffrage so that they could start affecting the political process. And last but not least, the value of labor and the worker rose in economic terms as the demand for labor increased. '*The worker had to be better fed and educated, in order to fulfil his role in the production process. His health had to be protected because it had obtained economic significance*' (ibid.: 236). In all three countries, but more so in the UK, enlightened industrialists saw this interest and started constructing better housing for their employees. In the UK, one specific circumstance contributing to that was the fact that entrepreneurs wanted an end to the cutthroat competition and tried to become more socially acceptable to the upper class. '*Just as England was nearing the pinnacle of its industrial strength, the British business community suffered a loss of self-confidence. Instead of seeking to expand its influence, it became co-opted by a resurgent aristocracy. Rather than to continue to emphasize industrial growth and international competitiveness, it now sought economic stability and social acceptability*' (Vogel, 1986: 243).

Managerialism?

The effect of the reference to the public interest was rather positive. The first land use controls in the Canadian province of Alberta - The Town Planning Act of 1913 - were instituted with support from the business community (Laux, 1990: 17). This support was based on economic considerations: improved living conditions for the working class was expected to result in a healthy and more productive workforce; planning would reduce the cost of housing and thus assist in keeping wages low (Laux, 1990: 17). But now that regulations had been adopted, what would be the role of the professionals who were to implement them? In all three countries, there was a hint of managerialism, although the 'technocratic movement' as Lintsen (1985: 163) calls it, was not as strong in the Netherlands nor in the UK (Pollit, 1993) as it was in the USA. Canada seems to have occupied a middleground. In either case however, the reformers did use managerial language. They placed themselves between manufacturers, whom they considered shortsighted, and the working class, whom they considered too militant. Instead, a managerial approach to regulation was their answer, 'they promoted the values and ideals of professionalism, scientific and technical expertise, administrative competence and neutrality, and efficiency in both business and government' (Vogel, 1986: 231).

If we look at the role that professionals saw for themselves, we can see two main points on their agenda. The first is the demarcation of a certain exclusive field of expertise and development of that expertise. It has been noted that the Dutch hygienists who had pleaded for public health legislation were not 'real' medical doctors and in fact had a wrong conception of what caused cholera[3]. Whereas their influence continued to be felt, in part through their very appealing story line 'light and air', the medical profession moved on to promote academic research and training for doctors and to set entry requirements to the profession. The same pattern is observable for Dutch

[3] For them, foul smells were a prime suspect.

engineers (see Disco, 1981) and planners (see De Ruijter, 1986) and for these professions in the two other countries. One expression of such professionalization is the foundation of organizations that united the professions such as the UK Institute for Town and Country Planning, the Canadian Institute of Planners and the Royal Institute for Engineers (KIVI) in the Netherlands. Such institutes started to control, to a varying degree, access to the relevant labor market by setting the curricula for academic training programs and worked to expand the market for the profession. This connects to the second main point of the agenda: the increase of the role of the professional in society. A Dutch journal for engineers in 1891 stated that the engineer is '*often the obvious choice to cushion the shock resulting from the collision of capital and labor and can, by exercising his influence, contribute directly or indirectly to the reconciliation of the different layers of our society*' (Disco, 1981: 727). Engineers themselves believed that scientific management of both the private and the public sector could help increase general welfare, while at the same time making a revolution unnecessary. The professionals who obtained important positions in the developing state bureaucracy (e.g. as head of planning departments) emphasized that government interference should relate to what they considered the dominant societal development of the time: modernization (De Ruijter, 1986: 63). Dutch engineers saw the elites of the time, who often had legal training or 'only practical experience', as 'dilentantes', who lacked any valid title to their high positions. Their professional organization complained that '*practical experience is as good or better a recommendation for occupying civil and state command positions than credentialled scientific knowledge*' (Disco, 1981: 725). Dutch planners were not satisfied either. '*Whereas everywhere else for the implementation of serious tasks only experts are called for, is housing with money of the Housing Act still the domain of organizations of lay people, controlled and as good as possible led by a far from almighty State Inspection, and under the supervision of sometimes capable, sometimes fully incapable council members*' (De Ruijter, 1986: 228). Rydin (1998) argues that the early UK planners had little developed theory relating to their activities, except perhaps 'geographical determinism'. This term relates to the belief that urban areas and their inhabitants could be 'improved' by careful expert design, which seems a rather paternalistic idea. But who then would control this process? No surprise here as '*This approach saw planning activity as essentially a technical process of design and drawing, separated in its operation from economic or political processes. The success of the activity depended on the technical ability of the planner, who could be considered an applied engineer*' (ibid.: 17-18).

The social position of the new experts in the three countries was not as high as it is now. The training of engineers in the Netherlands for instance occurred at a school (Delft) that was considered a middle-level education, explicitly *not* a university (which it became only after decades of lobbying). In the Netherlands, almost all engineers could only find meaningful employment in the railways and in the sphere of water management. The private sector had virtually no interest in engineers. The professional organization of engineers looked upon this situation with great concern. Changes were made to the curriculum to make the engineer more attractive for private employers. Interestingly, the way this occurred was by introducing business administration as a topic and more generally, increasing the awareness of economics among the students of engineering. This change implied that engineers, who were initially mainly concerned with technical quality, also obtained an eye for the economic quality of their solutions. Although this is admittedly speculative, it would seem that the birth of managerial discourse must partly be sought in the new curricula for professionals that were introduced at the time. Employment of Dutch engineers by the private sector consequently increased considerably (see Disco, 1981).

Internal and external pressures

Whereas the quotes above suggest an ambition to make the role of experts as large as possible, it should be acknowledged that the experts' ambitions did recognize certain boundaries. Nowhere in the three countries studied here, was there the ambition to fully take over the steering wheel of society even though there was a deep disdain for elected politicians (especially in Canada, where corruption was a serious issue) and strident attempts were made to present scientific management as value-neutral[4]. In the various countries, the boundaries of ambitions seem to have been largely related to property. This was partly a concern of the professionals themselves, partly a result of external pressure. Regarding the professionals themselves: in all three countries the newly appointed administrators saw it as their task to work with industry or to avoid confrontation with property owners. In the UK, the concern for cooperation with industry was so great that the first pollution control officers voluntarily decided to operate under high standards of secrecy, so that no company should feel threatened by being open towards them. In the Netherlands, the first 'extension plans' were carefully designed on the basis of the existing land ownership patterns, even at the expense of recognized public concerns such as traffic flow. The early

[4] Take the Dutch engineers. They took efforts to remove themselves from elected politicians as far as possible: '*if anything is outside the sphere of politics it is the applied science of the engineer*' (De Ruijter, 1986: 65).

Dutch land use planners were very shy of expropriation of property, a possibility that the Housing Act of 1901 did offer. Querido notes that from 1900 onwards, the reformers who had worked for greater public health controls were now becoming more influential within the state. They had initially been very critical of the state and had expressed critique 'in such a fashion that is not common for public officials'. Now that they were in influential positions, their critique became less vocal and their agenda less radical. Blowers (1980: 14) has observed a similar development in the UK.

The external pressure (so to speak) upon planners came from various fronts and was mediated in different ways in the three countries, with the courts playing an important role and with a qualitative difference between the public health type of regulations and land use controls. Regarding this difference, Alder and Wilkinson (1999: 235) make the case that even today, land use controls in the UK are seen as '*imposing a burden upon an "innocent" property owner for the public benefit*'. This in contrast with pollution controls, which do not affect rights traditionally associated with land use in the public eye, and are thus more easily accepted. The fact that pollution controls were enforced rather poorly may have contributed to that. In Canada, this has led to interesting situations where the courts were more active in blocking obnoxious development than the public regulators. It is argued that Canadian courts were refusing to modify substantive rules or rights in order to accommodate the demands of industry (Nedelsky; quoted in Schrecker, 1992: 85). Canadian courts were willing to make concessions to the 'public good' associated with industrialism, but then only in the form of compensatory damages. In some cases, the resistance against industrialization pitted the courts against provincial Legislatures. As late as 1949, a restrictive court decision was dissolved by the provincial legislature of Ontario because of this reason. The provincial Legislatures are said to have deliberately hollowed out the constraints on industrial development implied by the common law in Canada. In 1919, riparian rights (protection of water quality and quantity for downstream landowners) were expropriated by the Legislature of Nova Scotia and assigned to the provincial government. This occurred without compensation or recourse to the courts and was considered the culmination of '*a pattern of statutory removal of common law obstacles to water-related development common to many provinces*' (Schrecker, 1992: 86).

Different positions for planners

Turning the attention to land use controls, the shaping of institutional arrangements has been strongly affected by the courts. The results of this process are that land use planners in the UK are relatively free to take decisions along their professional norms, whereas Canadian and Dutch land use planners are relatively constricted. McAuslan (1980) observes that the UK courts were initially siding with landowners who resisted limitations on land use. This resistance had a certain effect, which was mainly procedural in character, as courts were striking down decisions by land use planners that had been taken without allowing for representations by the landowners. The UK courts were an active member of the conservative pluralist discourse coalition. They regarded '*Development control under the planning acts (. . .) imposed ab extra; they are not part of the normal bundle of rights of the landowner and as such are rightly subject to control in the interest of the landowner*' (McAuslan, 1980: 3). Note that the control in the interest of the landowner was of course to be exercised by the courts themselves. In placing themselves in that position, they had very little statutory leeway as the legislation was strongly managerial in spirit. The newly adopted legislation had conferred wide powers to administrators to do as they saw fit. It also provided precious little opportunity of redress as the assumption was that administrators were advancing the public interest.

Although one could be led to believe that the courts were only inserting some procedural fairness through their verdicts, this is incorrect. The verdicts they reached are described as wholly on the side of private parties. '*Judgements were couched solely in terms of the need to protect land owners against hasty government action with no regard being paid to the objectives of the legislation (. . .)*' (ibid.). A change came around 1905, when the UK courts started to accept that it was legitimate for administrators to advance the public interest, that they had to be assumed to act in good faith, and that administrators had a certain degree of democratic legitimacy, as their political masters were accountable to Parliament (ibid.: 5). McAuslan contends that this ideational turn by the courts stemmed from a new generation of judges who were versed in the utilitarian ideas of Jeremy Bentham and accepted state interference in private property. McAuslan contended that this has had lasting effects on land use planning in the UK. Land use planning in the UK became largely a discretionary activity with land use plans being only one material consideration for the administrators in concrete cases.

It is interesting to see how the position of planners greatly differed in Canada and the Netherlands. Turning to Canada first, the provinces there almost literally imported UK legislation in the first two decades of the 20th century. However, the strength of the reform movement and the distrust of elected politicians embedded in it, forbade the transfer of land use planning powers to local councils. Instead, many so-called planning boards were founded (Hodge, 1998: 131-132). These boards were distanced from local authorities and were controlled by appointed members, largely business representatives. The developments in Canada demonstrate the affinity between the managerial and conservative pluralist languages as both planners themselves and business interests were interested in planning. However, because of their control over planning boards, business interests were able to control planners to a great extent and prevented them from growing into a fully managerial role. Writing on Alberta, Laux (1990), notes that the UK-inspired first Albertan planning legislation, if fully implemented, would have implied great intrusion upon private property rights. Laux notes that the implications of the law were too drastic however in the context of local government with limited resources and the Albertan ideological climate. Not a single town-planning scheme was put into effect under this act. The Albertan economy fell into a recession that lasted from 1914 until the mid-1920s and the ideas on '*Grandiose development plans were set aside in favour of survival plans*' (ibid.: 19). Laux finds that each later version of the act focused less on planning schemes and less and less intrusion of private property was envisioned. Grant (1989: 13) argues that the transformation of Canadian planners from utopian dreamers to helpers of efficient government was complete after the great depression of the 1930s. The apparent control of property owners over planners had consequences for the way planning was and is approached in Canada. Already in the 1920s and 1930s, zoning (the separation of various functions that are considered incompatible) was the most popular form of planning in Canada and not comprehensive planning that was also legally possible. Grant explains the popularity of zoning from the fact that zoning ordinances protect the value of property (1989: 12)[5]. So large is the obsession with zoning that these days Canadians practically equate planning with zoning (ibid.).

In the Netherlands, planning also came to be equated with zoning. The process in which this occurred involved two discourse coalitions which tried to influence the development of land use planning legislation (Van der Valk, 1989: 486-508). The first consisted of landowners and their representatives in the bureaucracy. They stressed conservative pluralist language and did not agree to the limitations on development by the act. On the other side stood the engineers, architects and legal experts for whom the act did not go far enough. It seems the Dutch government oscillated between these two discourse coalitions as it increased at the same time both the legal importance of the expansion plan (later renamed zoning ordinance - 'bestemmingsplan') and the protection of private landowners in every amendment of planning law that was made until the Second World War. The courts, in combination with very active elected representatives, in the end

[5] Indeed, Laux (1990) notes that most zoning bylaws in Alberta focus on preserving the status quo, and are not a vehicle for ambitious planning efforts.

took away much of the discretion land use planners were accustomed to. An important decision was taken in 1921, when expansion plans started being described in the law as zoning ordinances with binding force. Subsequently, the courts imposed a duty to refuse building licenses in case of non-adherence to the bylaw in 1931. In order to soften the expected resistance from developers against such moves, procedural guarantees were introduced including the right to object to the province and the right to appeal to the Crown. It was the close supervision by these higher authorities that led to a large degree of judicialization. In the name of landowners, the discretion of municipalities was largely hampered as the demand of legal security for landowners became paramount. This implied that only very detailed land use plans were approved, which were very hard to prepare, would take much time to introduce, and result in much resistance. In addition, municipalities feared enormous claims for compensation. The consequence was that municipalities shied away from formally adopting land use plans. Instead, they started working with informal plans and used private law instruments such as concessions to get matters done.

Implications

The potential implications of the differences in the position of planners could be manifold, but would first of all tend to depend on the way planners in the UK exercise their discretion. While UK planners are said to be less hesitant in interfering in the private sector than Canadian planners (Cullingworth, 1987) the guidance they receive and the institutional make-up of decision procedures seem to reflect much conservative pluralist thought. The 'bias in favor of the developer' is a well-known story line in UK land use planning, and it refers to central government guidance requesting restraint from planners in the imposition of conditions on developers. Secondly it refers to the fact that authority and boundary rules grant the proponent of a change in land use special, exclusive rights, in the sphere of appeal and judicial review. This situation presents a bit of a paradox, as managerial discretion in the UK apparently leads to greater freedom of developers than exists in the Netherlands and the UK. This paradox is resolved if one considers that property owners have different demands. The developer seeks change, but the current landowner or occupants of nearby lands may well seek conservation of the existing situation. The principal choice then is between continuity and change. It is clear that UK planning law, even despite its managerial overtones, protects rights to develop and thus change whereas the Canadian and Dutch situation have evolved into one where land use planning controls protect the current landowners and thus continuity. The discretionary power of the UK planner then often consists of the authority to impose the burden of new development on those protesting it.

3.2.3 Institutional developments since 1940

Introduction

All three countries of interest here were involved in the Second World War and this war profoundly affected (particularly) the land use planning system, which had by no means been a great success until then in each country. The pollution control regime in all three countries also went through major changes after the Second World War. It appears that this was mainly a consequence of environmental disasters and emerging knowledge on the negative consequences of industry.

Land use planning

Both the economic (all three countries) and physical (UK, Netherlands) damage incurred by the war and the obvious success of wartime planning resulted in a desire to extend the practice of planning. In the UK, the hallmark Town and Country Planning Act of 1947 reflected the desire of 'public sector direction of resource allocation, mainly through controlling land uses' (Rydin, 1998: 25). Rights to develop were effectively bought up, and all development was to be subject to local planning control. This implied an extension of land use plans to the entire country. Similar developments are visible in the Netherlands and Canada. In Canada, the post-war federal government was more enthusiastic about planning than ever, and local authorities were required to have land use plans in place if they wanted federal housing subsidies. In the Netherlands, the reconstruction ('wederopbouw') was a highly technocratic and centrally-led effort, which effectively nullified the importance of pre-war land use plans because they were unsuitable for the massive amounts of new houses that were needed (see Van der Valk, 1989; Bosma and Wagenaar, 1995). When local planning control was reinstated in the 1950s however, the idea of local land use plans that cover the entire country became paramount and this idea was embedded in the Land use planning act ('Wet op de ruimtelijke ordening') of 1962.

The enthusiasm for land use planning seems to have been growing well into the 1970s. In the UK, the Labour government of 1974 was perhaps the better expert as it embarked on an attempt to start 'positive planning'. Positive planning implies that the authority to propose developments is taken out of the hands of the private sector. The underlying goal was progressive pluralist in nature: the party wanted to use the land use planning system to increase distributive equity. This concern with equity had been visible already in the TCPA of 1947, which had contained a 'betterment levy' intended to nullify 'windfall' profits from increases in land value related to land use planning (see Blowers, 1980: 115). The betterment levy had effectively slowed down and complicated the land market and had been repealed by the Conservatives in the 1950s. In the 1970s, Labour announced plans to reintroduce the betterment levy. The positive attitude towards planning was almost universal, witness the fact that even the 'Texas of Canada', Alberta, had a go at province-wide planning at some point. In the Netherlands, a layer of national land use plans concerning every sector of government activity was introduced in the 1970s.

The increased application, or at least popularity, of planning implied that the pretensions of planners grew. Speaking of the UK, Rydin (1998: 35-38) notes that planners, starting in the 1950s, recasted planning as a general form of decision-making, justified by 'procedural planning theory'. Procedural planning theory proposed optimal ways of taking decisions given certain goals, with the rational comprehensive model being the most ambitious element. The planner, under this theory, is a neutral arbitrator of interests who lays claim to land use and actually, to anything else for that matter. If anything, procedural planning theory implies a very strong managerial claim to neutrality, but at the same time carries a pluralistic flavor as competing interests must make their claim known to the planner so that the planner can decide, yet again, 'in the public interest'. The key role was thus for the planner. Along these lines an editorial in 'Plan Canada' in 1967 said: '*the fact remains that in the coming years the environment of the Canadian people, and its fitness for their use, will depend increasingly on our leadership, our ability, and our imagination*' (quote from Grant, 1989: 13-14).

In the Netherlands other government departments saw the increasing importance of the Department of Housing and Land Use Planning with suspicion and conflicts erupted over what would fall under the planning regime and what would not. The Ministry of Agriculture and the Ministry of Transport and Public Works were both particularly successful in keeping planners off their turf. At the same time however, certain areas of their own agendas were included at the height of land use planning, including ambitious land reclamation projects in the lake IJsselmeer. It has been argued that through the increase of the extent of land use planning, it was about everything but at the same time about nothing. The latter remark especially refers to the fact that planning activity lacks a normative core, which is reflective of the pluralist ideas embedded in procedural planning theory. The planner as a judge of other interests may of course not be biased. It is in this sense that the managerially sounding language of planners that refers to planning being in the public interest must be understood. They must position themselves as politically neutral in order to play the desired role of arbitrator. In the Netherlands this has implied the cutting of long standing ties with nature conservation interests. On the UK, Blowers comments that: '*planners recognise the necessity for politics. They acknowledge the need for political support and initiative to get things done but tend to disclaim any political inclinations themselves*' (1980: 3). Blowers notes that planners in the UK avoid the charge of being partial by claiming to serve the public interest, which is a non-existent notion in his opinion. '*In Britain there has been a tendency to evaluate issues in terms of utilitarian principles, which implies that gains achieved by some groups are at the expense of others. Choices have to be made, and any appeal to a notional public interest merely helps to obscure that fact*' (ibid.: 15). I take it that Blowers meant to indicate here that the public interest is a fallacy, not that the public interest is never used as an argument. In fact, my analysis, if anything, shows that the public interest has been used as an argument to *justify* difficult choices in the UK.

There are various points worth making in relation to the independent and unbiased status of planners. Some of these are in the line of a more general critique of pluralism. Take for instance the general problem with pluralist decision-making that not all interests are weighted in the end. This critique is reflected in Blowers' discussion of the UK planning system. According to him, planning '*tends to advantage those interests which are able to achieve access to the decision-makers*' (ibid.: 185). These interests include professional interest groups, well-off communities, and industry. What characterizes such groups, says Blowers, is their ability to describe their preferences in terms of the public interest. By implication, other groups do not have such capabilities and can not affect land use decisions. If this is correct then the critique that institutional arrangements with a pluralist flavor are too dependent on the ability to organize and motivate claims would seem to be correct. Secondly, it would imply that even though everyone uses the story line of the public interest (planners and interest groups alike), distributive justice is not achieved. Indeed, this is Blowers' argument as he says that in being responsive to the interests that express their concerns best, planners are in fact responsive to the existing power structure in society (this is akin to the critique of managerialism that was formulated in chapter 2). Note that both in the UK and the Netherlands, progressive governments in the first half of the 1970s were actively trying to make distributive justice their concern. In both countries 'land policy' or positive planning was envisioned but never accepted. In fact, one Cabinet in the Netherlands lost its Parliamentary support for a move in the direction of land policy[6]. The point of distributive justice connects to what I have already concluded: the land use planning systems of interest here are biased towards change (the UK) or to conservation (Canada and the Netherlands, with the Netherlands being an extreme case). Rather than just a reflection of 'existing power structures' I wish to observe that this difference has come about in interaction between property owners, administrators, elected politicians and the courts. The courts in particular, would be my argument, have been active in shaping the discursive space for the other parties. As already noted, courts in the UK accepted discretion of administrators on the basis of various story lines, including the presence of political control over administrators. In the Netherlands, the courts used a story line with a constitutional status, the one of 'rechtszekerheid' ('certainty of rights') to constrain administrators. In the process, the courts directly or indirectly defined where the public interest lay. In the Netherlands it was in the protection of vested rights, in the UK wherever administrators and their political masters thought it was. Political guidance to the administrators made clear that development must be made possible.

This points to another issue, which is the relationship between land use planners and elected politicians. Certain aspects of this relationship are shared between the three countries, others less so. To start with some of the differences, one can see that in Canada local government reasserted itself somewhat in the second half of the 20th century and started taking over some of the responsibilities that had been exercised by independent commissions and boards. Planning boards still exist, but the term is now used mainly in conjunction with bodies where appeals can be made against planning officers' decisions. Whereas the planning boards used to be partially manned by local councilors, this practice has ceased as the courts held that it did not meet the requirements of the Canadian Constitution Act of 1982 for unbiased procedures. Political control over development must thus be shaped in the form of instructions to development officers. As land use planning is mainly seen as a technical matter, this oversight is relatively minimal. Zoning takes place on the basis of '*the persistent belief that, at best, planning remains a fundamentally rational and technical endeavour, undertaken by trained professionals in pursuit of incontrovertible and generally shared public goals*' (Kiernan, 1990: 14). This observation must be seen in the context of

[6] Since then less ambitious legislation in that direction has been accepted in The Netherlands in the sense that under certain circumstances municipalities are given 'first choice' over other potential buyers at a 'reasonable price' if a piece of land is to be sold. This is part of the Law on preferential rights for municipalities ('Wet voorkeursrecht gemeenten').

the fact that land use planning in Canada is not as encompassing as it is in the UK or the Netherlands. It is observed that '*governments accepted planning inasmuch as it met their need to control the value of property and the character of neighbourhoods. As a result, planners became servants (...)*' (Gray, 1989: 13).

Land use planning was not for long an important concern for councilors in the UK either (Blowers, 1980). The reason, here too, was that land use planning was perceived to offer few opportunities to score politically and there was for a long time a certain consensus that economic development was good. Blowers observed in 1980 that this consensus was waning, thereby making planning a more politically attractive domain. In the Netherlands however land use planning appears to have known very active councilors since the early days. I have already mentioned the fact that the close supervision of amongst others elected politicians led to a strongly detailed and rigid system. Even as of today, representative bodies (provincial or municipal) are highly influential in planning. Especially at the provincial level, the Legislature ('Provinciale Staten') may explicitly ordain that the provincial executive can not vary certain aspects of the regional plans it approves of.

Is the relationship between planners and politicians somewhat affected by the political will of elected politicians, it is also affected by what the two parties involved bring to the table and this is largely the same in all three countries. Blowers observes that the power to make planning policy is formally put in the hands of elected representatives[7]. But the formal situation of supremacy of elected members is not realized in either country. Blowers' (1980) finding in the UK that power in reality is shared between high-ranking planning officials and a few powerful politicians rings true in two other countries as well. Planners are powerful vis-à-vis politicians because they have the ability to prepare decisions and have certain connections (e.g. to the national level) that are important in determining policy decisions. Especially the higher-ranking officials, e.g. the heads of the planning departments, have interactions with elected politicians and try to secure

[7] A qualification of this formal state of affairs is that bodies of elected representatives are relatively weak vis-à-vis the administration. It has often been observed that the UK and Canada are Westminster-type of democracies. This system favors clear majorities in elected bodies and implies a great degree of control of the executive branch over Parliament as the heads of the executive branch and the heads of the majority party are the same people. In the Westminster model, the legislative branch and the executive branch of government become practically indistinguishable from each other. The proportional representation system in The Netherlands implies less control over elected bodies, but Cabinets and executives tend to be based on strong majorities.

support for the proposals that the department makes. The power of the head, or other chief officers, is derived from his control over the department. His subordinates receive promotion through him and are therefore loyal to him, rather than to elected politicians. The chief officer determines what information and which options are presented to politicians. In case of conflict between this officer and elected politicians, the politician's position will be weakened, especially if there is a chance of appeal to a higher authority. Language plays an important role here. '*In general, the planning officer's appeal to "planning merits" or "principles" is sufficient to discourage politicians from challenging his recommendations unless there are obvious party considerations or outside pressures which urge them to do so.*' (ibid.: 27). Note that 'planning merits' and 'principles' are, just like the public interest, important story lines for planners. On the other hand, proposals do need political support in council. Here too, argues Blowers, are 'leaders' present who command the following of the members who belong to the same party. Because these leaders largely control the council, the interaction between senior officials and chief officers determines which policies are adopted and which ones not[8]. There are attempts to change the balance of control somewhat at the collective choice level. Specifically, some cases of corruption led to the Nolan report on Standards in Public Life of 1997. In the course of the preparation of that report, professional planning associations suggested that local councilors should be excluded from certain steps in the decision process. The Nolan committee rejected this suggestion, but still, the committee suggested that planning committee members would receive more training in planning law and procedures (Rydin, 1998: 171-172). In terms of my theoretical framework this would still imply an increase of managerial thinking as it is an advice to take the politics out of land use planning.

Land use planning that uses the story line of the public interest much (specifically the UK system) or is shaped by planners who think they should decide competing claims (in all three countries in the 1960s and 1970s) is confronted with some dilemmas haunting the managerial discourse. These dilemmas are related to the idea that a single public interest can be established. Especially problematic are situations wherein the competing interests also use the story line of the public interest to describe and defend what they want. Blowers encountered one case in his empirical work wherein opposing interests (a well-to-do community and an industry) both put forward normally accepted claims to the public interest. It appears that in each of the three countries, planners started encountering opposition to plans

[8] As stated, one party dominance of local councils is rare in The Netherlands. Political leaders may therefore come from different parties.

they advocated from groups claiming that planners were not propagating the public interest. This resistance was initially mainly felt in 'slum clearance' or 'city improvement' projects, where planners often proposed far reaching changes in the built-up environment. Later it moved towards projects with environmental implications such as roads. I will write some more on this in the section on the role of the community in decision processes. Environmentalists in all three countries (but less so in Canada) could argue that planning had failed the environment and thus the public interest. Certainly in the Netherlands, land use planners felt threatened to some degree about the upcoming pollution control legislation and much energy was devoted to defining the relationship between the two fields, with land use planning initially having the upperhand. In all three countries the demarcation between land use planning and pollution control is important, certainly when different authorities are at the steering wheel in these regimes (see e.g. Davoudi, 2000). Local councils tend to play a greater role in land use planning than in pollution control and are sometimes tempted to stretch the boundaries of their authority. In all three cases, the courts have played an important role in demarcation problems, commonly deciding that land use planning can only impose conditions for planning reasons[9]. Especially in the UK and Canada, central supervision over local authorities strongly reinforces this norm. More generally, local government is often seen as a service rendering government level in the UK and Canada (see e.g. Hodge, 1998). This explains perhaps why the most important reason for central supervision in the UK '*is to ensure that the interests of the nation as a whole will take precedence over the preferences of particular communities*', with the national interest '*almost invariably defined in economic terms*' (Vogel, 1986: 130).

To some extent the problems of demarcation have become less relevant. This is a consequence of planning theory developing in less pretentious

[9] In the UK, judicial review has resulted in the requirement that conditions must be related to 'planning purposes' and to the development in question. Planning purposes include the siting of buildings, their number, area, height, and mass. Design and external appearance, means of access, landscaping, impact on neighboring land, availability of infrastructure, traffic considerations and communications are also proper planning purposes (Moore, 1997: 193). Similar requirements apply in Canada and The Netherlands. In the Canadian province of Alberta for instance decisions in the planning sphere must be based on 'valid planning considerations'. The meaning of the concept is not always exactly clear, but regulations may certainly relate to issues such as lot size, yard space, building bulk and form, off-street parking, and aesthetic standards. Furthermore conditions must be supported by reasons and these must be substantiated (see Laux, 1990: 155-156, 297).

directions after a crisis in planning theory (see e.g. Rydin, 1998)[10]. Secondly, it is a consequence of the neo-liberal ideological climate in the 1980s and some of the 1990s that reduced the extent of the planning regime considerably. All state activity was suspect, but this was even more true for land use planning. Ironically, the drive to deregulate was directed quite centrally (mainly by Central Planning Guidance) and implemented in a managerial fashion. Comparative tables on the speed of land use decisions especially were a favorite means of central government. It should be noted however that neither in the UK, nor in the two other countries have hazardous waste facilities become exempt from land use planning controls. The sketched regime is thus still relevant.

Pollution control

After the Second World War, pollution control was initially far less a priority for the three countries than land use planning. Take for instance the Dutch central government guidance on the Nuisance Act to the municipalities implementing this act. It is stated that the introduction of much previously unknown industry would sometimes imply an impossibility to determine appropriate license conditions. Because of that '*it could be acceptable that a facility operates without a license*' (see Smits, 1995: 8). More generally, municipalities were kindly requested not to crack down too hard on companies which work without a license as closure of facilities (the only available sanction in the act) would '*harm rather than benefit*' industry (ibid.). The situation changed from the 1950s onwards, with the UK leading the way as a pioneer. Policy development in this field was driven by accidents and new scientific knowledge and thus somewhat less by ideology. Take for instance the introduction of stricter air pollution controls in the UK through the Clean Air Act in the 1950s. It was a response to the smog problem, which caused an estimated number of 4,000 deaths in London in 1952. The Act gave local authorities the power to declare smokeless zones within their jurisdictions. Somewhat later it also doubled the number of plants under the jurisdiction of the Alkali Inspectorate, which in effect meant that the most complex industrial sources of air pollution were regulated by central government. The subsequent improvements in local air quality greatly contributed to the sense that the existing institutional arrangements had adequately reacted and gave those within the system their sense of a 'proud record' (Hajer, 1995).

Like in the Netherlands and Canada, interest in environmental issues in the UK continued to surge in the 1960s and 1970s, leading to political responses in the forms of new legislation and policies on the one hand and

[10] This effect is somewhat countered by the fact that planners still tend to fall back on the comprehensive rational planning model (Rydin, 1998: 174).

newly created organizations on the other. Among the legislation involved there was a fair amount concerned with waste as I have already indicated. Here too there was an effect of the changed ideological climate in the 1980s and 1990s. A statement by Prime Minister Margaret Thatcher on the Falkland war indicates the UK government stance towards environmental issues in that period. '*It is exciting to have a real crisis on your hands, when you have spent half your political life dealing with humdrum issues like the environment*' (Rose, 1990: 314). Despite such statements, the pollution control regime was not nearly such a nuisance in the eye of the Conservatives as the planning system. In fact, international and national concern forced the Conservatives into a White Paper on the Environment in the early 1990s and Mrs. Thatcher then portrayed herself as an environmentalist.

Some of the issues that came forward regarding planning control deserve treatment for pollution control as well. Among these issues is the relation between elected representatives, the language used to justify decisions and the role of science. It should be noted that all these issues in pollution control are profoundly affected by, if possible, an even deeper-rooted idea that the policy field is technical in character. Whereas I have often read the statement that planners actually have little specific knowledge that a lay person could not understand, such statements are far less often made on environmental issues. Yet, the system of pollution control in the three countries does not emphasize technical quality of decisions, but the economic quality. This is an indication that regulation is not set up in line with the managerial discourse, but more in line with the conservative pluralist discourse. It must be noted that in all three countries experts receive political guidance at a very abstract level. Legislation in all three countries has tended to be 'enabling', meaning that authorities are created in the name of a certain purpose that is not defined with great precision (see e.g. Harrison and Hoberg, 1994: 10). Environmental decisions must often be 'in the public interest', and other examples of such relatively vague goals are easy to find: 'safety' must be guaranteed, conditions must result in the 'Best Available Technology Not Entailing Excessive Costs' (BATNEEC), emissions must be 'As Low As Reasonably Achievable' (ALARA), 'Best Practicable Means' (BPM), etc. Terms such as BATNEEC and ALARA are core principles of pollution control in all three countries. The dominant concern that these concepts express is that costs must be 'reasonable', 'not excessive' and options must be 'practicable' in other words not involve too many costs. It follows that the three systems of environmental regulation are not about eliminating environmental pollution and environmental risk, but about 'optimizing' it. This optimization occurs in a utilitarian fashion, that is, by comparing the benefits of reducing pollution to the costs of reduction and control (see Vogel, 1986; Hajer, 1995; Weale, 1997). Experts do the optimization work, not politicians. No representative body in any of the three countries has with any precision defined what the terms mean and leave that to the experts to determine. This fits the description of conservative pluralist institutional arrangements.

Perhaps the dominance of the conservative pluralist discourse explains why in all three countries there is so much cooperation between government and private experts. One common characteristic of pollution control in all three countries is that there is a strong degree of cooperation between administrators and experts employed by industry in this process (Harrison and Hoberg, 1994; Vogel, 1986; Hajer, 1995). Only in the Netherlands of the 1970s was such cooperation (and then again only at the national level) lacking and were relations between administrators and industry rather antagonistic (see Leroy, 1994). In that period, officials at the Ministry were relatively zealous and saw industry as their enemy. Their actions appeared to have been motivated by progressive pluralism as they set out to redress inequities caused by pollution.

The agenda of the Ministry changed in the 1980s so that the cooperative tradition of earlier decades was picked up again. Note that under a managerial discourse, private parties are not trusted to co-determine the definition of the public interest, but that the emphasis is on administrators in the service of the state. In other words, if this discourse is dominant, there would have not been any discursive space to justify the cooperation with industry. In reality however, the cooperation between state and private experts is strongly defended in all three countries. The forms in which this occurs differ somewhat, with the UK regulators stressing the need to be able to make case-by-case decisions for every company, and the Netherlands and Canada aiming for national consultations that lead to agreement over proper norms[11]. The main justification for the close working relationships tends to

[11] Canadian regulators have standards, but are said to often treat such formal standards as little more than a starting point for negotiations with individual firms (MacDonald, 1991). In this sense, Canada takes a middle position between The Netherlands and the UK, because UK regulators have been and still are averse towards uniform standards (except in the water field, see Lowe and Ward, 1998) and The Netherlands fully embraces standards. Although European regulations now force the UK to specify standards, UK regulators still prefer to decide license conditions on a case-by-case basis. There have been informal standards - 'presumptive limits' developed by HMIP regulators in the course of their work (Vogel, 1986: 76-79) - but such standards have generally been secret and great care has been taken to avoid that they would obtain binding force. It appears that UK firms have not generally pushed for the development of binding norms although

be that private market parties have a better information position than government bureaucrats (Nemetz, Stansbury and Thompson, 1986) - which sounds hardly managerial. The impression of administrators lacking resources appears to be correct as in all three countries, regulators are often not economically trained. Environment Canada for instance is described as an 'economically illiterate' department (see Doern and Conway, 1994). This sheds a strange light on the fact that the cooperation between regulators and private parties is often described as a negotiation process between the parties involved. As MacDonald has noted: bargaining is the essence of the environmental regulatory process as it is practiced in Canada (1980: 33). But if the parties at the negotiating table are not equal, it should be no surprise that the critique that is often expressed towards pluralist institutional arrangements - outcomes reflect bias in access and capability - is often expressed in all three countries. The Canadian system of regulation for instance is considered to be vulnerable to subversion by particular interests as there are few internal checks within the bureaucracy to guarantee a balanced decision and to avoid the exclusion of relevant information (Nemetz, Stansbury and Thompson, 1986)[12]. The system of cooperation has however not apparently undermined the credibility of the regulatory system in either country.

The conservative pluralist accent in the regulatory system is somewhat greater in the UK than in the Netherlands or Canada as the starting point of regulations in the UK is that activities must be approved (this is similar to the land use planning system). Take section 5(3) COPA 1974 for instance. On licensing, this section reads that '*it shall be the duty of the authority not to reject the application unless the authority is satisfied that its rejection is necessary for the purpose of preventing pollution of water or danger to public health*'. The legislation in the Netherlands and in Canada is framed differently, as they start with a general prohibition to undertake certain activities unless they are authorized by the administration. Take section 9(a) of the Ontario Environmental Protection Act. '*No person shall, except under*

---

they participated in the development of informal norms on the exact meaning of terms such as Best Practicable Means (Ball and Bell, 1995: 26). In The Netherlands, the courts have recognized standards developed in consultation with industry and have decided that licensing authorities may only deviate from them if there are convincing reasons such as local environmental circumstances.

12 The little empirical evidence from international comparisons that is available however, does not suggest that Canadian or UK regulators are 'weaker' than their American counterparts who operate under great openness (Harrison and Hoberg, 1994; Vogel, 1986). The difference between the three countries seems to be located in the starting point of the debate, in the aggregation rules, the role of science and the location of negotiating processes. These issues are interrelated.

and in accordance with a certificate of approval issued by the Director, construct, alter, extend or replace any plant, structure, equipment, apparatus, mechanisms, or thing that may discharge or from which may be discharged a contaminant into any part of the natural environment other than water'. Obviously, under such a rule, the approval to undertake a certain activity is something akin to an allowance of something that is actually forbidden. It can easily be seen that such different starting points lead to different aggregation rules for concrete license decisions. Specifically, the burden of proof - to demonstrate that conditions are warranted - in the UK is on government, whereas the burden of proof in Canada and the Netherlands - to demonstrate that certain conditions may not be warranted - will more easily be put on applicants[13]. Note the long existing resistance in the UK against the precautionary principle. Whereas initially outright rejecting the principle, the UK has formally adopted it but its status is still politically and legally equivocal (Jordan, 1998: 176). Canada has not been very receptive to the principle either, but the first provinces have started legislating it (Benedickon, 1997: 18-20). The Netherlands has embraced the concept for some time now - certainly on paper.

The role of science - important in relation to the burden of proof - seems to differ somewhat between the three countries. First of all, the UK positions itself, particularly in the context of European negotiations, along managerial lines. One British minister in the 1980s: '*I have been struck by how often we appear to be dealing with subjects not really on the basis of an objective assessment of environmental priorities but as the result of the changing fashions in pressures from outside. It is necessary in an area which should be science-based to put up pretty formidable hurdles and tests of scientific nature if we are to make rational priorities*' (see Weale, 1992: 80-81). There are apparently different 'types of science' too. Hajer (1995) has indicated that the British rely on 'research science' which mainly implies that heavy burdens of proof must be met before it is believed that a certain problem exists or a regulation that limits the freedom of market parties will be introduced. Take Pollution Paper number 11, dating from 1977, for instance. It indicates that: '*it is important to ensure that the standards being imposed do not rest on an unsound scientific justification or require disproportionate economic costs, since this would make it difficult later to introduce further*

[13] Since the early 1990s, this has shifted somewhat in the UK. HMIP guidance on Integrated Pollution Control for instance started indicating that when decisions are made on authorizations, operators must demonstrate which options they have considered when selecting the best techniques. And when the best technique is not put forward, the applicant must justify this by showing that the technology would entail excessive costs.

measures, however well founded' (Hajer, 1995)[14]. Although this managerial reference to science sounds pretty impressive, it must be noted that the UK government (at least under the Conservatives) was not very willing to spend money on developing such science. In the sphere of waste management for instance, research activity was effectively halted after 1980, pretty much to the disgust of UK expert bodies such as Harwell Laboratories[15]. It is indeed somewhat schizophrenic to indicate that policies will also be changed on the basis of scientific findings and then considerably reduce the fund for scientific research. The Dutch use of science, according to Hajer, is akin to 'regulatory science', which is more reflective of the precautionary principle and accepts somewhat lower standards of proof. In Canada, Harrison and Hoberg (1994) find that Canadian regulators are trusted a great deal by the public, which gives them broad discretion and room for the application of 'research science', or even very personal judgements of the costs and benefits involved. Canadian regulators greatly benefit from the large amount of research that is done in the US. The practice of environmental decision-making in that country is often considered highly antagonistic, with science being used as a weapon by contending parties. Harrison and Hoberg make the point that the extensive use of science in the US is rooted in the lack of trust in government regulators. Science is apparently the only authoritative source that people will accept, but science has become more and more corrupted as a source in the eyes of the public by the fact that apparently any claim can be scientifically supported. Thus, the US experience increasingly demonstrates an inability to solve conflicts by the use of science.

3.2.4 Review

The three institutions of interest in this paragraph: representatives, experts and private market parties are all highly relevant in the collective choice debate on regulations as the managerial and conservative pluralist discourse have dominated much of the debate since the inception of environmental controls. The basic position is that private parties are free to act as they please with Parliament having powers to reduce this freedom. At the start of industrialization, the political system was very much based on conservative pluralism, but also a version of it that saw little use for much government interference, which was not considered to be in the public interest. The definition of the public interest was changed in a complex

[14] A similar emphasis is to be found in the government's white paper on the environment in 1990 (Weale, 1992: 81).

[15] There are indications that the UK science on landfilling was far from complete too. The problem of landfill gas for instance was 'discovered' after (parts of) certain landfills exploded.

process that can not entirely be understood from the perspective of discourse alone. Various factors came together to change the dominant definition of the public interest that had existed until the mid 19th century. These factors included the increased value of labor both by labor shortages and unionization, the fact that poor public health also was a threat to people who were well to do, empathy with the laboring class, the insight that regulation could reduce competition somewhat or increase property values and the arrival of a professional class. The professional class played an important role in solving the tensions that were present at the time. Note that in all three countries the progressive pluralist discourse - specifically the concern over distributive justice - was initially quite strong amongst reformers, but had relatively little success. Only after they reframed their arguments in managerial terms they started succeeding in gaining support for their agenda. The magic formula was to introduce regulations and have them administered by neutral professionals which would maximize utility for all. The professionals saw a great role for themselves in the society that was emerging, but were kept in check by the other institutions in all three countries. This worked in various ways. Initially the regulations that were introduced implied a rather large degree of interference over private property and conferred quite a bit of discretion on part of administrators. In the Netherlands, the courts restrained this discretion, in Canada it was the planning boards that controlled the implementation of the legislation. In the UK, a certain degree of discretion was left intact but planners do receive guidance that sounds strongly conservative pluralist. The most fundamental difference in outcomes of this complex interplay seems to be that the UK system, despite its greater discretion for managers, appears to favor private initiative to change the existing situation more than the two other countries do. The managers are therefore clearly instructed to exercise restraint in their regulating tasks. In the Netherlands and Canada, zoning ordinances protect existing land uses, whereas the UK system protects the right to develop too. In pollution control, the most important difference appears to be related to the place of uniform standards, with the UK rejecting them and Canada and the Netherlands embracing them. Here too is the effect that UK regulators have greater discretion, which again must be used in a very restricted fashion so as not to overburden private development. Given the important role for scientists and scientific information it is quite remarkable that the landfilling of wastes was looked upon so differently in the three countries. In the UK landfilling of hazardous wastes along with other wastes (co-disposal) is practiced up till today, and supported by scientists (who do not have enough resources in their opinion). In Canada and the Netherlands, by contrast, co-disposal was stopped decades ago. The different approaches seem to reflect, more than an important role for science per se, the previously stated preferences of the UK government for research science and avoiding overburdening of industry, whereas the Canadian and Dutch policies reflect their preferences for regulatory science and a greater willingness to intervene in the economy. It is interesting that since the end of the 1990s, a research program of older landfills in the Netherlands, where co-disposal had taken place, seems to indicate relatively little problems of contamination[16]. The basis for the Dutch policy had been a fear of contamination of groundwater by leakage and non-degradable toxins. This did apparently not occur at the anticipated scale, giving the often-ridiculed UK point of view that mixing hazardous wastes with biodegradable wastes helps degrade toxins more credence.

3.3 The role of the courts

3.3.1 Introduction

This section is about the role of the courts in the regulatory process of the three countries involved. I address various important issues that determine the importance of this institution. In section 3.3.2, I analyze the general position of the courts in the governmental system of the three countries. The basics of this position are broadly similar, implying that the courts are supposed to defer to Parliament, but may be critical towards the administration. However, there has been quite some debate about judicial activism and section 3.3.3 discusses the extent to which courts are indeed very active in the three countries. In this respect, the Netherlands occupy a distinct position as they have 'administrative courts', which have traditionally been seen as part of the administration. This implies they have great freedom in deciding issues, a freedom they have not given up on now that administrative courts have been cut loose from the administration a decade ago or so. Section 3.3.4 is about the authorities of the courts once they have taken a decision. The courts tend to judge only individual cases that are brought before them. This implies an important limitation, which is that they are not in a good position to develop policy, as policy development requires a certain degree of oversight over all relevant cases. This implies that the authorities of courts are mostly negative: they can quash a decision, but often not replace it with their own. Having extensively discussed the authorities of the courts, it is also important to look at the parties who initiate court cases. The usefulness and usability of the courts for parties who seek 'justice' depends partly on the boundary rules that apply to the process of seeking access to the courts, which are discussed in section 3.3.5. In all three

[16] See 'Zelfreinigend vuil. Afvalscheiding is wellicht contraproductief', in: NRC-Handelsblad, 6 March, 1999.

countries, rules on 'standing' before the court have 'widened' considerably, but more so in the Netherlands than in Canada and, particularly, in the UK. Access to the courts in the UK is restricted in part because a rich practice of quasi-judicial procedures exists. Such procedures are not before a fully independent court, but an effort is made to hear all parties involved and to get a third party's view on the issue at hand. The most important quasi-judicial procedure in the UK is that of the public inquiry. In Canada, the practice of quasi-judicial procedures also exists, mainly in the form of appeal boards, where plaintiffs can argue their case against the administration. Quasi-judicial procedures are discussed in section 3.3.6. Finally, in section 3.3.7 another issue that is relevant from the perspective of plaintiffs is addressed: the issue of costs. Here, Canada stands out as both the federal and the provincial governments are working with 'intervenor funding', which implies they give plaintiffs money to argue their case. This phenomenon is unknown in the UK and the Netherlands.

3.3.2 The general place of the courts in government

The role of the courts is limited in all three countries studied here. There is a long tradition of deference to elected bodies, especially Parliament, in all three countries. As a correlate of the deference to Parliament (or other elected bodies), courts have been hesitant to interfere with decisions of administrators too. Here one sees the consequence of the interwovenness of experts and elected representatives resulting from the interplay between the pluralist and managerial discourses that I described in the previous section. There is however a certain development toward greater activism by the courts visible, be it to a differing degree. If present, such activism seems inspired by a continued process of norm development in the sphere of relations between the state and its citizens, which is often called 'judicialization'. There appear to be two reasons for judicialization. First the increased importance of (international) legislation that is above ordinary law gives the courts a means for judging government actions. In the UK and the Netherlands, European legislation (including the European Convention on Human Rights) has had a profound effect on this process - even going so far as to open up the possibility of judicial testing of national legislation. In the UK and Canada development towards 'constitution like' legislation such as the Human Rights Act in the UK, the Charter of Rights and Freedoms in Canada, and the Constitution Act for Canada also had or is predicted to have important consequences[17]. A second factor is the work of judges themselves.

[17] '*Many conservatives are terrified that it will wreak havoc with many of Britain's treasures institutions, giving too much power to interest groups like gays, minorities, criminals and women at the expense of government authority.*' in: 'As a

Starting from the most blatant abuses of government power and then moving towards the more subtle cases, the courts have developed certain norms for reasonable government conduct.

The Netherlands is an interesting case to see both factors at work. The country has had a constitution since the late 19th century, but that has not particularly increased activism of the courts. Conflicts that involved governments, even if property was involved, were seen as a no-go area for the courts for long. Indeed, when courts started being instituted for such cases, these were administrative courts, that were long seen as part of the administration. However, administrative courts did not by far cover all areas and private courts also started slowly developing norms for government behavior. This is a dangerous game for courts to play, as in the 19th century Dutch governments had claimed the right to stop court cases from proceeding, but in the 20th century such counter actions were not undertaken.

The practice of having administrative courts implies a stark contrast between the Netherlands and the UK and Canada where the courts were outside the administration. Yet it seems that the need for recourse within the administration is quite great, because both Canada and the UK have founded quasi-judicial procedures that provide appeal opportunities for certain parties and which are akin to the old Dutch practice of administrative courts. Quasi-judicial proceedings in the UK and Canada and Dutch administrative courts reflect a mixture of pluralist and managerial thought. It is recognized that there is a conflict between two parties and counsel will often represent these. On the other hand, the emphasis is on truth-finding and taking the right decision. The atmosphere is more informal than in court, and the chair of these proceedings has greater responsibility for probing the issues than an ordinary judge. It is essential to realize that administrative courts and quasi-judicial proceedings occur under the authority of an elected representative, usually the head of the relevant part of the administration (e.g. the SoS for the Environment). This person can, directly or indirectly, set the terms of the debate and can interpret the outcomes of the debate relatively freely, although ignoring the advice results in loss of credibility and is thus rare. However, the possibility is there, and must be seen as a sign of the dominance of the mixture of pluralist and managerial thought that is embedded in the institutional arrangements that were discussed in the previous section.

The Dutch practice of seeing administrative courts as part of the administration was shocked in the late 1980s when the European Court

---

British Bill of Rights arrives, courts brace for action', International Herald Tribune, 2 October 2000, p. 5.

condemned this system as offering no independent review of decisions. Since then, administrative courts have been placed outside the remit of the administration. Interestingly, the tradition of relatively deep testing of governments has continued however, which implies that the Netherlands have a far-reaching review system which is seen to be at odds with representative democracy.

The category of quasi-judicial proceedings is especially relevant in the UK, where an important avenue for environmental decision-making is formed by local inquiries, but also in Canada, where there is a tradition of administrative tribunals. In the UK, inquiries are called quasi-judicial procedures as they resemble court cases in many respects, even though they are not chaired by judges and are intended to be less formal through a lack of formal rules of behavior. Furthermore, inquiries do not result in a definitive decision, but merely an (important) input to the Secretary of State, who sets the terms of the debate and after the inquiry decides the issues. It is then possible to appeal this decision in the courts, albeit that pay-off rules, be it that pay-off rules in the UK are posing a serious hurdle for those who seek judicial review. In Canada, administrative tribunals are to provide recourse for parties who do not agree with bureaucrats' decisions. Examples include the Natural Resources Conservation Board in Alberta. Administrative tribunals in Canada tend to operate less formally than the courts, normally provide broad rules of standing, and are regarded as a part of the administration which explains their capability to 'retry' the issues before them as a 'trial de novo'. In some cases, their conclusions are the final decision of the administration, in others they are just an advice to the minister. Consequent judicial review of federal tribunals' decisions falls either to the Federal Court of Appeal or the Federal Court Trial Division. Provincial Superior Courts have jurisdiction to review in all cases other than those under the Federal Court's decision. It is argued that Canadian courts over the last two decades have tended to show increasing deference to the decisions of administrative tribunals (Tollefson, 1998). If the courts decide that deference is warranted, they will only interfere with patently unreasonable decisions.

In each country, the dominant belief appears to be that judges have no democratic accountability and must therefore be very careful in determining the boundaries of their own authorities. Despite that being the case, the courts (and inquiries for that matter) have played an important role in the interpretation of statutes, e.g. through the interpretation of the intention of lawmakers and of inconclusive concepts. It is largely the result of court verdicts that environmental groups, for instance, have gained access to the courts. This is because, traditionally, boundary rules for access to the courts have stressed ownership, for instance of land, as a requirement for access to the court. The process of widening of boundary rules to environmental groups often occurred gradually, with the first step being the admittance of groups who had landowners among them. The second step is then the acknowledgement that some environmental groups must be having an interest in the decision per se (thus without land ownership). The third step is the admittance of practically any group or person. The three countries here do differ in the amount of money available to bring court cases (negligible amounts in the Netherlands, considerable sums in the UK) but all tend to leave the costs of representation to the parties. Only in Canada has the concept of intervenor funding been of real substance in environmental issues.

Some specifics of the three countries

In the UK, the tradition of judicial deference to Parliament goes back to the Bill of Rights in 1689. Until that day courts had considered themselves capable of testing acts of Parliament against ‘common right and reason’. The doctrine of supremacy of statute law that followed the Bill of Rights is seen as self-imposed by the courts and as a consequence of the liaison between the courts and Parliament in their struggle to control the powers of the King (Norton, 1998: 408). However, courts in the UK are not subordinated to the administration and can interpret the meaning of laws and review the actions of the administration, as they did with the first planning legislation. The language in the law, however, does certainly not stimulate judicial review. Except for a very few situations, local land use plans but also SoS decisions under the TCPA 1990, ‘shall not be questioned in any legal proceedings whatsoever’ according to section 284 (1) of the TCPA.

Whereas the courts in the UK started to become more accepting of the need for regulation under administrative law in the early 20th century (see the previous section), the courts in Canada resisted for longer. Canadian courts continued to apply traditional nuisance law to public nuisances until 1930. It is interesting that whereas in the UK the courts were apparently protecting landowners (including industrialists) from restrictions on the free use of their property rights, in Canada the courts were unwilling to accept the infringement that industrial activity brought to the free enjoyment of property by adjacent landowners (Schrecker, 1992). In some cases, the resistance against industrialization pitted the courts against provincial legislatures. As late as 1949, a court decision was dissolved by the provincial legislature of Ontario because of this reason.

In the Netherlands, the distinction between private and administrative courts is important. Although private courts during the brief French rule in the early 19th century could be stopped from interfering with any administrative action, this system was done away with quickly after the French left. Matters that involved private property could from then on be heard exclusively by private courts. Because the courts were quite active in testing the legality of government action and the administration had in fact only a limited set of acts to work with, the French system was reintroduced from 1822 to 1844. Since then, the courts have been free to interpret their own jurisdiction, which was however restricted to issues that involved property (De Haan et al., 1986: 176-177). Subsequent Constitutions enabled Parliament to establish administrative courts, and then decide the division of labor between these new courts and private courts, but the attempts to do so failed. Van der Hoeven (1989) provides some clues as to why this was the case. He argues that the difference between implementation of the law and the acceptation of the law was long neglected in Dutch practice. This implied that actions by the bureaucracy were considered to be backed by Parliament. If a citizen wanted to appeal administrative decisions, he would therefore have to seek recourse with Parliament and not the court. This rather orthodox view of the administration ignored the fact that Parliament could hardly be expected to actively fulfil the role of overseer of the administration. So strong was the adherence to the orthodox view of democracy, that even the obvious abuse of government power was accepted as a consequence. Public authorities in the Netherlands at the end of the 19th century sometimes demanded that property owners relinquished property in return for construction permits. The standing doctrine among lawyers, administration and courts in that period was that such abuse should be held in check within the bureaucracy or by overview by representative bodies, but not by the courts. Apparently the judges themselves shared this view. There were some attempts to legislate for administrative courts in the beginning of the 20th century, but the debate was extremely confused and did not lead to approval of any legislation.

Progress towards the current situation of independent administrative courts stemmed in the end from various different sources. The first was the installation of checks on the administration by higher authorities and in the end by the State Council (seen as part of the administration). Secondly, appeal mechanisms were introduced in a limited number of areas such as employment of administrators. Finally, private courts started developing, from the 1920s onwards, a theory of unlawful state conduct and judging administrative actions. All three practices contributed to the development of norms regarding the exercise of discretion. Was administrative discretion initially seen as normatively desirable and not to be bound by any norms, gradually norms developed that bound the administration. Private courts developed more and more norms for reasonable government behavior, as did the administrative appeal courts, which for instance started testing the question whether certain decisions were really in the public interest or not. Checks on the administration were deliberately not intended to serve as recourse for ordinary citizens, but as a means to determine the public interest when several government bodies could not agree what the public interest was. To determine this interest however, several 'general principles of good governance' ('algemene beginselen van behoorlijk bestuur') were proposed in the 1950s. These principles were deliberately not presented as appeal grounds for the affected citizen, but as principles for administration (ibid.: 186-187). However, they did evolve into appeal grounds later. This was a phased process, with boundary rules initially allowing citizens to present their case to higher authorities on the basis of these principles (e.g. contest a municipal decision to the provincial executive) and the Minister deciding the issue in the end. The State Council played an important role in this because it advised the Minister (formally 'The Crown') on how to decide. When this system was found to not provide independent review of government decisions in 1989 through a verdict of the European Court of Human Rights[18], the system was overhauled, with the State Council now providing the final decision. Because the general principles of good governance have continued to receive statutory backing, they became the norm for State Council decisions and the decisions of lower administrative courts which were instituted after 1989.

3.3.3 Judicial activism?

Introduction

It is argued that since the 1960s, the UK courts have become more active in performing tests of administrative decisions, even dramatically so (Sunkin, 1994: 126). Johnson (1998) suggests that there are various reasons for this, including a growing influence of the European Union and personal factors such as the preferences of certain judges. Most profoundly however, the increased importance of the courts, according to him, must be seen as a reflection of the movement towards 'an articulate rights-based culture' where trust in representative bodies and the administration is less than traditionally was the case in the UK. In Canada, any role the courts play should be seen in the light of their own conception that '*their task is to review the legality not the wisdom of a decision or regulation*' (Tollefson, 1998: 241). Canadian courts traditionally did not attempt to review cases on

[18] European Court for Human Rights, 23 October 1985 [AB 1986-1].

the basis of the facts, as long as administrators operate within their jurisdiction and act on the principles of fundamental justice and due process. Even today, when the scope for judicial review on the basis of new legislation may have expanded somewhat, the courts have consistently continued to defer to legislators (Hessing and Howlett, 1997: 207-208). In the Netherlands, there is currently a certain amount of complaint about the judicialization of government. A special government commission was instituted in the late 1990s to study the phenomenon. An important scapegoat was the General act on administrative law of 1993 ('Algemene wet bestuursrecht') that was explicitly modeled on the vision of government and citizens being equals, with citizens having a certain right to expect proper behavior from government. The act profoundly affected some standing Dutch practices, including the ideology on compensation. Dutch administrative law was never overly generous towards landowners with regard to compensation in accepting damage to private property in the public sector and awards compensation only if private interests are 'unreasonably heavily' affected. This principle is referred to with the French phrase 'égalite avant les charges publiques', which implies that all must bear a little in the public interest (see Van Zundert, 2000). Zwerver (1975: 107) notes that the intention of the Dutch government in the subsequent versions of the Land Use Planning Act Compensation that compensation would never be awarded. Therefore, landowners could only expect compensation for unlawful government conduct. The 1993 act changed the situation in so far that it explicitly endorsed a doctrine that had started to emerge amongst judges in the 1980s. This doctrine holds that damage caused by a government decision, which can not reasonably be expected to be borne by a private party, must be compensated (see Van Zundert, 2000: 341). So whereas the criteria to award damages have not changed so much, the 1993 act is far more explicit in stating that compensation must be offered than was the case under the Land Use Planning Act. The indirect effect of the 1993 act on government practices has triggered a reaction. A discourse coalition consisting of the development industry (including a strong infrastructure lobby) and various elected politicians who saw their decisions reversed by the courts expressed concern over the inhibiting effects that the legislation had on the administration[19].

[19] Most notably the Queens' Commisioner for Noord-Holland, Van Kemenade. He chaired a working group that published a report 'Bestuur in het geding' - perhaps best translated as 'government in court', which called for restraint of the courts vis-à-vis the administration. This report was so influential that the national government studied the abolishment of appeal rights for certain groups (De Gier et al., 1999).

Grounds of review

In the UK, much of the overview of administrative bodies is done by the High Court. The main ground for review is that bodies must remain within the statutory powers given by the law (see Ball and Bell, 1995: 29-35). To step outside these boundaries would constitute a decision that is 'ultra vires'. There are three reasons why decisions could be declared ultra vires and these are the making of a decision without the power to do so, the improper exercise of discretion and procedural unfairness. The improper exercise of discretion is said to have occurred when irrelevant considerations have influenced decisions, when relevant considerations have not influenced decisions, and decisions that are so unreasonable that they could only have been made in bad faith. Procedures must be fair, and this is taken to imply that everyone is entitled to a fair hearing (to answer cases brought against him and to argue a case from their perspective) and that there is no bias in the decision process. The tests that apply to all administrative decisions in the UK also have relevance for planning decisions. In general, it can be said that the courts do not wish to interfere with issues of planning policy, and therefore, as long as the policies that are being applied are lawful, they will not intervene. The courts 'see their role as ensuring that decisions are made rationally in the light of all planning considerations' (Ball and Bell, 1995: 235). The way of ensuring this is to see that reasons are given for decisions and that all relevant issues are taken into account. The specific criteria for testing planning decisions are contained in 'Seddon Properties Ltd. vs. Secretary of State for the Environment'[20] and confirmed in 'Centre 21 vs. Secretary of State for the Environment'[21]. First, the principles are first, that the SoS must not act perversely. Perverse action is considered to exist when no reasonable person in the position of the SoS, properly directing himself of the relevant material, could have reached the decision that is being challenged. Second, the SoS must not take into account irrelevant material or fail to take into account what is relevant material. Third, the SoS must abide by statutory procedures and fourth, must not depart from the principles of natural justice (Moore, 1997: 325-326).

Similarly, but I would also say more far-reaching, grounds of review exist in Canada, where eight - long-standing - grounds of review have been codified in the Federal Court Act. These grounds are: substantive ultra vires, failure to consider relevant considerations or the considering of irrelevant considerations, unlawful fettering of discretion, real or apprehended bias, breach of procedural fairness, exercising discretion for improper purposes or in bad faith, unlawful sub-delegation of discretionary power and errors of

[20] ([1978] 42 P & CR 26).

[21] ([1986] JPL 915).

law/fact. The grounds for review in provincial courts are roughly similar but not codified. The meaning of these grounds can be found in various court verdicts (see Tollefson, 1998: 250-257). From the Halfway River First Nation case[22], it can be concluded that considerations that are specifically mentioned by the statute or implied by it, for instance the protection of a forest, are relevant and must be factored into the decision of an administrator. Unlawful fettering of discretion is assumed to have occurred when an official simply complies with the direction from a superior rather than making his or her own decision on the merits of the case. Apprehended bias is considered to have occurred when the decision-maker has not brought an open mind to the decision process and has pre-judged the issue. Procedural fairness is generally taken to mean that notice is given to parties and that an effective opportunity to be heard is provided. Errors of fact occur when there is absence or too limited presence of evidence for some of the findings of fact which have been taken into account by the decision-maker.

In the Netherlands, the change from administrative appeal within the state, which seems to fit mainly a managerial approach to decision-making, to independent administrative courts, has implied various changes. In the past, the appeal bodies could decide issues anew, as if they were the initial authority taking the decision. This implied they could step out of the bounds of the conflict presented to them and could in fact decide to change the decision in the opposite direction of what the appellant had requested ('reformatio in peius'). Some traces of this are still present as administrative appeals are decided in a relatively informal atmosphere. The emphasis is still on truth finding, which implies that the judge may engage in independent fact-finding and may see whether other grounds of appeal apply to certain contested conditions of a license. The practice that the State Council may request the advice of an expert official who works for the national government ('Ambtsbericht') has remained. However, legal experts have seriously questioned the independent position of such experts. The issue is important because the advice of the State Council tended to conform to the expert advice to a high degree (in the 70-80% range). In response to the critique an association has been founded with government subsidies (The Association for Advice on Administrative Courts, or 'STAB') that now advises the courts on technically complex issues. This occurs excluding of other experts, which has also raised concern with legal experts because it might create a monoculture of advising (Brenninkmeijer, in: Heldeweg, 1995). Lawyers are very critical of the continuing practice of deep probing by such experts and plead for restraint. Their idea is that the administration

[22] Halfway River First Nation v. British Columbia (Ministry of Forests) ([1997] BCJ no. 1494 (SC)).

should be left with a certain level of discretion and that the courts only decide the issues brought before them by the parties to the disagreement (see Heldeweg, 1995). This advice has been neglected.

The criteria that Dutch courts use in testing government decisions are partly codified, partly not. It is important to point out that concrete decisions and general regulations are considered separately, with the court having great deference to general regulations (most zoning ordinances will be considered general regulations). For concrete decisions, the starting point is lawful conduct of government ('rechtmatigheid'). This is taken to imply eight criteria (see Michiels, 1999). Decisions must be made weighing all directly involved interests, decisions must have consequences that are proportionate to the goals that are envisioned by the act, and government must act consistently (i.e. based on policy) and treat equal cases equally. Government organizations may not abuse authority (i.e. use it for different purposes than for which it was given), may not take decisions that have effects before they were taken unless these were foreseeable, government must act according to expectations it has created, and governments must motivate decisions properly.

3.3.4 The authority of the court

Beside the scope of judicial review, the usefulness of judicial review for litigants depends greatly on the actions that a court can take in response to unlawful administrative actions. In Canada, it is assumed that four possibilities exist: (1) the court may quash the administrative action (certiorari), (2) the court can prohibit the action (prohibition), (3) the court may require the authority to act (mandamus), and finally, (4) the court may clarify the legal situation by declaring existing duties, rights or powers (declaratory relief). What is clear from these options is that in no case the court can replace the judgement of the authority with its own. The situation in the UK is comparable. If the tests of a proper decision are not met, the court will generally quash the decision, but not replace it with its own. In the situation of the quashing of a decision by the SoS, the consequence is to leave the appeal outstanding, and the SoS must make a further lawful action. The action taken to remedy the situation will depend on the defect that caused the decision to be quashed, but it may imply that the SoS must deal with the matter anew ('de novo'). Only in the Netherlands (as a leftover of the past) are courts able to decide the issues themselves, but then only if their verdict leaves only one option open for the administration anyway.

3.3.5 The issue of standing: what are the boundary rules?

In the UK, boundary rules for requesting judicial review on planning decisions are in line with the managerial language, which stressed the necessity for freedom to act independently on the part of administrators. There is a recognition '*that public bodies are vulnerable to actions by busy bodies and meddlers and that some filter is necessary to prevent them being harassed by such people. In addition, the filter is seen as a safeguard against the courts being overflooded with irresponsible applications*' (see Moore, 1997: 333). A distinction is made between appeals on procedural and appeals on substantial grounds. In either case, the person must be aggrieved, which has received a rather limited explanation in the past, but is since the 1960s explained in a more expansive fashion (Moore, 1997: 324; Hough, 1992)[23]. Based on the idea that procedural errors are not as bad as substantive errors, those willing to appeal on procedural grounds must demonstrate not only that they are aggrieved, but also that they have suffered 'substantial prejudice' because of the decision. This explains that there are situations wherein appellants have not been allowed to launch an appeal against serious procedural errors (such as a refusal to hear them): the courts considered they did not suffer substantial prejudice (see Hough, 1992: 323-324).

It is interesting to note that in the case of planning permissions where the applicant made no appeal and where the SoS did not call attention to the matter, different rules of standing seem to apply. In such a case, any third party that wishes to challenge the planning permission may use the procedure provided by the Supreme Court Act (section 31) and the Rules of the Supreme Court (Ord. 53). The procedure consists of two stages. The party seeking judicial review ('the applicant') must first apply for a leave to apply for judicial review to the High Court. This leave is only granted when the Court is convinced that the applicant has an 'arguable case'. An arguable case is supposed to exist when the applicant has 'standing' and has acted promptly. For judicial review of planning decisions, to have standing, one must have 'a sufficient interest in the matter to which the application

[23] In Buxton vs. Minister of Housing & Local Government ([1961] 1 QB 278) it was reasoned that aggrieved persons are not those just simply dissatisfied with a decision, but those who are wrongfully deprived of something to which they are legally entitled. The appellant, who was a neighbor, was therefore denied review. However, there have been more liberal interpretations of the term, for instance in Bizony vs. Secretary of State for the Environment [1976] JPL 306, where a neighbor was accepted as an aggrieved case, and Turner vs. Secretary of State for the Environment (1973) 28 P&CR 123, where the chairman of a local amenity society was held to be an aggrieved person (see for all cases: Moore, 1997: 324).

relates'. The jurisprudence is not particularly clear on what this means. But UK courts consider various factors relevant: vindicating the rule of law, the importance of the issue raised, the likely absence of any other responsible challenger, and the nature of the relief sought (Moore, 1997: 334). Note that some of these criteria potentially imply a broader interpretation of boundary rules than the 'aggrieved party' criterion, and could thus imply access for environmental groups. Hough (1992: 326) however notes that conservation groups often meet 'a somewhat spartan' hospitality from the court. Courts seem to be more sympathetic to those with a close and enduring concern with the subject (e.g. local conservation societies) than to those with more general concerns. Here too, the second requirement for an absence of leave is prompt action. This requirement is met when the applicant has acted within three months after the ground of appeal has arisen[24].

Rules of standing before the courts in Canada had been expanding in the 1970s, 1980s and early 1990s, but the trend is now considered to be in its twilight phase (Tollefson, 1998: 260). To understand this development, one must keep in mind that Canadian courts - correlated to their deferential attitude - have always been concerned about overextension of judicial resources, something that may happen when standing rights become very liberal. Still, the Charter of Rights and Freedoms of 1982 has served as a catalyst in the already occurring development towards greater standing. Prior to 1982, people could seek declarations that legislation was invalid as long as they could show that they were directly affected, that they had a genuine interest in the validity of the legislation as a citizen, and that there was no other reasonable and effective way in which the issue could be brought before the court. With the Charter, a meta-yardstick for legislation and other administrative actions was introduced. The case of Finlay[25] served as an important step in clarifying the new situation. It was established that the courts have a discretion to award 'public interest standing' to challenge the exercise of administrative authority as well as general rules. But, "public interest standing" was limited to situations where no directly affected citizen might be expected to initiate litigation, which seems to reflect conservative pluralist thought as property owners are principally expected to serve the public good, not environmental groups. The purpose of the limitation in 'Finlay' was to screen out the 'mere busybody' (Tollefson, 1998). These

[24] There is a relation with the appeal procedure to the SoS for the Environment here. If a third party wishes to challenge a decision, he may want to wait until he is certain that there will be no appeal. On the other hand, it seems unlikely that a judge will pass a verdict in a matter that is still under appeal with the SoS for the Environment.

[25] Finlay v. Canada (Minister of Finance) ([1986] 2 SCR 607).

restrictions do not apply when a statute provides for appeal to the courts. Such statutes exist, especially in land use planning. Taking the Alberta Planning Act of 1977 as an example, a distinction was made between 'statutory appeal' to the Court of Appeal and 'statutory review' to the Court of the Queen's Bench (see Laux, 1990: 471-493 for a extensive description). The first appeal was for individual decisions, the second for appeals against zoning bylaws. Interestingly, the rules of standing are more liberal for the second type of review, which seems the reverse of the UK situation. In case of an individual decision (i.e. refusal of development permit), the appellant had to be 'genuinely and relevantly affected', had to act expeditiously, bring forward questions of law and jurisdiction, and had to have 'an arguable case' in the eyes of the Court of Appeal. By contrast, appeal against land use bylaws was open to all electors of the relevant municipality, be it that they had to be 'specially aggrieved' and should want to argue that the bylaw was an 'illegality'. Such illegalities exist mainly if the authority involved used the law incorrectly (excess of jurisdiction, no exercise of jurisdiction, and errors of law). The boundary rules in the second type of cases have been expressed rather liberally by the Albertan courts, and include interest groups.

In the Netherlands, rules of standing in planning law have been expressed quite restrictively in the first half of the 20th century, stressing access to the courts for landowners (see De Gier et al., 1999). The dominant thinking on the subject changed in the 1950s. The courts had already started liberalizing rules of standing to those with a clear interest in decisions (such as farmers who rent the land in question). When a state committee advising on a new planning law considered the issue, managerial concern for the quality of the decision in technical terms prevailed in the deliberations. The committee advised introducing the so called 'actio popularis' in planning law. This implied that any party could file objections about proposed land use plans with local council. This advice was followed and ended in the Land use planning law of 1962. Helpful in overcoming any possible concern about the effects were the positive experiences with the reconstruction legislation, which had also known an actio popularis, and local practice, which effectively defied official rules of standing by accepting arguments from anyone. Pretty much unforeseen, this also implied access to the courts, because the rules of standing in court were linked to the right to object to council decisions. A similar line of development is visible in pollution control legislation. The actio popularis has been somewhat contentious since its inception, especially by those who were concerned with the speed of procedures. Empirical research confirmed however that there was little effect of these liberal standing rules as most participants in environmental appeals have an accepted interest in the matter anyway, as the courts tend to accept pretty much any environmental organization as having an interest in environmental cases. Research in 1999 (De Gier et al.) showed that only in a very limited number of cases the actio popularis leads to additional participation (1% in a sample). Furthermore, the arguments brought forward do not seem much different to the ones brought forward by the parties with an interest. Therefore, little extra work was generated by these rules of standing. This relatively unprincipled argument seems to have convinced the government that things should remain as they are.

3.3.6 Quasi-judicial procedures

Introduction

As stated in the introduction, quasi-judicial procedures can also be very relevant. In the UK, the most relevant example in the context of hazardous waste decisions is the local inquiry. Such an inquiry is held if the SoS calls in cases or if an appeal is made against a decision made by the local authority and the appellant requests that the case is discussed at an inquiry. A local inquiry can be held both on issues related to planning permission and issues related to pollution control licenses. These inquiries are largely similar, except that inquiries on pollution control licenses are less open to the public. It is important to note that the majority of all appeals in the UK is not decided by an inquiry and that local inquiries are only used in certain special cases. Generally speaking, local inquiries are considered best suited to complex development proposals and to cases that generate substantial third-party representation (Moore, 1997: 312). Because hazardous waste proposals are likely to be among these, they are the most likely way of deciding on an appeal in the case of hazardous waste facilities. Such inquiries are heard by Planning Inspectors, figures employed by central government who often have a technical training and/or a military background (Barker and Couper, 1984). One remarkable thing about UK inquiries is that the inspector can actually advise awarding costs against an authority if its behavior is considered unreasonable[26]. In Canada, quasi-judicial procedures are omnipresent. If we take the example of Alberta again, there are various administrative tribunals that operate in a quasi-judicial way. These include the Municipal Government Board, the Environmental Appeals Board, the Energy and Utilities Board and the Natural Resources Conservation Board.

[26] See Circular 5/87, paragraph 6. It states that an authority must have acted unreasonably, with unnecessary expenses on behalf of the applicant as a consequence. Unreasonable behavior is considered to have occurred when decisions are not substantiated, or substantiated with reasons that are not precise, specific or relevant (Manchester vs. SoS for the Environment and Mercury Communications Ltd. (1998, JPEL 774) and Blyth Valley Borough Council vs. SoS for the Environment (1990, JPEL 904).

UK planning inquiries: arenas without rules

The character of UK planning inquiries has puzzled many. Inquiries at first glance give the impression of a judicial contest between two parties with a third party deciding a difference of opinion (which would have been a pluralistic approach to the decision-making process). Indeed, Gerrard (1994) describes the atmosphere during inquiries as conflictual. But that is not how inquiries are formally intended. Ball and Bell (1995: 247) write that an inquiry should be understood as 'not primarily a contest, but a forum in which all relevant information may be produced and tested so that the Inspector may make a rational decision'. It seems that there is a managerial rather than a pluralist logic underlying the inquiry process: the expert (the planning inspector) is to be better informed as a result of the inquiry. Consequently, inquiries have been labeled 'decision advice procedures' (Barker and Couper, 1984: 386) and not decision procedures.

This reading is confirmed by an explanation of the purpose of inquiries by a former SoS for the Environment, quoted by Vogel (1986: 109). He said that '*I believe that as a result of our inquiry processes we can make better informed decisions which fit the facts and which fit into national, regional and local objectives; second, our inquiry system enables us to hear and to have answered the legitimate anxieties of people who have the right to express their concern; third, the public examination of these large questions assists us in achieving a measure of consent when final decisions are made*'. Note that improvement of the decision (in managerial terms) within the framework of set objectives is the primary concern here. Secondly, the inquiry is about answering 'legitimate' anxieties of people, which sounds like a one-way process. The third point seems to be about the inquiry as a means of letting off steam for people who object to a proposal. They must be given the impression that they have had an opportunity to have their say so that they can be satisfied, even when the final outcome is not what they would have liked to see. It is in this sense that Barker and Couper (1984: 393) argue that inquiries are mainly about 'impressing'. The appellant, the local planning authority and the public must be convinced that the inquiry is run fairly and competently, so that the final decision becomes acceptable to them. It is observed that as an instrument, inquiries have had their ups and downs in terms of legitimacy. From that perspective the rules that guide the debate, are important.

For a long time, inquiries could rightly be called 'arenas without rules' as the procedures to be followed had developed very much by practice and consent and these rules were not formally written down at length. As 'gentlemen were gentlemen' these informal rules were followed without much protest (Dudley and Richardson, 1998: 736). In the 1970s however, inquiries came to be used by environmentalists to further their cause, most notably in the case of highway inquiries. There, the inquiries were used to question government policy on road construction itself, not just individual stretches of road and their implications (for which the inquiry was really meant). As there were little formal rules to prevent this from happening, 'issue expansion' took place and control of the debate slipped out of the government's hand (ibid.). Issue expansion is also observable in the field of waste management where attempts have been made to use inquiries to discuss issues such as the need for waste facilities and regulatory standards at inquiries (Petts and Eduljee, 1994). This is described as 'misuse' of the planning inquiry system and a debate 'after the event' (ibid.: 395).

Use of the inquiry process as a means to gain access to decision-making on wider issues demonstrate the potential conflict between communitarian, pluralist and managerial thinking. The pressure for increased participatory rights is said to come from a discourse coalition that consists of 'local and national bodies of residents, environmentalists and generally anti-development interests' (Barker and Couper, 1984: 378). Interestingly, the government itself stimulated the development of this discourse coalition, for instance by introducing the Civic Trust and the 1957 Civic Amenities Act. These acts resulted in the proliferation of local amenity groups, which consequently started to play an effective role in land use planning (Grove-White, 1991: 38). But arguments that are echoing the critique of communitarian thought are present in the UK debate. It is considered that there is '*a basic conflict between a process in which some people claim special rights to influence government and one where elected representatives speak for their electing constituency and oversee government in terms of the "common good". It can be argued that, in a democracy, activities of third parties that go beyond informing the decision-makers to having voice in that decision-making gives these people an improper line of influence beyond the ballot box and one that is not representative*' (Barker and Couper, 1984: 378).

The discourse coalition opposing use of inquiries as a participatory forum seems to have been used particularly by officials at the DoE. They motivate their resistance largely by reference to the administrative burden that inquiries impose (ibid.: 377). Indeed, since 1976, there have been attempts to better control the debate during inquiries and to let the managerial thinking prevail. These attempts have taken the shape of sets of rules that have come to regulate the inquiry procedure[27]. Devising such rules is a complicated task because the greater the number of rules, the greater the impression that the inquiry is a judicial process, which in turn may result in expectations concerning the degree to which the outcomes bind the SoS for the Environment (which they do not). The rules must avoid giving the impression that the inquiry process offers phony participation, but at the same time not create the expectation that the inspector or the SoS is bound by the outcomes. The complications of this task can be made clear by focusing on the practice of highway inquiries. There, after the challenge from the anti-road protesters had become fully clear, attempts were made to reincorporate them into the decision-making process and to shift inquiries back into their 'traditional modality'. Concessions were made to their complaints about the inquiry process in the late 1970s. More information was to be published prior to the inquiry, pre-inquiry meetings were introduced and permission for television cameras at the inquiries was given. These concessions were sufficient to maintain a temporary peace until the mid-1980s and decision-making returned to normality. The road lobby however, only grudgingly accepted the truce and managed to exploit the new political climate under Prime Minister Thatcher to reassert itself in the 1980s. The White Paper 'Roads for Prosperity' which came out in 1989 was a clear sign of this, as a doubling of investment in roads was announced. Environmentalists at major inquiries did not question the need for new roads. Instead, environmentalists were playing by the new inquiry rules, not disrupting the process and proposing various improvements to proposals such as tunnels. Such proposals won some favor with the Transport Department, but were then vetoed by the Treasury on the grounds of cost. The effect has been to largely discredit the inquiry process in the eyes of environmentalists, who consequently turned to 'direct action', including the occupation of road construction sites. The government has responded to this with an attempt to create new modes of consultation, such as the Standing Advisory Committee on Trunk Road Assessment, and an increased use (again) of pre-inquiry round tables (Dudley and Richardson, 1998: 737-740).

The Town and Country Planning Appeals (Determination by Inspectors) (Inquiries Procedure) Rules 1992 (SI 1992 no. 2039) reflect the discussion just described. The rules apply to cases decided by inspectors without further review by the SoS. The rules for inquiries in cases that are covered by the SoS are however not much different[28]. The inquiry rules have a dual

[27] E.g. Circular 10/88 DoE and 15/88 Welsh Office, 'Code of practice for hearings into planning appeals'.

[28] Hazardous waste disposal sites are not listed as a category of development for which the SoS for the Environment automatically covers jurisdiction.

purpose; on the one hand following them will imply that the rules of natural justice will not be infringed upon[29]. As such, these rules are in place to protect the interests of all parties present, and to see to it that they are all receiving a fair say. The second purpose however is '*to impose a discipline on the parties so that the inquiry is conducted as efficiently and effectively as possible, which in turn should lead to speedier decisions*' (Moore, 1997: 313). Therefore, it is assumed that anyone who can make a contribution to the determination of the appeal may appear at the inquiry, with the exception of parties who behave in a disruptive manner. These may be required to leave. The statutory parties are entitled to give evidence and to cross-examine persons giving evidence; other parties may only do so to he extent that is permitted by the inspector. It was concluded however that '*the inspector's discretion to refuse people who wish to go beyond these legal constraints is hobbled both by a need to behave in a way that is publicly legitimate and by the courts' requirement that everyone who has sought to be heard should enjoy "natural justice"*' (Barker and Couper, 1984: 377). The procedure, as far as not regulated by the rules, is at the discretion of the inspector.

After the inquiry, the inspector must take a decision or advise the SoS. Showing their discretionary leeway, the inspector and SoS are allowed by the rules to take into consideration any new evidence or matter which was not raised during the inquiry and which he considers material to the matter. In my opinion, this possibility supports the contention that there is a managerial discourse underlying the inquiry structure. Under a pluralist discourse, a third party would decide on the issues that are presented to him by the other parties and he must limit himself to that information. Clearly this is not the case with inquiries in England. Bryan Wynne pressed the issue when he argued that inquiries are '*political control with the benefits of judicial authority*'. His conclusion is based on the fact that the analysis of the debate is kept strictly confidential[30] and there is tidy control and termination of participation (see Vogel, 1986: 133). If this is correct, then the decision process is really political, but disguised as technical and judicial. It is apparently necessary for legitimacy of the decisions that the political aspect is obscured and replaced with the impression of objective (managerial) and unbiased (judicial) decision-making. Should the inspector (or SoS) take into account issues that were not raised at the inquiry, he must notify the statutory parties and afford them the opportunity to make

[29] One of the rules of natural justice is that the inspector may not listen to any party behind the back of another.

[30] This is incorrect in the sense that the inspector's report is a public document and does provide for an analysis of the debate.

representations concerning these matters. The inspector is required to notify the statutory parties and those parties who had requested so at the inquiry of his decision. The decision must be made in writing and must give the reasons for the decision; such reasons must be adequate and intelligible. This is however not taken to imply that every point raised at the inquiry must be addressed in the decision; the decision must give the 'salient' reasons on which it is based (see for a discussion of jurisprudence: Moore, 1997: 316-320).

Canadian tribunals

The various administrative tribunals in Canada naturally differ in purpose and in the way they operate. I have not found the same amount of in-depth analysis on the workings of these tribunals as exists in the UK. Just to give a flavor of the way such tribunals work, I will say something about the admittedly unrepresentative example of the Natural Resources Conservation Board in the province of Alberta, which was used to decide issues in one of my case studies. This board was set up by the Alberta government specifically for large developments that infringed upon natural areas and caused considerable controversy. Interestingly (and uniquely) this Board must approve of proposed developments before they can proceed to the ordinary arenas of decision-making, usually the Alberta Cabinet.

The procedures under which the Board operates can be found in the Rules of Practice of the Natural resources Conservation Board (345/91)[31]. The purpose of the board is - managerially it seems - defined as determining whether proposed projects are in the public interest. The public interest is somewhat loosely defined as lying in the triangle between social, economic and environmental considerations. This broad approach is another reason why the board is not very representative of other administrative tribunals, which tend to have a much narrower focus. The normal procedure of the NRCB consists of a pre-hearing and a hearing. Pre-hearings have been established to facilitate efficient and effective use of hearing time, a goal that is pervasive in the NRCB's material.

Despite the managerial emphasis on efficiency, the NRCB operates largely on a pluralistic logic. The NRCB acknowledges the existence of various perspectives and it actively seeks to hear such perspectives in determining whether a project is in the public interest. It uses pluralist language in explaining its purpose: '*the NRCB's responsibility is to conduct a fair and open process for reviewing applications*' (1996a: 1). The main instrument for the NRCB to decide issues is by using hearings. These hearings are seen as a clash of opinions, with cross-examination being

[31] They are neatly summarized in two leaflets that I have plundered for my purposes (NRCB, 1996a, 1996b).

stimulated. Such cross-examination may focus amongst others, on the soundness of evidence, the assumptions on which it is based and internal consistency. This extends to the qualifications, expertise, objectivity and knowledge of other persons who participate; such factors may be questioned. '*Cross-examination is an important component (. . .) because it allows participants to test conflicting evidence and competing arguments before the Board*' (NRCB, 1996a: 14-15). Presentations and questions must however remain within the scope of the hearing as decided by the NRCB and must be relating to issues within the mandate of the NRCB. The latter effectively excludes policy issues in most hearings. The boundary rules are somewhat a mixture of conservative pluralism and managerialism as the Board decided that 'persons directly affected by a proposed project, and other persons where the Board considers it necessary, shall be given the opportunity to contradict or explain information presented by the applicant and other interveners' (ibid.). In practice, status is given to anyone who has expressed an intent to file a submission to the hearing.

There is some progressive pluralism present as well in the sense that attempts are made to level the playing field through intervenor funding (see below) and through the requirement that all information to be presented to the Board is given to all those participating. After the hearing, the members of the Board, who are appointed by the Alberta Government, decide whether the project is in the public interest. If the project is not considered to be in the public interest, procedures may not proceed. This outcome is relatively rare however. More common is the decision that the project involved can be allowed to go ahead provided that certain conditions are met. The Board communicates such conditions to the responsible decision-makers (i.e. the Director of Approvals at Alberta Environment), but can not effectuate them. This creates the ironic effect that conditional approval by the NRCB sometimes becomes far less conditional further along in the process. Decisions of the board can not be appealed except for errors of law or jurisdiction. Such errors occur only if the board has decided outside its authority or has not upheld procedural fairness. Such an appeal would be to the Alberta Court of Appeal.

3.3.7 Costs of judicial and quasi-judicial proceedings

A topic that is not without relevance for both judicial proceedings and planning inquiries is the issue of resources and costs. Obviously, both inquiries and judicial reviews are costly. This is in the sense that they cause direct costs, for instance because in the UK, the former DoE had to employ around 200 planning inspectors to hear and decide appeals. It is also in an indirect sense, which is because investments are delayed and thus potentially profitable economic activities are not undertaken for some time. For now, I will neglect these issues, and focus on the costs for the participants in the appeal process[32]. Clearly, developers possess advantages over those who may oppose a certain development. The route to a local inquiry is through the appeal process, which is only triggered when the applicant chooses to do so. Although this is biased, this bias is defended in conservative pluralist terms. '*The applicant's appeal is against the limitation of his rights to develop property imposed by the state's control in these matters. By contrast, any third party individuals or organisations seeking to "appeal" against granted consents would in fact be calling for the removal or reduction of the owner's existing development rights (. . .)*' (Barker and Couper, 1984: 378).

Somebody - other than the developer - who wishes to make a case against the development is in a disadvantaged position. For a proposal to be called in, there must be signs of quite a high level of controversy, and even when an inquiry is held, the purpose tends to be a critical examination of limitations on development, not an examination of the development itself. The burden of proof to explain why a certain development should not be allowed is very much on them. It has been noted with concern that amenity organizations have limited resources and when they seek to influence land use decisions, they must therefore set clear priorities on which proposals they will try to influence and which ones not. The consequence may be that some proposals that potentially warrant public discussion according to many, actually pass through the system without such discussion[33]. Moreover, it is pointed out that knowledge plays an important role at inquiries, as successful opposition to a development requires that the opponents demonstrate that the proposal is not necessary or that an alternative site is preferable. To make such a case, the support of testimony by engineers and economic consultants

[32] The DoE, in line with the neo-liberal ideology of Prime Minister Thatcher, has spent much time and energy on studying both the direct and indirect costs associated with planning appeals and has generally tried to lower them (see for instance Department of the Environment, 1985). Some time ago it was concluded that '*far more administrative attention and analysis has been given to the speed of decision-taking on these smaller appeals than to the content and quality of decision making*' (Barker and Couper, 1984: 394). There are no indications that the concern over the speed of decision-making is limited to smaller developments, nor are there indications that this concern has waned over time.

[33] Examples mentioned in this vein are a nuclear power station near Lancaster, which was built close to a large population area, and an oil production platform at Loch Kilshorn. In the latter case, the energy of potential opposition was directed elsewhere (Vogel, 1986: 126).

is much needed. If credible witnesses do not refute claims by a developer, the inspector tends to believe him. As such witnesses are expensive, the issue of government support arises, because 'procedural' fairness may otherwise not translate into 'substantive' fairness. It has been argued that because of the lack of funds on behalf of amenity groups and the refusal of the UK government to supplement these, this is exactly what has happened (Barker, quoted in Vogel, 1986: 126).

The Netherlands do not offer intervenor funding to parties that are engaged in court cases, but there is a more general income-dependent scheme to assign free lawyers to people who can not afford one themselves. This scheme also applies to administrative courts. Furthermore, the fees for bringing a case are often quite low and on many occasions (certainly when the complaint was not outrageous), the defending public body is sentenced to payment of these fees to the plaintiff. In Canada, the parties seeking recourse from quasi-judicial boards do sometimes obtain 'intervenor funding', which implies that they can hire relevant expertise to argue their case. Intervenor funding applies to both the environmental impact assessment procedure and to cases before quasi-judicial bodies and is operated under broadly similar principles for both types of procedures. These principles are discussed under the heading of environmental assessment in section 3.4.5.

3.3.8 Some conclusions

Deference of the court to Parliament and restraint of judicial control vis-à-vis the administration that is overseen by democratically elected leaders are the norm in all three countries. Yet, paragraph 3.2 already demonstrated the enormous importance of the courts in developing an institutional fabric for land use and pollution control decisions. The Netherlands has an interesting tradition in the sense that the country had conceptual difficulty in assigning the courts a place in issues that involved private parties and the state. This difficulty was caused by a rather strong equation of administrative actions with the will of Parliament. Oversight of administrators should take place by elected representatives and there was considered to be no place for the courts, even in cases of blatant injustice. This problem was solved when private courts started interfering with actions of the administration and when administrative courts were founded. These latter courts were seen as part of the administration, which explains why there is a tradition of deep probing of government decisions. Together with the private courts, administrative courts developed an impressive amount of principles to guide administrative decision-making. Judicial activism has however clearly dissipated at the expense of administrative discretion, which went out the window on the basis of requirements for certainty. In the UK, the deference of the courts is greatest, whereas Canada takes a somewhat middle position. In all three countries however, courts do reserve the right to review administrative decisions (be it on the basis of different criteria) and also play an important role in developing the boundary rules. Quasi-judicial review in Canada and the UK must perhaps be seen as a system similar to the system that the Netherlands initially had (with a tribunal advising the Minister). The idea is to fight out a conflict in a pluralistic way within the confines of the administration, albeit under a process that looks independent and unbiased. Apparently, there is a need to obscure the rather political character of decisions. It is remarkable that the practice of quasi-judicial review has also affected actual judicial review, with the courts being more hesitant to critically examine cases that have been to quasi-judicial review.

3.4 The role of the community

3.4.1 Introduction

The concept of 'community' gets little mention in the legislation of any of the three countries, and is little found in the collective choice processes leading up to legislation. In the Netherlands, the environmental groups that sprung up in the 1970s were very much activity orientated, they consisted of people 'who wanted to do something for the environment' (Cramer, 1989). One of the activities was the spreading of new scientific insights into environmental issues and the organization of counter-expertise. Development of expertise and using this in decision-making procedures was, almost from the beginning and for almost all groups, the goal, which implies that the environment movement mainly sought the debate in managerial terms. Only here and there were these small groups who advocated communal living, but exactly because of the communitarian agenda, they do not participate in collective choice processes that lead to institutional design. Cramer (1989) argues that the environmental movement in the Netherlands radicalized itself somewhat in the period 1974-1980 with some elements entertaining 'ecosocialist' ideas of introducing central government, but this was again a minority. It is of interest that many activists in this period considered general knowledge of environmental issues and a good political vision sufficient to function (ibid.: 85). Scientists who had supported the environmental groups retreated around that time because they felt the groups became too political. Furthermore, the groups started building their own expertise. This in turn led to their acceptance and invitation by the Dutch government as negotiation partners in the 1980s. Ecological modernization became a term they used and this opened the door for closer cooperation with businesses. The pattern of the more communitarian parts of the ecological movement not participating in the collective choice debate is also visible in the UK and Canada (see e.g. Connelly and Smith, 1999: 68-102), as is the pattern of development of the environment movement. It should be noted however that in the UK a radicalization of the environment movement took place in the 1990s. The arena for action shifted from inquiries and formal proceedings to direct action, most notably in road development.

The only possible indication of an influence of the communitarian discourse therefore is that the respective governments have allowed a certain level of citizen participation. Citizen participation is in a sense the acid test for the various discourses at work at the collective choice level. Communitarians favor it greatly (although what they really favor is local self-rule), pluralists are mostly hesitant although progressive pluralists would offer liberal access to the courts and members of parliament. Managerialists finally would not favor citizen participation, except as a device that generates information for the real experts. The rest of this section focuses largely on the topic of citizen participation. In section 3.4.2 I discuss the level of citizen participation offered in the pollution control systems of the three countries and section 3.4.3 analyzes citizen participation in the land use planning systems involved. The most recent trend of participatory procedures is associated with the introduction of environmental impact assessment, a practice that is discussed in section 3.4.4. Section 3.4.5 finally, presents some conclusions. It is of interest therefore then that the roots of the increase of participatory rights are located in the 1950s, with the real blossoming of the desire to gain more participatory rights coming in the 1970s. The reasoning for increased rights in all countries has traditionally been pluralistic. The limited participatory rights that existed until the second half of the 20th century were largely confined to landowners and are, as I have shown, a result of judicial activism, demanding rights of representation for affected landowners. Such representation was often given to the elected representatives overseeing the work of administrators, who would then determine the reasonableness of the actions. If the result was affirmative, the decision was often final as the courts deferred to decisions of elected bodies (at least in theory).

A pattern that can be observed in all three countries is that boundary rules - for these relatively limited rights - became wider and wider. Take the Housing Act of 1901 in the Netherlands, for example. Whereas the rights to make representations concerning draft land use plans was reserved for decades to owners of the land embedded in the plan only, this started to include non-owners, mainly renters, in the 1950s (see De Gier et al., 1999: 5) and then evolved to access for anyone, the 'actio popularis'. The three countries differ somewhat in the degree to which this development has proceeded. In all three countries, the managerial argument that citizens can improve decisions has been important, but also the progressive pluralist argument that all parties with an interest must be given a say has too. Together these arguments have implied that rights of representation have been liberally distributed. Whereas such rights have been granted in all three countries, the prescriptions of the communitarian discourse, with its emphasis on localized decision-making and dialogue, have not really been followed. The UK government, particularly, appears to have been very hesitant in advancing a participatory agenda because it feared that business secrets would be leaked in participatory decision process, but even more because of the managerial fear that the administration would be overburdened. As I have already described for local inquiries, participation in the UK is something that must be wrestled from the authorities. Although the Netherlands and Canada seem to be a bit more forthcoming with participatory rights, such rights are 'safely' couched in a layer of managerial and pluralistic decision processes. In the Netherlands, the backlash against citizen participation, especially in the field of infrastructural works, has been enormous, most notably from elected politicians and administrators. Environmental assessment has been introduced in all three countries, and in the Netherlands and Canada, the procedure for environmental assessment allows citizens to better inform themselves of proposals and to participate. However, the terms in which this occurs are hardly communitarian, as the debate must be held in technical terms and reference to community interests is not very effective in influencing decisions. The practice of impact assessment in the UK differs from that in the other countries in that there is no independent review of the assessment. However, in Canada and the Netherlands, where there is such a review, the reviewers are largely drawn from the technical sciences, implying that scientific knowledge in a sense plays an even further increased role.

3.4.2 Citizen participation in the pollution control system

Participation in the UK

In the UK, waste licenses are only granted after planning permission is in place. In part because of that, in part because of the view that waste licensing is a technical decision, there is little citizen participation in this type of decision. Ball and Bell (1995: 374) for instance observe that the EPA 1990 simply excludes the public from waste decisions. '*One feature that is missing is any element of public participation, since an application does not have to be advertised. This situation arises because it is perceived that the grant of a waste management license is a technical question: the wider question of the appropriateness of the site will have been considered when planning permission was sought*'. Indeed, when one studies the provisions for citizen participation under EPA and COPA, one gets the impression of a decision-making process that consists of experts consulting with each other, while the general public is hardly mentioned. The Waste Regulation Authorities (and currently the Environment Agency) were authorized to add non-expert parties to the consultation process, but this was not mandatory. Important criteria in the approval process are the existence of a threat to water or a threat to public health. Whether this threat is acceptable or not is to be decided by the experts of the WRA itself, but they are obliged to consult with the water authorities, the waste collection authorities and with the Health and Safety Executive (HSE). When compared with COPA 1974, the EPA 1990 extended the set of statutory consultees somewhat. Under EPA, proposals that involved a Site of Specific Scientific Interest (SSSI), had to be presented to the Nature Conservancy Council. It is my impression that the presence of the Nature Conservancy Council as a statutory consultee is an enactment of a long-standing practice that is broader than what the law prescribes. Indeed, in each of the local councils that I have visited for my research, the list of consultees was much longer than the list to be found in the relevant legislation. All local environmental groups normally receive correspondence when applications for licenses are made. The critical comments made by Ball and Bell were thus not confirmed in these three authorities[34]. What seems significant however is that the weight of representations by these consultees differs, with a bias towards the water authorities (the NRA, now also dissolved into the EA). These authorities have a special place in the procedure as they may request the WRA not to issue the license. A successful application is unlikely when the water authorities oppose it (Ball and Bell, 1995: 374). Regarding the position of groups such as the Nature Conservancy Council, it is often noted that in the UK, important segments of the environmental movement play an active part in policy development. Environmental organizations often receive offers for a closer working relationship with a particular government agency. The dilemma facing the environmental organization is then to accept this offer, which according to Vogel (1986: 51) is the most effective means to affect policy, or to remain a relative outsider. The twist is that the privilege of consultation comes at a price: they are only given to 'respectable' organizations that behave 'responsibly'. This seems to imply mainly that they do not demand too drastic changes or employ extralegal means of action. It is noted that the value of the close working relationship between environmental organizations and governmental agencies is somewhat limited

[34] Cleveland County (no longer in existence), Newport (Wales), Rotherham MBC.

as environmental organizations are encouraged to develop such relations with governmental bodies that are peripheral to the major centers of decision-making. Some therefore speak of a 'phony insider status' and regard UK public officials as masters of the game of co-optation (see Vogel, 1986: 52).

Note that if citizens are expected to participate in hazardous waste decisions, it is likely that they should have access to the relevant information. UK regulators seem rather keen on secrecy, the reason being that they were afraid of losing the voluntary cooperation of factory officials if absolute secrecy was not guaranteed (Jordan, 1993: 408). The language used in public defending the situation was however largely managerial. The chief alkali inspector for instance said in 1972 that openness of data was not something to advance. '*You would get amateur environmental experts and university scientists playing around with them. People can become scared of figures, they can get the wind up*' (Vogel, 1986: 92). Pollution Paper 11, dating back to 1977 added that regulators who communicated on standards with the public should bear in mind '*the risk of gross misinterpretation of data and the need to avoid disclosure of truly confidential information*' (Hajer, 1995).

A discourse coalition consisting of the Royal Commission on Environmental Pollution, Members of Parliament and certain parts of the environmental movement has attacked the secrecy surrounding pollution control ever since the early 1970s. Their activities had some effect. COPA 1974 for instance provided for public registers of licenses. However, the implementation of these (relatively toothless) provisions has been delayed on the basis of the reason that implementation was too costly, which seems a managerial concern. Petts and Eduljee (1994: 395) sum up the negative consequences of the relatively closed system of decision-making in the UK. '*In the UK, there has been a lack of discussion and assessment in the public domain on the need for waste facilities, in terms of capacity, type, and the optimum siting of facilities so as to minimise inequities in risk bearing*'. Note the progressive pluralist concern with equity here. There are certain extra-legal developments in the UK, which might influence the degree of citizen participation in a positive way. Among these is the establishment of a nationwide network, Communities Against Toxics, that supports local groups that wish to participate in the decision process and feeds them with information. Secondly, there have been experiments with various alternative ways of developing waste plans, for instance by the use of community advisory forums such as in Hampshire (see Petts, 1999).

Participation in Canada

In Canada, or more specifically the province of Alberta, proponents of hazardous waste facilities had to obtain two licenses under both the Clean Air Act and the Clean Water Act. One license was needed to construct the facility and one to actually operate it. Operating licenses were only issued after initial tests of the operations at the facility. The procedures to be followed were described in Cabinet regulations, which made the licensing system essentially closed to the public. License conditions were determined by negotiations between government and the proponent. There were no public registries to provide proposal information to the public, and no public notice was given when license applications received (Barringer, 1990: 91-92). The Cabinet regulations for license procedures were essentially only concerned with the information that was to be supplied to the Director of Approvals[35]. No requirements were in place regarding citizen participation in the process. The Director essentially made the decisions. A quite typical formulation describing the decision process was: 'Where the Director of Standards and Approvals is satisfied that a license should be issued, he shall issue a license in accordance with the Act'[36].

The Alberta Environmental Protection and Enhancement Act that was adopted in 1992 (in force since 1993) brought a change in the direction of a more participatory system of decision-making. With the AEPEA came many new regulations, including the Activities Designation Regulation (211/96) and Environmental Assessment (Mandatory and Exempted Activities) Regulation (111/93). The regulations require that a public register for all projects is maintained and public notification of the Director's decisions in the assessment process. From the perspective of citizen participation such regulations may not be revolutionary, but still provide a small improvement. It seems that such improvements were largely made because of the activities of environmental groups, specifically the Environmental Law Centre from Edmonton. This group had access to the committee that prepared the act and had a participatory agenda (see MacWilliam, 1996: 3; Tingley, 1994: 33).

The actio popularis in the Netherlands

In the Netherlands, participation rights are granted to anyone willing to participate. The system that was introduced in 1952 was that of the 'staged actio popularis' (De Gier et al., 1999: 10). Actio popularis implies that anyone has the right to make representations to the decision-makers. It is

[35] See Alberta Regulations 33/73 (Clean Air Regulations), 34/73 and 216/75 (Clean air (General) Regulations), 35/73 (Clean Water (General) Regulations, and 36/73 (Clean Water (Industrial Plants) Regulations).

[36] See Alberta Regulations 33/73, § 10.

called 'staged' because the licensing process is seen as consisting of stages, including representations (the Dutch tended to say objections) to elected representatives, to a higher administrative body, and then judicial review by the courts. The system was such that participation in the early stages, in fact at the stage where there is a notification of an application for a license, led to a right to continued participation at later stages. Citizens were however not entitled to change the scope of their arguments after the first response, except when changes were made in the decisions. There was little motivation for this system (it is a copy of the system in land use planning, see the motivation there) but it was defended in the 1970s by the working group Duk. This working group used managerial language to motivate the system: the ideas of the population 'should directly and immediately be available for government policy' and participation would allow for 'the common solution of the problems' (De Gier et al., 1999: 11). More stringent boundary rules were rejected on the basis that the quality of argumentation was more important than somebody's status, that sieving out parties that do not have a right to participate would not work anyway and finally, that misuse of such rights was not to be expected. Note the relatively unique situation that managerial concern for technically optimal decisions led to an increase of citizen participation. This development must then be seen as an explicit embrace of the increased skills on part of the public. The government followed this advice (which was mainly intended to harmonize various laws with each other) and added a principled reason. The administration must not take decisions before it has heard citizens 'because there are issues involved that are fundamentally important to everyone' (De Gier, et al., 1999: 12). In the end however, the administration decides.

3.4.3 Citizen participation in land use planning

Canadian planners: the cheerleading public

In Canada, one constant element of planners' thought until the 1960s was the idea that they stood above the general public and elected politicians. Although planners in the early parts of the 20th century had been convinced of the need for public support, it appears that such support was much more expected in the form of cheerleading than interference. Planners used their supposedly unique capability to determine the public interest as a means of largely excluding the public from decision-making. '*As "experts" and "professional" planners they saw themselves as able to determine the needs and interests of the community*'. Once the public would come to understand this, it would support planners more than it actually did at that time (Grant, 1989: 12). This situation turned to an almost hostile attitude towards the general public after the Second World War. Planners began to describe those who participated actively in debates as 'special interests'. While the planners' goals was to determine the public needs and the public interest, the members of the public were considered 'self-interested' and 'irrational' and their input was to be avoided. Equal criticism was launched at elected politicians.

In the late 1960s and 1970s, Canadian planners found themselves in the middle of protests against urban renewal, high rise redevelopment and urban expressways. These developments had been accepted up until that time, but started to meet with resistance. These protests, and the demand for participation and democracy with which they were accompanied, affected the theory and practice of Canadian planning. Planning had been seen as an essentially rational exercise aimed at the promotion of the public interest, meaning growth, technical progress and efficiency. Public participation however highlighted that values within communities were rather diverse and forced planners to accept that they could not determine a single public interest. In response, planners began to devise ways in which citizens could participate, and tried to help to incorporate participatory democracy in local governance. But Grant says they came up against the boundaries of local government and their own profession. Not only had the local politicians difficulty in relating to the concerns of an amorphous public or anti-development protestors. Planners also lacked the skills to facilitate public involvement or evaluate the results, and the public held unrealistically high expectations about the degree of control that it would obtain over development decisions. '*Participation exercises seldom progressed beyond what Arnstein (1969) would characterize as various levels of tokenism or co optation. Critics argue that governments made no real commitment to public participation, but instead used it as a screen to mask the lack of democracy in decision-making*' (Grant, 1989: 15). By the 1980s, the 'participatory era' in Canada was over, even though planning ideology and legislation are said to have retained much of the rhetoric of that era. The right of the public to be heard and consulted is enshrined in the planning legislation of most provinces[37]. However, planners are said to have retreated from their

[37] There are cases in Canada where official requirements for notification of rezoning had been met, but common law principles were considered to have been violated and rezoning decisions were quashed (Laux, 1990: 202). Under the common law, procedural obligations have generally depended on the type of decision. No procedural obligations applied where a legislative (general policy) decision is taken, limited obligations applied where an administrative decision (regarding a finite group of individuals) is taken and strict obligations were in place where quasi-judicial decisions (regarding a specific case) were taken. Currently, the Charter of Rights requires 'fair procedure' for every government decision, which is taken to imply rigid obligations in the case of concrete, individual decisions and less rigid

advocacy of citizen participation and now '*provide opportunities for residents to participate in the preparation or area and municipal plans while recognizing that the most important development decisions often take place in contexts where the public has little input*' (ibid.: 16).

Grant effectively argues that some politicians made up an important part of the discourse coalition against citizen participation, stressing that citizen involvement is inefficient, unrepresentative and lacks the legitimacy of elected governments. Thus, developers, politicians and experts are still firmly in control, not the public (ibid.: 15). It is commonly understood in Canada that planning contains a policy element and needs to be judged in the interest of the municipality at large. The implication is that a pluralist, adversary, approach to hearings that are obligatory is rejected. According to Laux, '*it is not expected to permit the type of seemingly endless debate that may be characteristic of the judicial process*'. A local council '*need only afford interested parties at a public hearing a reasonable opportunity to state their views. It will not be faulted if it does not permit the parties to make point and counterpoint, to cross-examine one another*' (1990: 211)[38]. What strengthens the role of citizens however is that the principles of fair hearing and impartiality that are enshrined in the Charter of Rights are interpreted quite broadly. Unlike the situation in the UK, the council may receive no evidentiary material outside the hearing and lobbying of one of the members of the council is not permissible. The first principle even applies to a council that seeks advice from its staff, which results in fresh evidence. Affected persons must be made privy to that information and must be given an opportunity to respond (Laux, 1990: 215). The principle of impartiality does not go as far as prohibiting the hearing of rezoning by councilors who have taken a political view of certain developments. The courts have acknowledged that councilors are elected politicians and must therefore have normative views of the issues at hand. The consequence is that councilors must demonstrate a willingness to entertain views from all sides, and must act so that a reasonable person watching the hearing will not think that they have closed their minds (Laux, 1990: 225)[39].

---

obligations in case of general regulations. In either case, procedures must be unbiased and impartial. This requirement has affected procedures, for instance because members of a local council may not sit on appeal boards that decide planning appeals.

38 In fact, it is doubtful whether the requirements go much further than being allowed to express one's concerns. The city of Edmonton, as can be gleaned from Kwasniak (1998) and Laux (1990: 212), applies the rule that presentations to the council concerning rezoning may last five minutes at most. This rule was upheld in court.

39 See the discussion on reasonable apprehension of bias, earlier in this chapter.

The UK: a veneer of openness?

On the UK planning system, McAuslan notes that 'the participation ideology' had virtually no influence on the legislative framework. This ideology, inspired by the work of John Stuart Mill, contends that the law must provide rights of participation by virtue of more abstract principles of justice and democracy. This implies that all those who have, for whatever reason, an interest or concern in a proposed development should have a right to participate in the decision. The ideology denies the land owner a special place in the decision-making process, but it also denies '*that the public interest can be identified and acted upon by public servants on the basis of their own views and assumptions as to what is right and wrong. Public servants should act after full public debate (and by public debate is meant a debate in which the general public can take a direct part) and always subject to continuous consultation with the public*' (McAuslan, 1980: 5). Obviously, this 'ideology' differs from communitarian thought in that it recognizes public servants and does not emphasize the local community. But even this more moderate agenda has not been achieved. McAuslan contends that the few rights of participation that ordinary citizens have in UK planning law are often conditionally phrased, that is, subject to approval by local councils, which are not really stimulated to extend rights of participation liberally because of cost considerations. His conclusion is that: '*the legislation, despite a veneer of openness and participation, is, when carefully analysed, a good example of the ideology of public interest at work; not merely are the public authorities to make the decisions on the substantive question of grant of permission or no; they are to make decisions on the questions of publicity or no; actual as opposed to formal regard to be had to representations, and the only clear duty they have, unclouded by subjective phraseology, except in two cases, is the duty to consult with certain other public authorities in respect of certain types of application*' (McAuslan, 1980: 14-15). This would seem to suggest the managerial bias that I also found in procedures under the pollution control system.

The Netherlands: quality of the argument

In the Netherlands, participation rights have been distributed on a relatively large scale too. In the Land use planning act of 1962 and subsequent revision in 1985, the right to make representations on land use plans or changes therein was given to anyone. The reasoning behind this move was managerial, as it was based on the argument that the '*quality of the argument, not of the participant*' (De Gier et al., 1999: 5) should determine boundary rules. The Commission Van den Bergh, which prepared the act, indicated that '*so much as possible should the idea be followed, that objections can not be filed by those, who are already interested parties under private law, but by anyone, who considers he has a reasonable ground to think that the plan is inappropriate. In this way can, in the process of judging a plan, all advantages and disadvantages be brought forward (. . .)*' (ibid.). I am calling this a managerial line of thinking because it is based on the assumption that decision-making is the weighing of the pros and cons and there is little realization of value conflicts. There was a certain concern within Parliament over the costs associated with the implications of the 'actio popularis' but such arguments could be overcome by pointing at the limited use that was made of the participatory rights under the reconstruction act. It was also pointed out that it already was standing practice in the land use planning to have arguments of those not formally entitled to participation to be considered by administrators anyway. A rather principal discussion was thus resolved by resorting to rather pragmatic arguments. This continued to be the case as in 1985 the act was revised and concerns were again expressed over the liberal boundary rules. Again there was research in place to demonstrate that little use was made of participatory rights by those who would have not been qualified otherwise.

3.4.4 Environmental assessment: participatory or management tool

High hopes in the UK

It might seem strange to discuss the item of environmental assessment in the context of community involvement in hazardous waste decisions. However, I think this can be justified when one considers the hope that is often expressed on the potential meaning of environmental impact assessment for citizen participation. Take for instance the situation in the UK. It is said that there has always been close scrutiny and the necessity to gather a vast amount of information on waste facilities. Yet, '*It is fair to say that until recently the public was largely excluded from the technical debate unless they had the financial resources to instruct experts to act on their behalf. If the environmental assessment process operates satisfactorily in disseminating information in intelligible terms, then the public may be able to play a more effective role in these inquiries*' (Ball and Bell, 1995: 267). Whether or not this hopeful view materialized depends on various questions of course. For certain, in the UK citizen participation is not located at the front-end of the decision process, but at the back-end. This becomes clear when one looks at the requirements aimed at the developer who needs to prepare an impact statement. Such a developer is obliged to closely consult with the planning authority and the statutory consultees. However, consultation of others is not required, only regarded as 'good practice'. In discussing UK central government guidance on environmental assessment, it is said that the UK is taking a restrictive view on participation, and most importantly stresses the role of the assessment as a means of allaying fears and informing the public (Petts and Eduljee, 1994: 404). The arrangements for the process of environmental assessment seem rather informal and the issue of citizen participation is left very much to the good taste of the developer.

Canada: experts versus experts

The situation in Canada is somewhat different to that in the UK. Although the requirements for environmental assessment there stress managerial aspects, there is some progressive pluralist thought present as well. The first indications of a Canadian interest in environmental assessment are to be found in the early 1970s. Canada at that time had limited experience with environmental assessment and looked to the American experience to learn. Although the Canadians were impressed by the ambitious American approach, they were not at all attracted by the amount of litigation that had come to surround American environmental assessment. A Canadian interdepartmental committee of deputy ministers, advising on environmental assessment to Cabinet, promoted a very cautious approach to assessment in order to avoid some of the problems surfacing in the US. Essentially what most of the 'mandarins' in the Ottawa bureaucracy wanted was a not too ambitious process that could be managed within the confines of bureaucracy itself, thus environmental assessment was to be a bureaucratic decision-making tool (Doern and Conway, 1994: 193-194). The federal cabinet largely accepted this advice and created the rather cautious Environmental Assessment Review Process (EARP) in 1973 by secondary legislation.

The basic principle of this scheme was one of self-assessment by federal government departments that wanted to initiate projects that could potentially have significant environmental implications. Panels of experts were to play the greatest role in these assessments, and little value was attached to the involvement of the general public, which would only be involved (through hearings) in cases where the initiating departments themselves considered the potential environmental effects significant. Most Canadian provinces followed the federal lead and introduced provincial assessment regulations in the course of the 1970s. These were based on the same lines of thinking as the federal process[40]. Despite the success of the

[40] E.g., in Manitoba, the practice of environmental assessment dates back to 1975, when an Environmental Assessment and Review Process was adopted as Cabinet policy. Environmental assessment received statutory status with the Environment Act of 1988. The act mainly links environmental assessment to the process of licensing and authorizes the Director of Approvals to determine the type and extent

very participative Berger Inquiry (1974-1977), the 'most important environmental assessment in Canadian history' (Northey and Tilleman, 1998), there was no support for greater citizen participation in the assessment process. Assessment was seen as a management tool for the government - broadening the considerations for decisions to include environmental reasons. The view of assessment as a management tool is reflected in the fact that boundary rules did not require independent review panels. Most members of such panels were therefore government officials working for the departments that wanted to undertake projects. Under the Guidelines Order, there were two levels of assessment: the first was the basic level, called screening. The second was comprehensive study, which only took place for the larger projects that were included in the comprehensive study list. Northey and Tilleman (1998: 199) indicate that self-directed screening applied to 95 percent of all projects, while self-directed comprehensive review took place in 5 percent of the cases. Their assertion is that public review applied to less than 1 percent of the cases. Finally, it is important to know that review panels did not take decisions, but only reported to the initiating minister who then decided. In the words of one judge: '*The environmental impact assessment (. . .) is not meant to satisfy mere academic curiosity but to help a Minister in the exercise of a duty to intervene and act positively with respect to the execution or completion of a project*'[41]. Ministerial decision-making is still a basic tenet of assessment in Canada. But currently there is one exception to this rule. In Alberta, for a limited range of projects, environmental assessment is undertaken under the direction of the Natural Resources Conservation Board. Approval by the board is needed for a project to go ahead. In other cases however, the results of the impact assessment are merely but one input into the decision of the minister[42].

---

of assessment required for projects. It is the Clean Environment Commission (or at least some of its membership) that holds hearings into environmental assessments. Whether hearings will be held or not is determined by the Director of Manitoba Environment or the Minister of the Environment. Public concerns about proposals for development can be a reason for such hearings. For a comparison of federal and provincial environmental assessment systems, see Doyle and Sadler (1996).

41 Excerpt from Carrier-Sekani Tribal Council v. Canada (Minister of the Environment) and Alcan Aluminum Limited v. Save the Bulkley Society et al. [(1992), 8 CELR (NS) 158, at 175-6 (TC TD)].

42 This has led to odd situations. In particular, one contentious large pulp mill (Daishowa) had indicated that it would not be able to meet the Alberta pulp mill standards of 1988, but was still given permission to construct their plant (Johnston and Gismondi, 1989: 19).

Environmental assessment was a relatively ineffective exercise in the mid-1980s, with the federal and provincial bureaucracy mainly finding ways to get around it (Tilleman, 1995: 377-378). '*Ontario's Environmental Assessment Act is an attractive piece of paper. You can wrap gifts in it, start fires with it, and mail it to your friends. When the Legislature adopted it in 1975, there was even speculation the act could be used to safeguard the environment*' (Quote from the Globe and Mail, in: Cullingworth, 1987: 396). Unexpected change in the federal assessment process came in 1984 when the outgoing Liberal Cabinet approved an Order on environmental assessment[43]. The explicit purpose was to codify the existing approach in an administrative code of practice, but as time ran out, the Cabinet hastily approved a much more far-reaching text than that. The text ended the practice of self-assessment, broadened public access to the panel review stage, introduced the possibility of public review in cases of public concern, extended the scope of review to technical, environmental and social effects of the proposal, and opened the possibility for a discussion on need for a proposal[44].

Despite these changes, environmental assessment could still not be seen as stimulating citizen participation. This becomes clear when one looks at a report by the Federal Environmental Assessment Review Office (FEARO). 'Orientation for senior managers on EARP policy and objectives 19' was a manual that was produced around 1990. It was intended to convince other departments of the usefulness of environmental assessment and a participatory approach to such assessments. Tilleman (1995) quotes from this document[45] and the quotes indicate that FEARO recommends public participation in bureaucratic decision-making mainly on the basis of the advantages participation has for the bureaucracy itself and for preventing errors in decisions. For instance, the manual indicates that public input to departmental decision-making will improve the public image of the department, and will make the department less susceptible to opposition (Tilleman, 1995: 343). Furthermore, participation is, '*not without some truth*' (!), perceived as a '*long, costly, risky and foolhardy process*' that creates '*needs that otherwise would not exist*' and forces proponents into making '*undue concessions or changes*'. But, argues the manual, such risks

43 Guidelines Order (SOR/84-467).

44 A critique of environmental assessment in Canada is that discussion of the need for a project is in fact often precluded (Rowson, 1997: 336). This was for instance the case in the hearings on a nuclear fuel waste disposal facility. The effect in this particular case was to undermine the credibility and legitimacy of the project proponents (Kuhn, 1997: 40).

45 Unfortunately, this report was not available at any of the (many) locations I visited during my stay in Canada, and therefore I cannot be certain the excerpts are indicative for the entire FEARO report.

must be considered in the light of the risks associated with not consulting the public (Tilleman, 1995: 347), which are apparently even worse. Such risks are defined within a managerial line of thinking. They include widespread opposition and total rejection of '*a project that is poorly understood by those having to live with it*', which indicates that consultation is mainly an education event. Another risk is '*not knowing enough about the environment where the project is intended to be implemented*' and therefore lack of participation may lead to important errors in project design. In other words, consultation is defended on the basis that it may contribute to better understanding of the issues on part of the public, which then has the limited role of adding facts or data to the bureaucratic equation. That the decision-making process is very much about (democratic) values remains very much outside the scope of these comments. The impression from the above is reinforced by a remark in the manual that is obviously meant to reassure the senior officials of other departments. The manual states that (for citizens) '*in fact, the real sore points are often secondary in importance and can be resolved without endangering the project*' (Tilleman, 1995: 348). The expectation is thus that a fundamental discussion of the need of a certain project does not occur. I am not certain whether the manual is indicative of the attitude of FEARO officials towards citizen participation or not. Yet it is clear that these officials felt the need to talk about citizen participation in a managerial fashion as they addressed their colleagues from other departments. In this sense, the manual provides a strong indication that the managerial discourse structures environmental decision-making within the Canadian bureaucracy[46].

Provincial attempts to provide for citizen participation did not go as far as the federal requirements. In Alberta for instance, proponents of projects were free to decide how they would consult with the 'affected' members of the public. They often used one-on-one meetings as a means to approach the public (see Alberta Environment, 1977: 9-14). There was much critique of this practice; it was criticized as preventing public education on the environmental aspects of the proposal (Johnstone and Gismondi, 1989: 18). A review committee consisting of officials, citizen and company representatives studied environmental assessment in Alberta in 1980. It concluded that proponents were '*often unclear on what the government and the community expects for an adequate public participation program*'. Also: '*citizens and public interest groups are often put into a reactive and negative*

[46] Dr Tilleman (personal correspondence) wrote me the following: 'on the motives of officials emphasizing participation, I do not believe it is primarily to defend their bureaucratic or political stability. Instead, I believe the agency officials (hopefully) believe greater public participation results in faster decision-making (less errors) and diminishes the risk of future challenges (e.g. court).

stance by not being able to adequately participate in the development of a project' (Alberta Environment, 1980: 11).

In Canada, there had always been much resistance within the federal government against a legislated environmental assessment process - especially from bureaucrats - because it was expected that a legislated process would harm flexibility. This resistance waned after some important court cases wherein the EARP regulations were found to be binding on government anyway[47]. Legislation thus arose in the form of the Canadian Environmental Assessment Act (CEAA)[48]. A draft was released in 1990, but it took until June 1992 before the law received royal assent. Indicating its controversial status, the CEAA and four accompanying regulations were not put into force until January 1995. On the basis of the Act, the Federal Environmental Assessment Review Office was replaced by the Environmental Assessment Agency (EAA), which became independent from Environment Canada (Northey and Tilleman, 1998). One issue of debate when the act was introduced was whether assessment should be extended to policies and general regulations. The government shied away from this possibility, as did many provinces. More generally speaking, the development of the act was accompanied by an explicit reconsideration of the process as it had existed until then, with a discourse coalition of business interests, most provinces, and the many experts consulted, stressing that hearings should remain informal and some provinces and certain ngo's favoring a quasi-judicial approach (Study Group on EIA Hearing Procedures, 1988: 33).

Because one of my case studies was decided under the old federal environmental assessment guidelines (EARP), I shall pay some attention to that process. The information rules for environmental assessment under the Guidelines Order appear to have been quite liberal, that is, if the assessment was a public review process. The panel is made responsible for running an information program that is to make the public aware of its activities. The proponent is made responsible for a public information campaign to explain the proposal under review. The panel is to ensure that the public has access to '*all relevant information that any member of the public may request*' (§ 28 Guidelines Order). The Guidelines Order (§ 29) furthermore specifies that all information submitted to a panel must become public and that the panel must

47 The Rafferty Alameda (1989, 1990) and Oldman River (1992) cases both focused on the question whether the federal environmental assessment review process should apply to major dam projects, where the provinces (in this case Saskatchewan and Alberta) had already done provincial assessment. See for the second case: Friends of the Oldman River Society v. Canada (Minister of Transport) [(1992, 7 CELR (NS) 1, at 21-52 (SCC)].

48 SC 1992, c. 37.

allow sufficient time to examine and comment on the information before a public hearing. Most information will come in the form of the Environmental Impact Assessment. The EIS normally contains five elements: (1) a description of the proposal, (2) an indication of the need for the proposal and considered alternatives, (3) a description of the present environment, resource use and social patterns, (4) a description of potential impacts from the proposal, and (5) an indication of the measures to reduce or avoid adverse impacts (FEARO, 1987: 3). It is said that the discussion of alternatives for the project was especially important for an accurate environmental assessment, but that this was a weak point, and still is under the current CEAA (Northey and Tilleman, 1998: 218-220). This is because the question of which alternatives are available for a project hinges on the definition of its purpose and this can be tinkered with. The problem also exists for provincial assessment. In Alberta, opponents of various pulp- and paper mill projects in the late 1980s and early 1990s point at the fact that assessment for such mills did not include the actual cutting operations necessary to feed the mills. Therefore, the implications of the concession to clear-cut 73,000 square kilometers of forest for the ALPAC plant were never assessed. Neither were the wildlife impacts of other huge projects (Johnstone and Gismondi, 1989: 18).

In order to assess the proposal, federal panels were to hold public meetings (or hearings), that fall in either of two categories. The first category are special meetings to seek public input on issues requiring further study during the review ('scoping sessions') or meetings to receive comments on draft guidelines for the preparation of an EIS. The second kind of hearing is to provide the principal opportunity for public comment once an EIS has been submitted to the panel. Whichever type of meeting is to be held, the Canadian legislators clearly envisioned a non-adversarial process and rejected pluralist logic. This can be seen from statements that the meetings are to be conducted '*in a non-judicial and informal but structured manner*' (Guidelines Order, § 27). The FEARO (1985) developed the Environmental Assessment Panel Procedures and Rules, which serve to structure the debate during hearings. The rules were managerial in character because they stressed truth finding, not the act of participation in itself. This can be gleaned from the rules that the panel chair may '*limit questioning where participants have substantially similar interests in the project*' (§ 20) and in general '*may limit the questions asked and may limit participants in presenting arguments or making submissions*' (§ 21).

It thus seems that the practice of environmental assessment in Canada has been dominated by managerial thought. However, Canada currently exhibits the most progressive pluralistic ideas on support for parties involved in quasi-judicial proceedings. Current guidance from the Canadian federal government emphasizes citizen participation as an important issue: hearings are an opportunity to learn for both the proponent and citizens. It states: '*The public is an important source of local and traditional knowledge about a project's physical site and likely environmental effects. Through public participation activities, project proponents can obtain this information, better understand and respond to pubic concerns, and inform people about decisions*' (CEAA, 1994). This formulation is somewhat difficult to interpret from my theoretical perspective. In essence the statement appears managerial, but my prediction was that a managerial decision process would not emphasise local knowledge, which is clearly done here. There is thus a trace of the communitarian language in Canadian guidance. Most clearly however, is the fact that progressive pluralist thought underlies the new act. In order to enable citizens to participate in the decision process, the possibility of 'intervenor funding' was introduced, implying that citizens are paid for the costs they incur in participating. However, a managerial concern for both the level of costs associated with intervenor funding and decision quality in technical terms qualified this possibility somewhat. '*The work to be funded must relate directly to the environmental assessment of the project in question, and should provide the panel or mediator with new information not presented by the proponent or other participants*' (CEAA, 1994). Whether or not intervenors will provide knowledge that can not be obtained from the proponent or others seems hard to establish in advance, except perhaps base upon the reputation of participants in a specific field. The dilemma here is that participant funding is only available to a limited extent, and that providing such funds can attract groups that perceive them as a way of funding their activities. Norm Rubin of Energy Probe acknowledged this dilemma in an interview with me. He suggested that some groups become involved in particular issues because of this funding, while their actual interest may have lain elsewhere. It also appeared that there was some level of competition between environmental groups for intervenor funding[49]. This may be caused by the provision that intervenor funding is not granted if another participant already provides a certain type of information. Once an environmental group has received funding it seems logical that another environmental group may not add a significant degree of information to that. Intervenor funding is also present in the two provinces that are relevant for my case studies, Alberta and Manitoba. Parenteau (1988: 60-61), who evaluated environmental assessment procedures, notes that the entrance of experts hired by proponents had the effect of driving ordinary citizens out of the decision process and result in negotiation-type of arrangements between

[49] Interview, Toronto, 13 July 1999.

experts and local authorities. '*Private technocracy may privilege negotiation over mediation. Strengthened by its specialized knowledge, subject to the proponent's economic power of initiative, private technocracy is called upon to directly negotiate conditions of integration. This negotiation directly addresses local authorities, offering general economic advantages in return for a specific impact*'. Under intervenor funding, it can be expected that the negotiation process is becoming one between experts of three instead of two parties.

The Netherlands: an independent Commission for EIA

The Netherlands experimented with environmental impact assessment before actually introducing it. The experiments were on a voluntary basis and largely involved governmental projects; they lasted from 1981 to 1987. In that last year environmental assessment was formally introduced, and this move was partly motivated by a desire to enhance citizen participation in environmental decision-making. The government of the day indicated: 'openness of the relevant data at an early stage gives the participation by citizens more content; this could prevent that objections in a late stage, when investments have already been made, decisions have to be reversed or strongly amended' (Brouwer and Tompot, 1995: 302). Dutch environmental assessment procedures are about on the same level as Canadian procedures when it comes to citizen participation. Like in Canada, the guidelines for environmental assessment are determined after hearing the public.

The issue of expertise has however been solved somewhat differently here than in Canada. The Netherlands have instituted a Commission for Environmental Impact Assessment and this body hires external experts (working groups) to supervise specific assessments, acting on behalf of the Commission. Its role is largely advisory, but at the same time quite influential. The Commission, or rather the specific working group that is appointed for a certain assessment, advises the licensing authority on the guidelines for environmental assessment. The working group also comments on the progress that the proponent of a certain project is making in terms of the assessment process. If the working group considers this progress too small, the assessment document will generally not be approved. So instead of intervenor funding, an independent commission is supposed to guarantee the quality of the assessments that are made. In the end however, the decision rests with public authorities. A high public official from the Department of the Environment (Evers, 1986: 176) has explained the choices by indicating that environmental assessment is a 'management tool'. The manager in question is then the licensing public authority. Decision quality is thus an important concern, but this does not obtain purely managerial tenets. In fact, the official straightforwardly rejected the idea of 'technocrats in power'. '*I do understand the people who say that it should not be possible that members of the local executive, members of local councils, members of the provincial executive, Members of Parliament, etc., who are the responsible authority and who do not care at all about the environment take decisions "just like that". But that is where we have a principal issue: in the Netherlands we live in a certain type of democracy where we have agreed which each other that dumb decisions can be made, as long as they are made by a majority: this includes spoiling the environment.*' (ibid.). Evers justified the lack of intervenor funding, which was called for very often since then - especially from the side of government expert bodies such as the Council for Environmental and Nature Research (RMNO) - on the basis of environmental assessment being what he called a 'management tool'. Of concern was a dialogue between proponent, authorities and the public, a process. This somewhat vague motivation was supported by the contention that in the process, errors would be filtered out anyway, if not by the public, then by the Commission for EIA. During the Parliamentary discussions at the start of the experiments with EIA, a fund for expert assistance was called for by the smaller left wing parties, but the government resisted this claim in part on the basis of the fact that the environmental movement was already receiving government subsidies[50]. The ideal of dialogue in environmental assessment procedures does not seem to emerge too often in practice in the Netherlands. After about 15 years of practice, one of the permanent secretaries of the Commission for EIA remarked that: '*unfortunately it happens far too often that somebody from the room brings forward an issue and then is verbally completely wiped out by the full table of decision-makers. Decision-makers always have a headstart in terms of information.*' (Brouwer and Tompot, 1995: 311).

3.4.5 Some conclusions on citizen participation

The conservative pluralist- and managerial roots of citizen participation

Given the fact that I had associated citizen participation with the communitarian discourse, I find it remarkable that the roots of the UK, Dutch and Canadian reforms in the direction of increased citizen participation stem from elsewhere. Already I have observed the managerial motivation for increased citizen participation in the Netherlands. McAuslan explains that the discussion on increased participation in the UK land use planning system really originates in the conservative pluralist discourse[51].

[50] See 'Tweede Kamer der Staten Generaal, Vergaderjaar 1980-1981, nr. 16814'.

[51] It seems that Rydin (1998: 31) does not concur with this vision. She quotes the rate of development and the changes it was bringing as the reasons for people to want to

The first discussion on public participation took place in 1957, when the report of the Committee on Administrative Tribunals and Enquiries (the Franks Report) was published. This (government initiated) committee was concerned to redress the balance between the public interest ideology and the private interest ideology, as the public interest ideology was considered too dominant at that time and infringement of property rights were taking place too often. How could the landowner be better protected from the busybodies of government? In trying to answer that question, the committee identified the criteria it thought proper for a good administration and they arrived at openness, fairness and impartiality. The intention was that planning law should require government officials to apply these principles in their dealings with landowners solely. However, as these attributes could not 'rightfully' be confined to property owners, the report has had a far more profound influence than the committee's members would have thought. McAuslan discusses the consequent changes in law and policy as a result of a clash between adherents to the public interest ideology and the participation ideology, with the already described result, which can be read as a victory for managerialism. The pattern of rights for citizen participation as an outgrowth of the rights of property owners is also visible in Canada. By 1929, Alberta's planning legislation already had some rudimentary forms of citizen participation, which should largely be seen in the light of a private interest ideology. In that province, local authorities were required to publish announcements indicating where citizens could inspect drafts of zoning bylaws. Interested persons could make written representations to the local council, which then decided upon possible objections. Also, rights of appeal were conferred to aggrieved persons, a term that was interpreted quite narrowly. In the Netherlands similar provisions existed, with landowners being the only ones that could make representations to local councils and appeal extension plans to the courts for decades (the first half of the 20th century).

Citizen participation and neo-liberalism

It is interesting how conservative pluralist thought was recently used in the sphere of the discussion on citizen participation in the UK, more than in the two other countries. One serious part of the Conservative governments' agenda in the 1980s and 1990s was to push for more accountability of public bodies towards the public, reflected for instance in the 1991 Citizen's Charter. Government bodies were to be as accountable as companies in the market for whom accountability was a matter of survival. Private firms do their business, and to stay in business, they must be responsive to their

---

get involved. This explanation is more in line with the reasons that Grant (1989) gives for increased demand of citizen participation in Canada. See above.

customers. So the analogy to private business and the demand for greater accountability does not so much recognize the category of the citizen, but replaces it with the category of the customer, or indeed the (democratic) consumer. The public sector makes a product, and its customers should be satisfied. This 'customer' wants an efficient, flexible and speedy government and to achieve this, parameters and standards must be set against which the performance could be measured. These parameters and standards became an instrument of management in the public sector, and I think it is in this sense, that policy also became more managerial as Jordan has noted (1993). To given an example, the speed with which local authorities processed planning appeals was measured and comparative tables were published by central government, thereby putting pressure on these authorities to quicken procedures. The Conservatives also put pressure on local authorities to change their (traditional) role of service provider to one of strategic management, that is director of desired developments. Reflecting the affinity between conservative pluralism and managerialism, the Conservative agenda for public services has been called neo-Taylorian in character (Pollitt, 1993: 15). In general, the Conservative program resulted in an enormous reduction of local control over decision-making. This is reflected, for instance, in the fact that by the mid-1980s 60 per cent of local expenditure was covered by local taxes but in 1999 less than 20 per cent (Connelly and Smith, 1999: 285). Some are concerned about the compatibility of strong financial control from central government and democratic accountability at that level. Concerns are also expressed about the implementation of goals that have less value to private markets. In the field of waste management, waste collection has been contracted out, and this move has often been criticized for undermining recycling and waste minimization (ibid.).

An absence of the communitarian voice

The communitarian language is largely absent from the collective choice debate in all three countries. This is partly a consequence of the fact that the more communitarian-inclined parts of the environment movement do not see the state as a logical discussion partner. Other parts of the environment movement seem more concerned with developing counter expertise or with strictly political games. The first reaction is the dominant one, supported in Canada by the possibility of intervenor funding. In either case, the ordinary citizen does have some opportunities to influence decisions, but he must do so in a context that is largely framed by the mixture of conservative pluralist and managerial language that is found throughout the pollution controls systems of interest here. Take environmental assessment as an example. It is concerned with generating experts' information on potential environmental implications of proposed developments in all three countries. The essential characteristics of the regulatory system - including the managerial and conservative pluralist language embedded in it - are not changed. Citizen participation is a mere input into this system, with little real power attached to it. Where actors in the collective choice discussion advocate citizen participation (take the FEARO in Canada), they tend to do so in managerial terms, indicating strong structuration by that discourse. In practice, dialogue in environmental assessment hearings does not tend to occur.

3.5 Conclusions

Introduction

The findings of this chapter will be discussed under the heading of various questions. The first one relates to the balance of institutions assigned with the task of making hazardous waste decisions. Which institution does what in which country? The second question relates to the presence of various discourses in the various countries. Have the three discourses been used at all, and if so by whom and to what effect? The third and related question is the one on the theoretical framework introduced in chapter 2. There I had indicated that there is a certain degree of debate about the question whether language captures people and then makes them act, or that people have certain purposes and then choose a language they like. So is the collective choice debate in the three countries best explained as a contest of interests or a contest of language systems? Another issue of interest is whether the expected linkages between discourses and institutions have really occurred in practice or not.

The institutional arrangements in place in the three countries

My conclusion is that all the institutions which could possibly be relevant to hazardous waste decisions according to chapter 2, are indeed recognized in the relevant legislation in all three countries. The exception is the community, which as such receives virtually no recognition in any piece of legislation in all three countries. The basic institution for societal decision-making in all three countries is the market, with the courts deciding conflicts between market parties and the state providing little services except classic collective goods. In all three countries there initially was an intimate connection between property ownership and political representation. Only property owners could vote, run for office, and pay for court cases if their interests were affected. In the second half of the 19th century, the first 'real' attempts to address the consequences of industrialism are to be found. Regulation was introduced that authorized administrators to oversee private decision-making and curtail some of the freedoms that had existed until then. The right of the state to curtail freedom was resisted, especially in land use planning. The courts were sympathetic to landowners and developed certain norms for government criteria. It is interesting to note that the administrators themselves often despised politicians yet were placed under the control of elected bodies. They benefited greatly from this association because the argument that administrators were implementing the political will, made them less vulnerable to challenges from the courts that remained somewhat deferential. Political control however also implied that the agenda of some administrators for greater expert control was not achieved. Instead, administrators were used mainly for activities that mainly supported the continued growth of the economy (e.g. zoning in Canada). This has implied strong restrictions on the discretion of administrators in the Netherlands and Canada, and strong guidance in the UK. The courts have continued to play a certain role in the development of the relevant institutional arrangements. Particularly the continuing development of norms that guide administrative decision-making is relevant. It seems that the courts have become more sensitive to the distinction between administration and elected representatives and the impossibility of elected representatives to oversee the administration. This implies a greater space for the testing of administrative decisions, which is taken up especially in the Netherlands and Canada.

The presence of discourses at the collective choice level

Some conclusions can be drawn about the presence of various discourses in the three countries. The first is that communitarian discourse is largely absent. Although it is often argued that the environment movement used communitarian discourse to mobilize resistance against unwanted projects in the 1970s (see the previous chapter), communitarian argumentation does not appear to have permeated the collective choice level debate. The explanation is that the relatively small parts of the environment movement that have a communitarian agenda in the three countries, do not regard the state (the legislative branch at least) as a discussion partner. The purpose is rather to bypass existing institutions. Consequently, there is no discourse coalition for a greater role of the community in the collective choice debates. The development of citizen participation - a communitarian favorite - in all three countries demonstrates this dynamic. The development is mainly supported by an extension of conservative pluralist thinking (acknowledging the rights of property owners to defend their land), by progressive pluralist language (strengthening representative democracy) and by managerial thought (particularly the idea that citizens can improve administrative decisions by providing new information). This has however implied that citizens are 'allowed' to participate if they can demonstrate ownership in the environment (the UK and Canada) and must argue along the lines of the language employed by the administrators themselves, which is a mixture of (mainly conservative) pluralist and managerial thought. In all three countries concern has been uttered about the fairness of this approach, mainly from the side of academics, progressive politicians and environment groups. This critique however is also not inspired by communitarian ideas. In fact, the emphasis is pluralistic: critics argue that the counter-power of environmental groups should strengthened. These ideas have only gained ground in Canada however, where it has led to the introduction of intervenor funding. There are indications that the effect of intervenor funding is indeed pluralistic. Instead of increasing the possibility of the ordinary public to participate, there is evidence that intervenor funding has even further increased the structuration of the debate by managerial and pluralist thinking by turning hearings into a contest between experts. Furthermore, groups with a broadly similar agenda start competing for the funds.

The managerial discourse was present in the collective choice debates in all three countries. Especially the newly emerging professional classes (engineers, planners, and hygienists) in the 19th century had great belief in science to the extent that they did not believe in true conflicts of interests. The story lines of the public interest and increased utility for all fused well with ongoing developments in all three countries. Businesses started seeing benefits in regulation as it would reduce the steep competition somewhat, protect existing property values, and at the same time provide for an improved workforce. In political circles new ideas about regulation took hold, which specifically stipulated that the public interest would be served by greater government interference. The emerging trade unions and left wing parties finally also greatly believed in scientific management, although not to the extent that scientific managers should completely take over the steering wheel of society. Only in Canada did the aversion of the reformers against elected representatives become clearly visible in the institutional make-up of the decision processes. Planning boards there were placed outside the remit of local authorities, but then again these planning boards were placed under control of the business community. These days, administrators are nowhere claiming that the rule over society must be taken over by experts. The last real heyday of managerial thought was in the 1970s, when planners considered planning the model for decision-making in all fields and attempted to place themselves above other parties as arbitrators of the public interest. Note that as far as their agenda succeeded, this was mainly because the more progressive political parties were in control and these parties held some belief in state-managed economic development. Currently most professionals have accepted the existence of a certain level of political control and the need for free enterprise. However, it must be noted that administrators do demonstrate various lasting tendencies. They tend to maximize claims that issues should be under their control (see the Nolan Committee discussion in the UK). Furthermore, they acknowledge that they operate in a political environment, yet claim not to have any political inclination whatsoever; this being made possible by their utilitarian outlook on society, which implies that developments are acceptable if they increase utility for the collective society.

The pluralist discourse is very present in the debates at the collective choice level in all three countries. The conservative variant seems to provide the dominant reasoning for the relation between the market and the state. The market being the basic institution, the threshold requirement for government interference in all three countries tends to be the increase of utility in the eyes of experts. If regulations are introduced, there is a tendency to have administrators and private market parties consult with each other about the appropriate level of regulations. Political guidance tends to be rather vague, but where present it stresses a balance between environmental and economic considerations, which is indicative of conservative pluralism. It is interesting that elected bodies in all three countries are more concerned with utility and less with distributive equity. This may have largely to do with the fact that land use planning and pollution control were long seen as a-political, mainly technical, policy fields. This would explain why even in the relatively progressive decades after the Second World War the efforts that were made to get more redistributive policies in these two policy fields stopped short of positive planning. However, it also seems to reflect the deep conviction that the market is the best institution for making decisions. If one takes a look at the relatively few attempts to make land use planning a redistributive policy field (e.g. the Labour government in 1974) one cannot escape the impression that societal resistance was enormous as there simply was not the belief that government should take over all rights of development. Indeed, if anything, the UK land use planning system reflects the deeply seated norm that development must be allowed as much as possible. The courts in all three countries work under pluralist assumptions, with the parties involved fighting it out, albeit that the Dutch courts behave much more as if they were part of the administration (which they were for a long time) and quasi-judicial institutions are also more bent on a managerial approach to decision-making. Even in the Netherlands however, the constellation of administrators backed by democratically elected leaders has long made the courts refrain themselves from overly-active testing of government decisions. The boundary rules for access to the courts seem to demonstrate a turn from conservative pluralism to an extended form of conservative pluralism or even progressive pluralism. What I mean by that is that boundary rules initially stressed ownership in gaining access to the courts, then stretched this to ownership in the environment, whereas in the Netherlands boundary rules provide hardly any limitation at all. In Canada finally, there is a system of assisting those participating in quasi-judicial proceedings, which reflects the progressive pluralist discourse.

Some reflections on the discourse approach

In the final part of this chapter I want to present some conclusions on my theoretical framework. The first group of conclusions relates to certain unexpected consequences of certain discourses, the second more to the practice of discourse analysis itself. To start with the first group, I have found some remarkable patterns that do not really fit to some of the assumptions I had posited in chapter 2. Take for instance the issue of citizen participation, which I had equated with the communitarian discourse. What I found in reality is that this discourse is lacking at the collective choice level debate, but that citizen participation rights have certainly been expanded. The reasoning appears to have been conservative pluralistic and managerial thought. Especially the courts but also elected representatives have decided that environmental groups (or even anyone) have 'ownership' in the environment, implying that they have standing in decision procedures. Also, the quality of decisions, in the managerial sense, has been an important impetus for increased citizen participation rights. Another conclusion in the first group is that the use of the 'public interest' as a story line is not at all confined to those adhering to the managerial discourse. In fact, many groups claim to further the public interest and this is much acknowledged by granting standing in court to 'public interest groups' such as environment organizations. The result, unavoidably, is the move towards a more pluralistic view of decision-making. Note that planning theory moved from 'geographical determinism' to 'procedural planning', which is indicative of this move among planners. In the process, the term 'the public interest' becomes less and less appealing, because everyone can claim to be acting in the public interest. However, not using this term is not an option, because the alternative: an appeal to a 'special interest'. sounds markedly negative. Despite the constant fencing with the term public interest, it does not seem that the 'real' definition of the public interest has changed much. Economic utility is still a dominant concern in all three countries. Environmentalists may make any claim they like that they represent the public interest, in the end such claims are weighed against other claims using aggregation rules such as 'BATNEEC' and 'ALARA'. The consequence is that a strongly economic interpretation of the public interest ensues, and that technical and economic quality are really the types of quality sought in environmental decisions. The third and final conclusion in the first group relates to the second one. I have observed that even though a certain discourse - specifically conservative pluralism- may be very relevant in the collective choice level debate, this may still have very different consequences. The three countries of interest here all seem concerned with property rights, but the exact property rights of concern are not the same ones. In the UK, the rights of property owners to freely use their property as they see fit is engrained in the legislation. In the Netherlands, and to a lesser extent Canada, on the other hand: there is a far greater concern with the protection of what is there already. This difference is largely a consequence of the preferences of the courts, in combination with elected representatives and business interests. The consequence is a certain rigidity in regulations in the Netherlands and Canada and flexibility in the UK.

Finally, turning to some observations about discourse analysis itself, it appears that interests are very important in the selection of a certain discourse. Note for instance the efforts of professional groups to demarcate their profession by setting boundary rules and at the same time enhancing the standing of the profession through professionalization and lobbying for a certain territory. Experts undoubtedly believe in the story line that science is a good guide for policymaking, but a certain attention to one's own position can not be denied. As the frequent use of the concept of the public interest already suggests, certain story lines are very attractive and provide the speaker with a certain authority. Not just the public interest and science are attractive story lines, so are neutrality, lack of bias, and democracy. The fact that administrators can claim to act on behalf of democratically-elected leaders gives them great leeway vis-à-vis the courts, even though the actual meaning of democratic supervision is very limited, and even though administrators often claim not to be political at all. Take the science upon which the UK bases its policies as an example, that is far from complete or even up to date, but still regarded as very impressive. And what of the 'neutral' courts that in all three countries greatly affected the leeway for administrators and the protection of certain classes of property holders. Quasi-judicial proceedings are set up in such a way that they appear court-like. The goal is to mimic the impression associated with courts: neutrality and fairness. In reality, quasi-judicial proceedings have their terms of reference determined by others than the parties to the proceedings and result only in an input to a political decision-making process that, if possible, takes shelter behind the neutral advice of the quasi-judicial body that heard the case. What this all implies then is that parties do use discourse as far as it furthers their interests. It is remarkable that some concepts, mainly the ones associated with the managerial discourse, appear to have greater attraction to all parties than other concepts. It is thus crowded in the room where neutrality, the public interest, and science reside but there is room left in the place where politics, values, and clashes of interests reside. If anything, this demonstrates that modernization, or at least the impression of it, is what the collective choice debates are about. In the next chapters we will see whether the same is true at the concrete choice level.

Chapter 4

The UK: Political control with the benefits of judicial authority?

This chapter describes three cases that I have studied in the UK. Although it is often suggested that the UK population shows more deference to public authority than the people in other Western democracies, there is evidence here of a very strong desire to actively resist the applications. Such resistance is shared by the local authorities in each of the studied cases and results in refusal of planning permission. Such refusal is largely motivated in planning terms, whereas in two of the three cases there were clear indications that refusal was based on different concerns. In one case, refusal was partly defended on the basis of the argument that the proposal was against the local interest. After refusal of planning permission, the decision process is thrown into the ambit of national control. In each case a long public inquiry was held, which was really intended to probe the quality of the proposals in managerial terms. During the inquiries, experts were pitted against each other, with the action group in one case providing expertise that surpassed the expertise of the other parties. The action groups do not represent entire communities. And although they use a certain level of communitarian discourse to stir up opposition, they resort to managerial language during the inquiries. The use of this language is enhanced by the possibility of cost awards against those who deviate from it, and by the fact that any other than technical arguments do not carry weight at an inquiry. The cases suggest that the UK government is not afraid to go against the wishes of the local population and to override resistance. It appears that the procedure at the inquiry and the strong perception of objectivity on part of the planning inspectors results in an acceptance of the outcomes of the process at the local level. There is very little use of the courts as an avenue for decision-making.

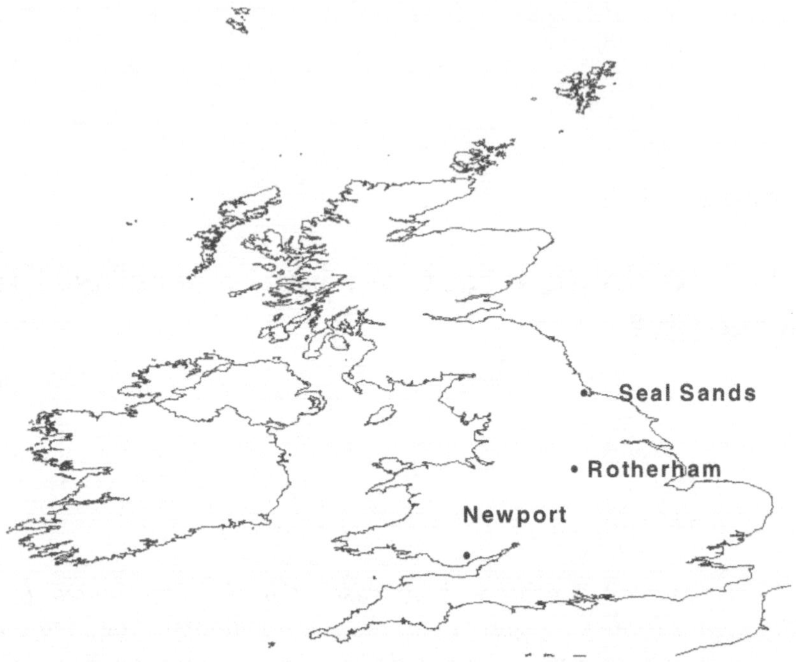

Figure 1. Places visited in the United Kingdom

4.1 Seal sands (Cleveland): linked incinerator proposals

4.1.1 Introduction

The first UK case I shall describe[1] involves the proposal for a merchant hazardous waste incinerator by a company named Ocean Environmental Limited. The proposal is set against the background of an active interest that the UK private sector started to take in hazardous waste incineration in the second half of the 1980s. At that time, burning and dumping of wastes at sea was to be phased out and recycling and waste reduction was still a relatively moot option implying that land-based incineration was considered an attractive waste disposal option. Furthermore, various waste disposal companies had become publicly traded firms on the stock exchange and had

[1] Helpful comments to earlier versions of this draft were received from Mr. D. Walsh (Leader of the Council, Redcar and Cleveland Borough) and Mr. B. Middleton (Cory Environmental Ltd.).

received quite a capital injection from that. The Northeast of England (specifically Humberside and Teesside) is an area where these dynamics were present. The area was home to large industrial companies that were producing massive amounts of hazardous wastes and had up till then relied on old-fashioned disposal techniques. Various firms, from the region itself and from elsewhere, saw waste incineration as a potential growth market and thought they should perhaps construct facilities in the northeastern part of the country. Among the newcomers to the waste disposal market were some of the newly privatized water companies, which thought they could combine hazardous waste incineration with incineration of treatment sludges. Ocean Environmental Limited however was one of the first firms to develop a concrete proposal for a hazardous waste incinerator.

4.1.2 Internal preparations, first contacts with authorities

Ocean Environmental long avoided publicity about its plans, mainly because the company did not want to alert competitors about its activities, but also because it wanted to be well prepared once it would go public, so that its plans could withstand scrutiny. The company performed an analysis of the regional waste market in the Northeast by interviewing the major companies in the area. On the basis of these interviews, Ocean assessed that it could capture about 30,000 tons per annum. Given this amount, the company considered a rotary kiln incinerator with secondary combustion chamber the appropriate technology for its projected facility. The desire to develop a plan that could withstand public scrutiny was taken very seriously. Although there was no official requirement under UK regulations to perform an environmental assessment at the time, the company decided to contract a consultancy firm to perform such an assessment anyhow.

The assessment report, or Environmental Statement (ES) as it is called in the UK, that later resulted from this exercise suggested that Ocean had looked at the entire North of the UK to find appropriate sites. The ES presented a list of 23 sites and indicated that these sites had been tested against nine criteria, amongst which certain business criteria. It seems that these business criteria carried a lot of weight because the four sites that were particularly suitable according to the ES were almost all in Cleveland County, where much waste was generated. At the time the ES appeared, there had been various meetings with the local authorities in the area, particularly the Cleveland County planning department. The department had provided certain data to the company in the context of the assessment process and planners had (non-bindingly) informed Ocean Environmental that there appeared to be no planning objections towards the plans. The county planners indicated that planning policies were being changed to put less emphasis on heavy industry, but the site of most interest to Ocean (at a location called Seal Sands) would remain designated as an area for the

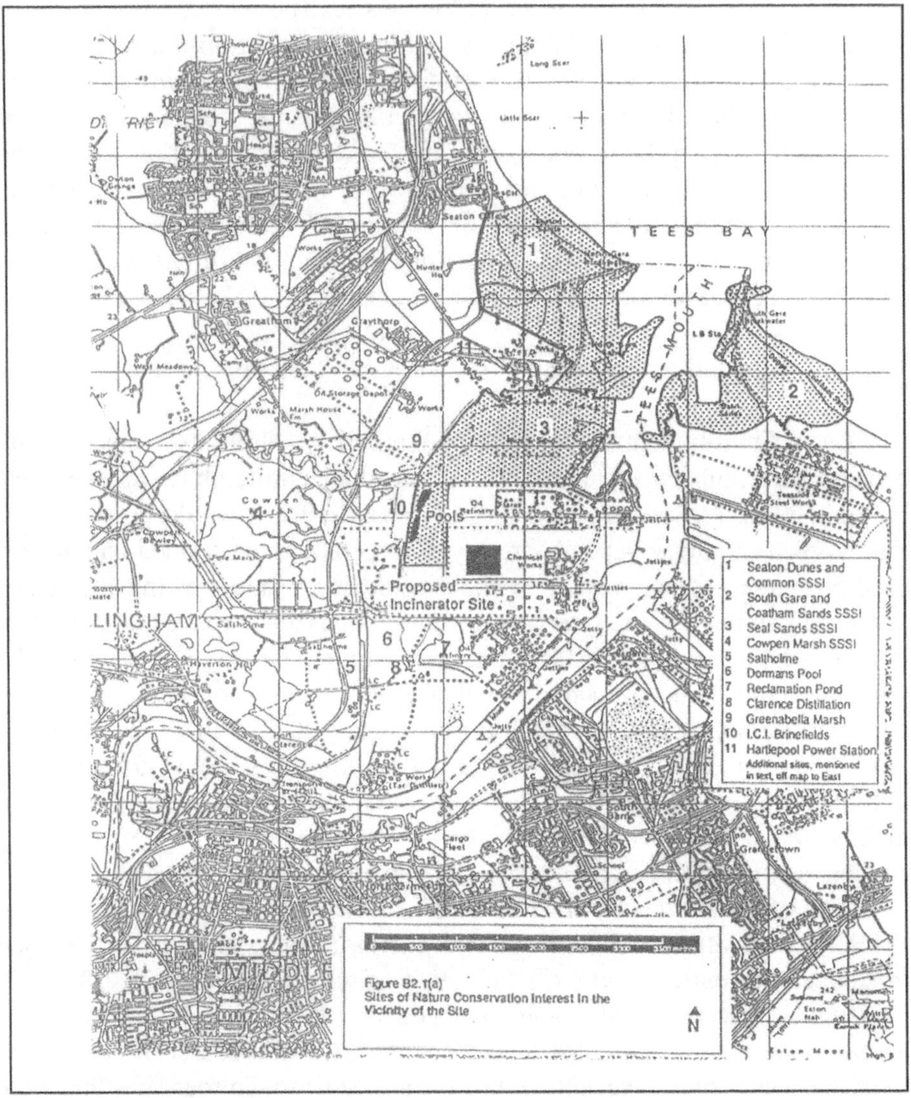

Figure 2. Location of Ocean Environmental's proposed Incinerator. The location is near Middlesbrough and Billingham and is depicted by the black square. Adapted from the ES.

chemical and oil industry. The Tees and Hartlepool Port Authority, which owned the site, did not object to Ocean's plans, and was willing to sell the land to Ocean Environmental. Figure 4.1 depicts the site that was eventually chosen.

4.1.3 Official procedures

Planning application

The first step in gaining public approval for Ocean's plans with the Seal Sands site was to get planning permission from Cleveland County. Ocean Environmental submitted an application for planning permission on 30 December 1988, after completion of its environmental assessment report. The application was treated as the planning legislation prescribed, and thus the application was publicized and sent to various bodies for comment. Given the complex nature of the facility, the county requested that Ocean allow an extension of the determination period, which the company agreed to.

Mixed responses from the statutory consultees

The replies from the various consultees to the application were essentially positive, especially from expert bodies, such as Her Majesty's Inspectorate of Pollution (HMIP), the Highway Officer, the County Surveyor and Engineer and the Country Side Commission. HMIP found the plans 'eminently acceptable' and other expert bodies had indicated some concerns initially, but these were adequately addressed in consultations with Ocean Environmental. The only statutory consultees who did not respond positively were the borough councils within the county (Middlesbrough, Stockton, Langbaurgh, and Hartlepool). These local authorities had contracted various experts to study the environmental statement (ES). Despite the fact that their advisors found few flaws in the ES and indicated that Ocean's plans were sound, the boroughs advised the county that they opposed the plans. The reasons given varied, but included the lack of a strategic plan for hazardous wastes at the county level. Also mentioned was the possibility that the facility would attract further unwelcome development and lead to an increase in pollution, and that the system of environmental regulations was not reliable enough to guarantee the predicted emission levels.

Decision-making within the county bureaucracy

Within the county bureaucracy, the discussion on the proposal largely focused on its compatibility with county policies. Both in the sphere of land use planning and waste policies there appeared to be little objections. The existing and emerging county structure plans designated the site that Ocean Environmental had chosen as 'special industry', a use that allows hazardous waste incineration under national planning guidance. The guidance in the county's fairly recent waste disposal plan (adopted in 1987) was perhaps somewhat more ambiguous. The intention of the plan was to continue the practice of waste exportation to other jurisdictions. However, the plan also indicated a shortfall in local disposal capacity and suggested that the private sector should take initiatives to create long term national capacity for waste disposal. It is doubtful that the county had envisioned *itself* as the proper location for such national facilities. However, because of the invitation to the private sector, Ocean Environmental's application could be seen as being in accordance with county policies.

The waste disposal plan contained a procedure for decision-making in case of any concrete proposal. The statutory consultees and the advice from the Cleveland Pollution Control Group, a consultative body of environmental officials active in the region, would determine the acceptability of the proposal. As both groups gave no reason to question the acceptability of the proposals, county officials advised the county council that the application should be approved under certain conditions. This conclusion was contained in a report from the Director of Economic Development and Planning that was presented to the planning committee of the council on 19 June 1989. In the report, the director indicated that Ocean Environmental was willing to voluntarily accept a Planning Agreement with the county that would prohibit waste importation, provide for continuous monitoring by the local authorities on site (not just by HMIP), and would lead to the establishment of a liaison group consisting of local citizens.

The response of the County Council

By the time the application for planning permission was received, it was already clear that the proposal would be politically controversial. Ocean Environmental had informally approached some key council members for a meeting. During this meeting, the council members had voiced strong objections to Ocean's plans. Immediately after the meeting, the councilors made Ocean's proposals public, thereby disrupting the company's low profile approach.

The reasons why the councilors in question objected to the plans included their concern over the potential health effects of waste incineration and a feeling that there was no proper framework in place for judging the proposal. Councilors feared that without such a framework Cleveland county, or even the entire Northeast, would be flooded with wastes and was to become the dustbin of the UK. The latter feeling was strengthened by the fact that various other proposals for waste incinerators were now being made or announced. Especially a second proposal for a combined hazardous waste and sewage sludge incinerator by International Technology Europe (ITE) was a cause for concern. Although Portrack was part of the Cleveland county, Cleveland county council was not the planning authority in this case. This was because central government had given planning authority to the Teesside Development Corporation, with the explicit purpose of stimulating economic development. The Teesside Development Corporation was seen as the creature of a Conservative government in the Labour-led Northeast. Besides, the members of Cleveland county council had little trust in the TDC since it had advertized its Portrack site as '*an unique opportunity to set up chemicals manufacture on a greenfield site unhampered by environmental objections*'[2].

Despite the fact that some councilors had already expressed objections to the proposals informally, the county council did still spend much time seriously deliberating it. Because the issue was very complex, one meeting of the planning committee on the 19 June was not sufficient to decide the planning application. The committee requested additional information, amongst others information on other local authorities' experiences with hazardous waste incinerators, waste arisings in the area and emission levels of the proposed incinerator in comparative terms. The information on experiences of local authorities elsewhere was specifically required because councilors had little faith in the pollution control system that regulated such facilities. The county bureaucracy quickly collected the requested information and wrote a report on the findings which essentially confirmed the need and safety of the operation that Ocean was proposing, although one Welsh borough had indicated it experienced considerable problems with an incinerator. Further meetings were held, including a three-day public meeting. The council was of a mind to refuse the application, but was especially concerned about what would happen if Ocean subsequently appealed and throw the application in the lap of central government. Ocean Environmental had been forthcoming with guarantees that would strictly speaking not be required by law (e.g. the citizen liaison committee) and the councilors feared that these guarantees would not survive the appeal process.

The three-day hearing, which featured both opponents and proponents of the proposals, was concluded by a meeting behind closed doors between county planning officials and members of the planning committee. The Director of Economic Development and Planning outlined the dilemma that the council was facing. Like the council he had his doubts about Ocean's plans. He agreed that the system of pollution control in the UK was not reliable and that there was insufficient data on the potential health effects of incinerators. However, he reminded the members of the boundaries of the authority's responsibilities under planning law. In a note to the council, he wrote that the problems that had been raised by councilors and opponents were '*outside the direct control of the County Council as local planning*

2 'Council's no over waste job plans', Evening Gazette, 12 June 1989, p. 5.

authority'[3]. The Director indicated that those who were responsible for safety, health and environmental issues were all of the opinion that Ocean's plans were acceptable. Therefore he recommended approval of the planning application, also in light of the many planning conditions that Ocean had been prepared to accept voluntarily. After hearing the Director's submission, the planning committee decided to postpone the decision and transfer the case to the full council. David Walsh, who was chair of the planning committee at that time, told me that the committee was faced with a dilemma. The committee was not convinced of the need for the waste incinerator, but was keenly aware of central government's direction which forbade them to take need into account in taking its decision[4].

It was clear that the council was walking a thin line between the public opinion in the county that seemed to have turned against the proposals at least to some extent, central government guidance and the advice of the bureaucracy. The leader of the council, now in charge of thinking up a proper approach, had to find a solution to this dilemma. In the days leading up to the meeting of the full council, he drafted various resolutions that would become the basis of the council's debate. The main resolution argued that planning permission should be refused because there was uncertainty over the health effects of the proposals. Secondly, Ocean Environmental's proposal was described as being premature because of various reasons. These included the 'facts' that the county's waste disposal plan was not up to date, and that the pollution control authorities were too weak to monitor the plant effectively. The resolution was unanimously adopted by the council, as were various additional resolutions, including one resisting involvement of the private sector in hazardous waste treatment and one indicating the believe that the proposal would be detrimental to new investment in the area. Finally, the council adopted a resolution that called for a (very rare if not unique) Planning Inquiry Commission, which would study the issue of hazardous waste treatment facilities and their location on a national scale. The latter is interesting because it implied that the county refused to develop a hazardous waste policy at the regional scale, which is what the Teesside Development Corporation, also starting to have doubts about the proliferation of hazardous waste facilities in the region, had proposed. The reasons that the Cleveland county council gave for refusing a regional approach included the argument that it would take too long to develop such an approach to serve as a basis for Ocean Environmental's applications and that the public in Cleveland had waited long enough for a decision. Note that

3 Underlining in the original. From the 'Second report from the Director of Economic Development and Planning to the Committee: Conclusions and recommendations'. Found in the Cleveland County Files at Stockton Borough Council.

4 Interview, Redcar, 6 January 2000.

Cleveland's objections to a regional approach make sense, but that at the same time similar arguments (and to a greater extent) apply to the county's proposal for a national study of the problematique. In other words, the county's refusal of a regional plan for hazardous waste should perhaps mostly be seen in the light of the fact that a national approach would perhaps shift the emphasis away from the county, whereas this was less likely for a regional approach. Indeed, this is what the local newspaper thought to be the purpose of Cleveland's actions and the paper actually lambasted the council for it.

Ocean Environmental appeals the refusal of planning permission

Ocean Environmental appealed the county's decision two months later. It was argued that there was a need for the proposal and that most of the county's objections could have been covered by a planning agreement between the two parties. Ocean requested that a public inquiry be held, which implied that a planning inspector from central government would hear the case and that other parties could be present as well. The Planning Inspectorate proceeded with the appeal on the assumption that there would indeed be a planning inquiry, but events overtook this intention. The county's push for a national Planning Inquiry Commission was stifled by two other refusals for planning permission that soon came in. The Teesside Development Corporation decided to refuse ITE's proposal for the incinerator at Portrack, and around the same time, ITE had another proposal refused on Tyneside, near Newcastle.

Both refusals gave momentum to the case that national issues were involved, and Cleveland was trying to lobby the Secretary of State to indeed have a Planning Inquiry Commission established. Perhaps as an indication of the relatively powerless position of local authorities in the UK, a different solution was announced by the Secretary of State in a Parliamentary debate, even before he had spoken to representatives of Cleveland or the other communities involved. After questions from various Members of Parliament during a session on 22 January 1990, the Under-Secretary of State announced that the appeals of the proposing companies in these three cases would be decided together in a so-called 'linked inquiry'. The decision was motivated by the argument that the three proposals were overlapping in this sense that they would compete for the same waste streams, whereas allowing all three would result in over-capacity. The Under-Secretary announced that there would be a general session on all three applications that would focus on the need for local disposal capacity and environmental effects. In addition, there would be three regional sessions that would cover site specific issues including land use planning. A team of four planning inspectors, assisted by a technical assessor who would help them in their assessment of the technical appropriateness of the proposals, would hear the inquiry. Somewhat prejudging the outcomes of the general session, the Under-Secretary of State indicated that the government felt that there was indeed a need for additional waste facilities in the Northeast as the region was a net exporter of wastes.

The local reaction to the decision was furious; the leader of the Cleveland council indicated that the government had refused to discuss the more fundamental issues and had pre-empted the inquiry debate by indicating that there was indeed a need for a facility. Ocean was not too happy with the government's decision either because a linkage of the three cases implied that it would loose its advantage. Ocean had been ahead of ITE in procedural terms, but now their proposals were treated together. There had already been a timetable for Ocean's appeal, but this was now withdrawn. In May 1990, the Secretary of State formally called in the proposals, which implies that the inspectors would not decide the cases on their own, but would only advise the Secretary of State, who would then determine the case.

Preparations for the inquiry

The inspectors held pre-inquiry meetings in June, July and August 1990. There, many procedural issues were discussed, including the timetable. In terms of substance it quickly became clear that the procedures would be highly pluralistic, not so say litigious, in character. Even basic facts such as the amount of undeveloped land at Seal Sands, climatic conditions on the site, distance of the plant to the nearest nature reserves, etc., etc. were presented as issues that should be discussed at the inquiry. One of the more important concepts that had to be defined was the term 'local waste policy'. This was because Cleveland, in the period leading up to the pre-inquiry, had collected new data on waste arisings which indicated that there was no 'local' need for an incinerator; that is, not in Cleveland county itself. Other parties however contended that the discussion was about the whole Northeast and not about particular counties. The inspector concurred with the second point of view, with the implication that waste arisings in the entire Northeast should be debated and compared with disposal capacity.

Already during the pre-inquiry meetings, the inspectors started forcing the parties to start negotiating conditions that they felt should apply if the plans were to go ahead. This requirement is common practice in the UK and is intended to speed up the decision process later. In the case of the proposal for Seal Sands, however, the county and Ocean Environmental severely disagreed, mainly because Ocean was now taking a less forthcoming attitude and would no longer accept some conditions that had previously been agreed between county officials and the company. The liaison committee and county monitoring of plant performance were items that had been agreed to before, but which the company now withdrew. The inspectors got involved in such debates to some extent, for instance by indicating that certain conditions the county was proposing could not be justified. However some issues remained unresolved, which implied that the Secretary of State should decide them later.

Inquiry and post-inquiry

During the inquiry, Cleveland's case centered around the uncertainties regarding the application, mainly those related to the waste market and technologies, to the detrimental effect of the proposal on the regeneration plans for the area, to public opposition, and lack of trust in Ocean as an operator of such plants. The county's barrister succeeded in making Ocean reveal the assumptions underlying its assessment of the local waste market and successfully attacked them, but this had a limited effect however because the definition of what was the local market had been changed from involving only Cleveland to the whole Northeast of the UK. The barrister for the county embarrassed Ocean by pointing out that it had wrongly added up waste projections in its ES, resulting in an overstatement. In return, Ocean embarrassed the county by revealing that the county itself had already been burning hazardous wastes in its own municipal waste incinerator without informing the public and apparently without raising very fundamental objections. Opponents used a study by the county's health department to support their contention that the incinerator would have detrimental health effects. Unfortunately for them, this report suffered from a lack of scientific credibility as it was felt it did not prove a causal relation between elevated levels of pollution and public health. The opponents succeeded however in generating considerable debate over the air dispersion models that had been used to assess the health impacts of the plant.

The inquiry ended in March 1991, and the inspectors submitted an advice to the Secretary of State eight months later. Their report was not made public, but its crucial recommendation was to allow Ocean's proposal and to refuse both ITE's proposals. The Secretary of State took another year to decide the case, and so, the decision and the inspectors' reports were made public in November 1992.

On the basis of the inquiry, the inspectors were convinced of the need for a facility in the area. They based this conclusion largely on the fact that both ITE and Ocean had seen a market for an incinerator. The inspectors also concluded that uncertainty about the health implications of the incinerators could not be a reason for refusal of planning permission. They indicated that the demand for absolute certainty about health implications from some of the opponents was impossible to achieve. On the other hand, concluded the inspectors, would it be wise to consider any remaining uncertainties as a factor weighing against the proposal. It was necessary to adopt a conservative approach to siting incinerators near residential areas. But this, concluded the inspectors, is exactly what Ocean had done in its environmental assessment process by studying the emission of 50 substances from the incinerator and applying very wide safety margins in assessing their health impacts. The inspectors concluded that remaining emissions were small and would be marginal when compared to already existing emission levels in the area. The inspectors also concluded that Ocean's proposals were as good as possible with present technology and that they fitted in the county's planning policies, which was not the case for the two other proposals made by ITE, which therefore looked less acceptable in planning terms. Some had expressed fear that the plant would interfere with the operation of a nuclear power station about three kilometers from the site, but the inspectors concluded that such an idea was virtually incredible. They did not believe either that the plant would be detrimental to the county's image because '*To be seen to deal responsibly with one's own industrial waste could be part of a positive image, in our judgement*'[5]. Because of these considerations, and because a great number of conditions would be attached, the inspectors recommended the Secretary of State to allow Ocean's proposals.

Secretary of State decides

The Secretary of State followed the advice of the inspectors and granted the appeal. In doing so, the SoS deleted some of the conditions that the inspectors had proposed, such as a ban on the import of wastes and certain emission limits. According to the Secretary of State, these were not issues that could be regulated under the planning system. Of course the decision of the Secretary of State caused a huge outcry in Cleveland when it was announced, but remarkably most of the bitterness was caused by the fact that the Secretary of State had not taken over all the recommendations of the inspectors. It seems that many local politicians felt that this was foul play.

With the decision of the SoS, Round 1 of the decision process had been completed, and planning permission granted. Normally, the next step would have been a decision on licences under the pollution control system. This step was, however, never taken. By the time this permission was given, the market for wastes had largely already changed, and in the direction that had been foreseen by groups as STINC and by the county. Extra treatment capacity had been installed elsewhere and demand for waste incineration had decreased (e.g. solvents were no longer burnt in incinerators but in cement kilns). Therefore Ocean Environmental has not actually realize its facility to

[5] Ibid. p. 39

date, although the company (now named CORY Environmental) did renew planning permission every five years since the original permission was granted. Because such renewals may only be refused in very exceptional situations and are really a formality, neither the county (now reorganized into various boroughs), nor anyone else, objected.

4.1.4 The role of various institutions and discourses

Introduction

The decision process on Ocean Environmental's proposals was largely structured in accordance with the relevant legislation, with a few notable exceptions and a few attempts to change the rules of the game. The most important deviations from the normal decision process are the fact that the company performed environmental assessment of its plans whereas this was not legally required yet, that the company was forthcoming with guarantees to the county, and that three proposals were linked. The possibility of a linked inquiry is not really a deviation from the law, as planning legislation provides for this possibility, but even that being the case, the possibility is hardly ever used and therefore still unique. One consequence of the fact that the inquiries were linked was that the issue of need came open for discussion, something that is often simply left to the market. Another consequence was that Ocean Environmental and ITE were pitted against each other, as the implicit assumption underlying the linked inquiries was that only one proposal should go ahead given the overlap between the three plans. This created a certain pluralistic dimension during the inquiry, with the two proponents attempting to discredit each other.

There were some attempts to go beyond the legal framework that seem to have failed. The county especially wanted to turn hazardous waste siting into an issue that would be addressed nationally, with central government developing a 'positive' plan for the location of treatment facilities. This managerially sounding proposal fell on deaf ears in Whitehall however. After it had refused planning permission, the county also tried to stir up local opposition by using communitarian language. There was indeed active opposition to the proposal(s) from a local action group, but this group was not so much instated as a consequence of pleas by the county, but an initiative from certain activists. They specifically targeted ITE's proposal for Portrack, but Ocean Environmental's proposals as well. Their motivation was somewhat communitarian-sounding as the leaders described the proposals as being against the interest of the community and they claimed to have waited with their opposition until they had the feeling that there was wide opposition to the plans. The action group did not however use communitarian language during the debates on planning permission. Instead, they presented data on the need for waste treatment and health effects mainly. Although they did a remarkable job in collecting evidence on such items, their actions seem to have had little effect in the end and the planning inspectors that heard the inquiry were not convinced.

The role of private parties, experts, and elected representatives

The initiative for the proposal in this case rested with Ocean Environmental, which is a private market party. The company saw a business opportunity, and as can be expected from a private market party, the desire to make a profit and to compete with other parties influenced its approach to the decision process. This influence is particularly clear from the fact that the company tried to keep its plans out of publicity so that competitors would not find out about them. I have noticed that this concern is realistic to some extent, because the county files show that quickly after the company applied for planning permission, competitors came to study the application. Taking the commercial motivation into account, it is remarkable that the company performed environmental assessment of its proposed facility as it was not legally required to at the time. The motivation for performing such assessment seems partly a genuine interest in designing a facility that is technically optimal and partly a wise strategic move. Concerning the second motive, having an ES to submit along with an application for planning permission will increase the chance that the proposal makes a reliable impression. The ES itself seems to reflect this desire particularly when it describes the actual site selection process. The report describes how the company has compared a range of potential locations on the basis of a host of criteria and thereby identified four 'ideal' locations. Especially the fact that 'business considerations' are among the criteria, and the fact that three out of four of the suitable locations are in one county, does seem to give some reason for thought about the weight of the business considerations in the entire equation. However, such questions were hardly asked during the decision process, and the impression of a rigid, if not managerial, site selection process became a strategic advantage for the company.

The plans needed approval from the county (and later central government), and this to some degree provides an opportunity to look at the relation between government experts and elected representatives. The county council, especially the planning committee, had dealt with hazardous waste issues several times in 1988. The experience of the county included the impossibility of stopping hazardous waste storage on agricultural land and requests of various companies to start treating and storing hazardous wastes that were only grudgingly accepted. The county council had expressed its concern over these developments, which it felt would turn the county into the dustbin of the UK. At the conclusion of the debate on one of these proposals, one councilor had said that no more wastes should be allowed to enter the county from elsewhere in the UK and it seems most councilors agreed. Apparently, this conclusion did not translate into a desire to change the county's waste disposal plan. But it does still help explain why, when Ocean arranged for an informal meeting with the council in November 1988 to present its plans and to get the council 'on board', the meeting had the opposite effect. Councilors were shocked to hear how far the company had already proceeded without their knowledge. Ocean's plans were leaked to the press and the leader of the council publicly attacked the company for sneak tactics and being secretive. He said: '*The company has already commenced behind-the-scenes investigations. (. . .) I am extremely concerned that plans are being drawn up to build a chemical waste incinerator capable of treating toxic waste products - from this country and abroad - without telling people what they are doing. (. . .) I would call upon Ocean Environment Management Ltd. to let the people of Cleveland know what type of development is planned and for a public debate to be opened on this issue*'[6]. Especially the suggestion that Ocean would import wastes from abroad or other parts of the UK caused quite some unrest in the county and local Members of Parliament were quick to jump on the bandwagon and declare their resistance. The company initially stuck to its strategy of not publicly commenting on its plans until it had formally applied for planning permission. However, the negative publicity forced Ocean Environmental to abandon this strategy and hold public information sessions, which did not weaken the impression of sneak tactics however, and to some extent actually intensified it, because they were badly coordinated.

It seems fair to say that the proposal put a strain on the relationship between council and county officials. The officials continued to process the application for planning permission under the procedures prescribed by the waste disposal plan and UK planning legislation. These procedures stressed the role of expert bodies and attached great value to the technical quality of the proposals. Most of the consultees that responded to the application, especially the ones that were most important (e.g. HMIP), reacted positively to Ocean Environmental's plans. Only the Boroughs were opposed to the proposal, but they gave little valid 'technical' arguments. Therefore, county officers could not fault Ocean's plans and they advised the council to approve the proposals and sign the Planning Agreement. This conclusion was not at all to the liking of the planning committee or full council. The planning committee however, used to working under the planning legislation

6 'Come clean order to a waste firm!', Evening Gazette, 3 November 1988, p. 5.

and recognizing its weak position if it went against the planning officers, did not dare reject the application and transferred the decision to full council. Having received little argument from their officials to reject the application, the leader of the council thought out a strategy that would still defeat the proposal. Key parts of that strategy was an offensive, managerial argumentation, which suggested that the authority to make proposals for hazardous waste facilities should be taken away from the private sector and be vested in central government. Another was an emphasis on the aggregation rules that the proposal should 100 per cent certainly not have negative health implications. The council also played a card that can be called communitarian to some degree. The fact that there was local opposition to the plans was made into an argument against the proposal and the leader of the council started explicitly framing the discussion as a battle between a shrewd private developer and the community. '*Over the next few weeks and months we will see a massive effort by those who stand to make a great deal of money, to impose their will on the people of Cleveland*'. The leader of the council called for '*a massive demonstration of people power*' because '*as the county council we cannot fight this battle alone*'[7]. The council's position led to an implosion of the relations between council and high-ranking officers. Although none of the people I spoke to wanted to elaborate on this, I have been told on various occasions that the Director of Economic Development and Planning changed jobs because of this decision.

The council took its case for central government planning on hazardous waste facilities to the SoS with the help of local MPs. However, the council was snubbed as the SoS accepted suggestions for a linked inquiry from the MP from Tynseside, the other area involved, before a delegation from Cleveland could even speak to the SoS. Positive planning of hazardous waste facilities by central government, as Cleveland desired was not feasible under the then current legislation, which is centered on private initiative to make proposals. With the linked inquiry, the SoS had usurped the maximum possible leeway under the planning law for comparing various private proposals with each other. The fact that a linkage was made is, in itself, an indication that central government felt there was an exceptional situation. What may have influenced this perception somewhat is the fact that some of the MPs from the Northeast were marginal government seats, implying that the few Conservative MPs from the area were very concerned they might loose their seat if central government tried to force the incinerators on the counties involved. There is some evidence to suggest that this was indeed a consideration for the Conservative government of the day. Various people I have spoken to told me that the timing of the Secretary of State's decision

[7] All quotes from: 'Experts ready for burner fray!', Evening Gazette, 11 September 1990, p. 12.

should be seen in the light of the general elections of 1992. Indeed, the decision was announced some months after the general elections that year, whereas the Inspectors' advice had been lying on the government's desk for quite some time already. Interestingly, Ocean Environmental had already transformed its option to lease into an actual lease of the land before the SoS announced his decision, which suggests that the company already knew the decision of the Secretary of State beforehand. This impression was confirmed by some of my interviewees.

Whatever the reason for the linked inquiry, it does seem that the comparative perspective was beneficial from the perspective of decision quality. Just the bare fact that three proposals were compared helped even the lay person to sort out some of the strengths and weaknesses of the various proposals. For instance, already at the start of the inquiry, the local papers were able to predict that Ocean Environmental's proposals would be approved. One paper wrote that the company's plans stood out because they had been better prepared in terms of site selection and centered around proven technology. ITE on the other hand had not studied various alternative locations for its facility and its proposals implied that technologies would be combined in a previously unknown way. The rigid impression that Ocean's ES made thus had the intended effect of giving the company a strategic advantage. This applied even to the issue of site selection, which was one of the more vulnerable parts of the ES in my opinion. This issue did become a topic of debate, but not for Ocean's proposal. The Teesside Development Corporation presented information on Canadian hazardous waste site selection criteria to the inquiry and argued that ITE had used very few of them. The TDC thereby shed doubt on the technical quality of ITE's proposals. No one cared to compare Ocean's site selection criteria to the Canadian standards, although I am quite certain that the results would not have been much more positive.

The quality of the proposals in managerial terms (mainly technical suitability) became one of the dominant focuses of the inquiry. The report of the Technical Assessor especially, which discussed the technical quality of the three proposals, was influential and the inspectors referred to his conclusions quite often in their advice. The Technical Assessor, an official from the Department of the Environment, expressed doubts about ITE's plans to mix hazardous wastes and sewage sludge, whereas he considered Ocean's proposals to be state of the art. The companies involved were also affected by the comparative nature of the inquiry. They clearly had the impression that they were competing with each other for one planning licence and part of their agenda was to discredit the competitor's proposal. They started commenting upon each others' proposals from day one and especially Ocean Environmental was effective in undermining ITE's case.

Ocean Environmental did for instance point out that various alternative outlets for sewage sludge existed in the area.

The general character of the inquiry is best described as a quasi-judicial process with both managerial and pluralistic overtones. It was pluralistic in the sense that all parties at the inquiry made the strongest possible case for their position, contesting just about every argument the opponents would make. The inquiry centered around the 'statements of case' that all the parties had made and involved a great deal of cross examination of expert witnesses by opponents. Especially the local authorities and proponents had invested large sums in good legal representations and had hired very expensive top barristers to aid their case. The inquiry was managerial in this sense that expertise and scientific findings were the main 'weapon' of all parties at the inquiry. Public interest was large at first, but as the debate became more technical, only a small audience kept following the debate, apart from the main parties.

The role of the community

Apart from the various dealings with hazardous waste proposals, the population of the county had become sensitive to waste issues through discussions on a possible radioactive waste repository in the area in 1983, which was aborted in the end. An action group named BAND had fought the nuclear waste repository. When some of the former BAND members heard about the hazardous waste proposals, they linked up with members of the local branch of Friends of the Earth and started organizing the opposition under the name Stop Toxic Waste Incinerators In Cleveland (abbreviated to 'STINC'). This group would later become one of the 'major parties' of the inquiry.

It took a while before STINC was formed. In fact, the founding of the group took place only in March 1989, months after the planning application to Cleveland County for Ocean's incinerator had been made. The motivation for this seemingly late reaction was given to me by one of the earliest members of STINC. He used communitarian language to explain it: '*I didn't immediately get involved when the incinerator was proposed because I needed to make sure that it wasn't just my egg. You have to be sure that your arguments are right. Why would I oppose something like this? You need to get information, it is not an area that you are familiar about. You also listen to the community, what you hear (. . .) It took about six months to gather enough steam to react*'[8]. Indeed, before STINC was formally founded, a petition was set up to assess local opposition. The result was a claimed total of 12,000 signatures under the petition, indicating a sufficient level of

[8] Interview with Mr. F. Jones, Billingham, 15 December 1999.

disagreement in the eyes of the organizers (but still a small segment of the population).

A meeting to actually found STINC was attended by about 60 people, 30 of which became members of the organization. Many of those who attended expressed concerns about the proliferation of waste incinerators in the area and STINC set out to stop both the ITE-proposal at Portrack and Ocean Environmental's proposal at Seal Sands. The group came to advance a range of arguments. One argument was that there was no real need for toxic waste incinerators since the local industry was starting waste minimization activities. Indeed, ICI, with a major local branch, announced plans to stop dumping 250,000 tons of hazardous wastes per year at sea by June 1989 and to embark on a waste minimization plan. STINC sent questionnaires to the major companies in the area to assess the local waste market. They claimed a response of 72% and calculated that no waste incinerators would be needed in the area, not even the relatively small size of the one that was being proposed by Ocean (30,000 tons a year capacity). A second line of argumentation was based on the firsthand experience that various members had with waste treatment through their jobs in the chemical industry. On the basis of their experience, some of the members argued that the private sector could not be trusted to safely deal with hazardous wastes. Other members of STINC were, however, convinced that it would be safer if chemical firms dealt with their own wastes, because they knew better what disposal technologies were appropriate. A third argument was akin to what the county council said. STINC indicated that the UK should first look at its waste strategy before local communities would be forced to accept incinerators. In a communitarian vein, the group pleaded against any pressure from central government in the decision process, and for local decision-making. Thus STINC invited '*London to allow us the opportunity to pursue our democratic process*'[9].

The county and STINC did not have very intensive contacts until the county refused planning permission and they started sharing the same agenda. In the time leading up to the inquiry, they started actively working together. The county decided to support STINC with GBP 5,000 and the two organizations agreed on a certain division of labor, which implied that STINC would focus on health implications of the proposals (besides the already described arguments). STINC's presentations on the topic emphasized that public health in Cleveland was already under stress from the existing level of pollution and that adding pollution would worsen that situation. They called the incinerators 'open-air laboratories with humans as guinea pigs'. Note that the emphasis on potential negative health

[9] 'We must not open these floodgates', Letter to the editor by F. Jones, Evening Gazette, 23 June 1989.

implications from the facility implied a shift from communitarian thought towards managerial thought. Suddenly STINC had to start debating air pollution dispersion models and similar issues like that.

The role of the courts

The proposal did not go to court, which is in itself an indication that the inquiry fulfilled a certain role in achieving conclusion to the debate. If we take a second look at the inquiry that was held in this specific case, a few items of particular interest stand out. The first is that the issue of need for the facilities, normally an issue for proponent to decide, but now topic of debate, was somewhat difficult to discuss because the data that the two companies had collected on the waste market were not made public. The inspectors tried to get the two companies involved to reveal the data, but both companies long resisted this pressure by indicating the commercial confidentiality of their data. They were willing to provide only relatively generic information about their market assessment. The inspectors accepted this argument, even though this acceptance rendered much of the debate on the need for the facility even more useless than the statement of the SoS that there was a need for the facility had already done. Even though much inquiry time was spent on the issue, the inspectors in their report demonstrated reverence to the private sector and used conservative pluralist language to motivate their advice. '*To forecast the role of incineration in this changing situation is not our role but credence must be given to the market's interpretation of the effect of those influences*'[10]. This argumentation is somewhat unsatisfying because if correct, then why should the topic have been discussed at the inquiry at all, and why had three proposals been pitted against each other?

A second noteworthy issue was that participation by ordinary people and action groups was severely inhibited. Many organizations and persons, 486 in total, had expressed a desire to address the inquiry and the timetable was changed to accommodate the large number of parties and witnesses they would be calling. The inquiry began in September 1990 and lasted until April 1991. The sheer length of the inquiry period (seven months) made it impossible to follow the entire discussion, which seems unavoidable. But ordinary citizens were also inhibited by the amount of documentation supplied. Although inquiry rules are in place to protect the interests of such groups, these were not well enforced by the inquiry inspectors. To give an impression: ITE submitted 40 proofs of evidence (3,000 pages of material) in the three weeks prior to the inquiry, after the official submission deadlines. In fact, the company submitted another 700 pages of evidence in the eighth week of the inquiry. One opposition group consequently withdrew from the inquiry, claiming that by allowing this behavior, the inspectors had

[10] Inspectors' Report, volume VI, p. 4.

turned the inquiry into an unfair procedure. This group claimed that the procedure that had been followed was a breach of the European environmental assessment directive, which guaranteed participation. In fact, at some point some of the participants informed the European Commission of their opinion that the UK was not correctly applying EU law. The European Commission consequently notified the UK government that it was studying the case.

The inspectors were forced to respond to such challenges and made a statement to the inquiry concerning the matter. They argued that the UK environmental assessment guidelines were a proper translation of European law. Secondly they argued that, although ITE's actions were a borderline case, nobody had been deprived of their right to make comments because the inquiry had lasted so long that sufficient time had elapsed after each submission. Furthermore the information that had been submitted too late did not drastically change the character of the proposals in the eyes of the inspectors. Although the European Commission kept a file on the case it seems that this specific complaint was only part of a larger file used to pressure the UK to alter its environmental assessment process. The complaint was not followed up on later.

The third issue of interest points at one way in which the decision-making rules embedded in the law are actually enforced. UK planning legislation allows appellants to request an award of costs against the planning authority. Such an award of cost will be given when it is felt that the behavior of the planning authority was so that the proponent has incurred unnecessary costs. Ocean Environmental used its authority in this respect and requested award of costs on the basis that Cleveland had '*erroneously put the burden of proof on the applicant, whereas national policy plainly and repeatedly places the onus on the authority to demonstrate clearly why the development cannot be permitted*'[11]. Ocean complained that the county had gone against the advice of their own planners, which, according to the company, implied that the county had decided the application on the basis of emotions and not rational arguments. This seemingly strong and managerially-sounding complaint did not convince the inspectors. In their reply, they dissected an 'orthodox view' of the planning system and a 'broad view'. According to the inspectors, '*it is unsurprising that the members took a broader view than the more orthodox application of conventional planning factors by the officers. It was reasonable for the members to do this as long as the wider factors were still legitimate and significant planning considerations*'[12]. Because the inquiry had given the inspectors the impression that there were indeed various uncertainties associated with

[11] Inspectors' report, volume VII, p. 3
[12] Ibid, p. 7

Ocean's proposals, and because the inspectors considered them valid planning considerations, they believed that costs should not be awarded. However, the inspectors also wrote that in some cases the county had gone too far in introducing non-planning factors in the proceedings, for instance by asserting that the pollution control regime was not reliable. However, argued the inspectors, such discussions had not incurred extra costs because the effectiveness of the pollution control regime would have been discussed anyhow with HMIP, a statement that seems to ignore the fact that there is a significant cost difference between ordinary meetings and inquiry meetings. The Secretary of State went along with the inspectors' reasoning, however, and did not award costs against Cleveland. The fact that the issue was brought up after all and seriously discussed after that did send an important signal to the authority.

4.1.5 Some reflections

When judging the decisions made in this particular case, we can observe that especially the role of private market parties, experts, and elected politicians was great, a situation that broadly conforms with the way the UK legislation envisions such decision processes. The fact that the site selection performed by Ocean Environmental and its assessment of the waste market were not really questioned underlines the importance of private market parties. Experts were specifically relevant at the stage of environmental assessment and preparation of the licence decision. The design for the facility that was proposed by Ocean Environmental's experts was not really altered throughout the entire debate and received much praise from most other experts. These experts included the Technical Assessor assisting the planning inspectors, most of the statutory consultees, and the county officials. Despite that being the case, the county council voted against the plans, because amongst others it feared to be covered by a flood of proposals for incinerators. The refusal of planning permission subsequently necessitated an involvement of central government and indeed the SoS took over the decision process. As there were indeed various proposals for incinerators on the table at that time, the argument of a flood of incinerators made sense and three inquiries were suddenly linked. Although the inquiries were to discuss the need for hazardous waste treatment facilities, the SoS effectively answered the question, even before the inquiry started. This severely limited the inspectors' moving space. I find it somewhat surprising that they did not refer to the SoS's statement in their advice, but chose to use the argument that private firms could best assess this market. On the other hand, the linkage of inquiries did work very well from the perspective of a comparison of the technical quality of the three proposals at stake. It does seem that the impact of the Technical Assessor's opinion on this issue largely determined the advice of the inspectors and consequently the decision of the SoS.

In terms of discourse, much of the decision process was framed in managerial language as the main questions that could be asked under the law were questions about the technical quality of the proposal. It does seem that the councilors (and some in the action group STINC) saw the proposal mainly as a threat to the community, but arguments based on this assumption were not advanced very much in the formal procedures, and - when used - did not really have a great impact on the decision process. Take the county's argument that the incinerator would be detrimental to the county's image. The answer of the inspectors that responsible waste management could actually contribute to a positive image does sound a little bit like an easy way out of this discussion. And indeed, how would it be possible to engage in a debate on such relatively hard to measure implications of the proposal, which could hardly be seen as a 'valid planning consideration'? The fact that various experts stood against each other gives a certain pluralist flavor to the decision process, and it does seem that the inquiry covered just about every technical detail of the various proposals with experts contesting a great many issues.

4.2 Newport: an attempt to reject the logic of the planning system

4.2.1 Introduction

The second UK case I describe[13] involves a proposal for a treatment plant of (largely) liquid hazardous wastes in Newport, South Wales (UK), near the end of the 1980s. The proposal was made by BF Environmental Services, a full daughter of BFI that was active mainly in the Midlands at the time. There are various parallels with the previous case. Here too, the local reaction, partly inspired by previous negative experiences with the waste industry, was negative. Planning permission was denied and an inquiry was held into the proposal. Although the central government did decide to grant planning permission, this did not solve the disagreement and Newport borough council, the responsible planning authority, used its powers as a Waste Regulation Authority to once again deny the facility approval, thereby necessitating another inquiry.

[13] Helpful comments to an earlier version of this description were received from Professor J. Petts (University of Birmingham).

Figure 3. Proposed site for BFES's plant in Newport, South Wales (adapted from the ES).

One difference with the previous case is that the authority undertook only moderate effort to conceal the political motivation for its decisions, emphasizing that the proposal would be against the interest of the community. In this sense the decision-making in this case can be seen as an attempt to reject the logic of the planning system. As was the case with Ocean Environmental's proposal, the proponent tried to have the local authority 'punished' for refusing permission on grounds that are not planning reasons. Costs were indeed awarded against the authority and this specific decision lead to a certain amount of litigation in the courts, which were otherwise unused as an avenue for decision-making in this case.

4.2.2 The development of the proposal: private initiative

The case study starts near the end of the 1980s, when BF Environmental Services (BFES) saw a potential market for a new waste treatment facility. The company anticipated a tightening of UK regulations and realized there was as yet limited existing treatment capacity present. The company had initially focused on the whole of the UK, but because privatized water companies were active in developing treatment capacity in most regions, there were relatively few niches in the waste treatment market. Because Welsh Water was not interested in treating wastes but in developing tourism, Wales was an exception. Hence the company decided to concentrate on South Wales. After analysis of local waste disposal plans and counts of vehicles carrying chemicals on the road, BFES assessed that there would be a potential market in the Southwest of the UK to the order of 4.5 million gallons of waste a month. Consequent contacts with public officials in various counties, in combination with the study of local waste and land use plans then resulted in the selection of a site on an existing industrial estate in Newport (Gwent), called Stephenson Industrial Estate. The relevant local land use plan designated the site for industrial use and the owner was willing to sell the land to BFES. Besides, there was sufficient treatment capacity for the plant residue at the Nash treatment works nearby and the site had excellent road access as it was near important motorways to the West (London) and to the North (Birmingham).

After BFES had set its eyes on the site, it commenced consultations with public officials of the responsible planning authority, Newport Borough Council. Even though BFES had been led to the site in part by the work of their colleagues in the Economic Development department, Newport's planners were somewhat hesitant about the proposals. They said they would need expert advice on the proposals as they were not familiar with the techniques proposed. From a land use planning perspective however, they expressed reservations as the facility would generate traffic, have a certain visual impact, and would possibly be too close to the urban parts of Newport. Therefore they suggested an alternative site in Newport, which was however not of interest to the company. In fact, even before the company knew it would receive support from the local council, it bought the site in question. Part of the reason for doing so was to show Newport Borough Council that BFES's intentions with the site were serious[14].

As was the case with Ocean Environmental, BFES did not want to go public with their ideas immediately. The company foresaw potential resistance and hired a communications expert to advise on the approach towards the local population. As a consequence, a communication strategy

[14] Interview with Mr. T. Butterfield (formerly with BFES), Coventry, 5 January 2000.

was developed that was based on a trickle-down philosophy as the so-called 'opinion formers' would be approached first in order to win them over. The first activity was to brief public officials and the local council. The idea was that they could then start to play a role in defending the ideas to the public, the second stage of the information campaign. However, the strategy failed at the first step and resulted in very negative feedback as local councilors adamantly opposed the proposal. The second step, informing the public, was only taken after local councilors had already broken the story to the press. The initial response of the local press seemed relatively neutral, but to the detriment of BFES, it was pointed out that this would be the fourth type of operation in the area, which implied an (unwelcome for BFES) association with other local plants that were suffering from problems. Among the other facilities in the area was a treatment plant in a small village outside Newport, which had experienced a fire that necessitated the evacuation of an entire part of the village in the year before BFES's proposal. Even more controversial was the RECHEM incinerator at Pontypool that had led to health concerns in the area. BFES's case was not helped when the local paper, tipped-off by Greenpeace, published a story on the legal problems that BFES's parent company (BFI) was having in the USA. BFES had initially used the experience and knowledge of its parent company as a selling point, but suddenly found this connection a liability. BFES could not respond very rapidly to these accusations because it was taken by surprise and the communication lines with the USA were long.

4.2.3 Official procedures start, decisions made

Environmental impact assessment and licence applications

Meanwhile, BFES was working on the legally required environmental (impact) statement (ES). This document was completed by June 1990 and then submitted to Newport Borough Council along with two planning applications for the same facility. The reason why two applications were submitted was strategic. If Newport would be slow in deciding the applications, an appeal could be filed for deemed refusal of one application, whereas the second could be left sitting in the local arena. In case Newport should refuse planning permission for the second application, the appeal process would have started up already, saving the company time. From the perspective of Newport, the double application did not make a very friendly impression of course. The ES portrayed the site selection process that the company had followed as a rigorous and systemic effort. It spoke of various sites that had been considered, offered a list of criteria by which the sites had been tested, and suggested that the chosen site was really the best site in objective terms. On the other hand, the ES offered little opportunity to check which sites had been considered, how they scored on the site selection criteria, and how the various criteria had been gauged in deciding the appropriate site in the end. Other sites than the selected one seem to have been written off.

Responses to the applications

Somewhat similar to the previous case, the reactions that followed when the application was publicized seem to reveal a divide between experts and non-experts. The expert bodies, such as HMIP and Newport's Environmental Services Department, all commented that the plans were acceptable. The only expert body that objected was the County Planning Department. Their objection was based on the lack of a county wide plan for waste disposal, which seems an odd argument as the lack of such a plan was largely to blame on the local authorities in the county themselves and could therefore hardly be a reason to refuse planning permission. Outside the realm of expert advisors there was resistance. Many consultees, including the firms near the selected site, rejected the site. These firms were afraid of the risks involved with the plant and feared a negative image upon their own image. Some firms suggested they would relocate from Newport if the plans went ahead.

Because Newport lacked relevant expertise in the area of waste treatment, their planners opted to appoint Loughborough University as their consultants with the task to assess the validity of the ES. Their report appeared months later and concluded that the site was well chosen and that no significant negative environmental impacts of the plant were to be expected. Various conditions should however be met before the plant could be operated safely, according to the University, including a hazard and operability (HAZOP) study. The results were discussed with BFES and they promised to follow the recommendations. The Newport bureaucracy now agreed to the plant.

Newport Borough Council decides

The application was brought before the Newport planning committee on 22 November 1990. The basis for the committee meeting was a report written by the Director of Planning, which made a case for granting permission. The planning committee meeting started with presentations by the Chief Executive and the Director of Planning. The Loughborough experts were at hand to answer questions. When invited to ask questions, the councilors voiced great fears, including some related to waste importation and waste incineration. Various councilors were dissatisfied with information available to the Council, which gave them little reason to vote against the proposal. Like in the case of Ocean Environmental, the planning committee members had difficulty in deciding the application because they really wanted to reject it, but were void of arguments. It was felt that the issue should be brought before the full council. This indeed happened and the full council decided unanimously to reject the application.

4.2.4 Appeal and inquiry

Not unexpectedly, BFES appealed the refusal of permission to central government, i.c. the Secretary of State for Wales (SoS). The strategic approach of filing two applications had worked in this sense that Newport's decision had been announced after the statutory deadline had passed and an appeal for deemed refusal was already underway. Therefore, an inquiry was already being planned even before Newport had decided the application and the planning inquiry took place relatively quickly after Newport refused it. A planning inspector would hear the inquiry, but the application was 'called in', implying that the inspector would only write an advice to the SoS who would then decide the matter for himself.

The inquiry started early October 1991 and was, in comparison with the inquiry on Ocean's proposals for the Seal Sands site, a relatively short one and it lasted even shorter than foreseen. One reason appears to be the fact that Newport was very reluctant to discuss the issues in the terms embedded in UK planning legislation. An important part of the Borough's argumentation was that the council, as the local representative body, could not be forced to accept the proposals even though there was little wrong with them in terms of the aspects considered relevant by law. The council felt that the planning system was not the appropriate forum for deciding this application. Despite such fundamental arguments being made, most inquiry time was spent on the issue of safety, which was also something that the council had cited as a reason for refusal of planning permission. BFES felt it need not go into the arguments about the inappropriateness of the planning system too much, but did point out that the council had ignored the relevant legislation and thus decided the application on the wrong grounds. Like Ocean Environmental, BFES applied for an award of costs against the planning authority on that basis. Regarding the issue of risk the company largely countered the arguments of opponents on the basis of data from its Environmental Statement. The company's experts essentially noted that the types of waste treated at the plant (described as polluted water) were not inflammable or explosive and that the building would be fully enclosed. Because of this, the '*toxic concern is too weak to cause concern off site*'[15].

At the inquiry, the Inspector forced the parties to work together on a quantitative risk assessment. Under the supervision of the Inspector, both sides came to agree that the worst case scenario that had been used by the

[15] BFES Inquiry, Proof of evidence A 9, by BFES-expert M. Vince.

council, the unintended mixing of incompatible waste streams and subsequent escape of toxic fumes, was not only very unlikely, but also involved little serious harm if it occurred. The risk of the worst case scenario occurring was estimated at $1x10^{-9}$, whereas $1x10^{-6}$ was the generally accepted maximum risk level. The effects on the surrounding population were calculated on the basis of air dispersion models, which predicted a certain concentration of toxics in the air after the dreaded incident. Only if extremely stringent standards for toxics concentrations in air were applied (those for working conditions, ergo practically constant exposure) would these temporary concentrations of pollutants in the air be too high and then only in a limited area. Judged against two, more relevant standards (specifically designed for short-term exposure), the concentrations were considered completely unproblematic.

Given these outcomes and given the fact that the inspector was not allowed to entertain arguments outside the remit of the law, the result of the inquiry was fairly predictable. Indeed, the inspector submitted a report to the Secretary of State fairly quickly, and he fully rejected Newport's case. Furthermore, the inspector agreed with BFES that Newport had acted unreasonably and therefore awarded costs against the authority (these costs could be up to several million of pounds). The SoS took over this advice completely. Thus, planning permission was granted, be it under a range of conditions, and Newport was to pay damages to BFES. The latter caused quite a bit of excitement in Newport, because a local businessman threatened to sue the members of the council and make them personally pay for these costs (this never materialized however).

4.2.5 A new round and a new inquiry

The decision process is over?

Despite its clear defeat, Newport Borough Council decided that it would not give up its resistance. The borough had yet another weapon against BFES, which was its power as a Waste Regulation Authority (WRA). In what followed, Newport's resistance against the plant was largely played out by using the ambiguity in one of the planning conditions that had been imposed on BFES by the Secretary of State. This condition indicated, according to the borough, that the safeguarding systems of the plant had to be approved by the borough before the plant could be commissioned. According to BFES, the condition implied that the borough had to agree to every single aspect of a scheme for commissioning the facility, something that is normally only required after issuance of a waste disposal licence and after construction has been completed because the actual operations of the facility can then be assessed.

BFES had applied for a waste disposal licence on 18 October 1993 and under the relevant legislation, the application should have been decided within four months. By February 1994, there were no indications that a decision was imminent. Therefore, the company appealed to the Secretary of State against a deemed refusal. In response yet another inquiry was planned, and was started by September 1994. Demonstrating its defiant attitude, Newport postponed the issuance of the licence as long as possible, quoting the absence of a scheme for commissioning. When the inquiry approached, the borough suddenly published a draft version of the waste disposal licence. Unfortunately, Newport had 'forgotten' to consult with the statutory consultees on it, which implies it had a somewhat ambiguous official status. Along with the draft operating plan of the facility that was prepared by BFES, the draft licence still became the focus of the inquiry. This time, the inquiry was not open to the public because of the sensitivity of the data submitted to the inquiry from a market perspective. Other parties were allowed to provide written statements, an opportunity that was indeed used to some extent. However, the submissions from the general public played virtually no role in the decision in the end, as the inspector found they were outside the realm of waste legislation and pertained to planning issues mainly. The inquiry was once again a relatively brief one and the inspector soon followed with his report to the Secretary of State. The inspector recommended granting the licence under certain conditions, including a completed hazard and operability (HAZOP) study. The Secretary of State, who had again called in the issue, had obviously learned from the first inquiry. By November 1994, he took the unusual step of writing Newport that he was 'minded' to grant the licence and replace Newport's decisions with his own. In doing so, he put pressure on Newport to grant the licence itself so as to keep a certain amount of control.

But, the pressure was not just on Newport, there was great pressure on BFES as well. The company had started to actually construct the facility and hiring personnel. These investments, funded by the mother company, were yet to result in any returns and there was pressure to start generating profits. Besides, the waste regulation regime was at the brink of being overhauled. This implied that the responsibilities of the local authorities were about to be transferred to the new Environment Agency. BFES feared much additional delay if its application were to be decided by this new entity, which was mainly occupied with the task of setting itself up and was not yet functioning properly. Given the fact that both parties were now under pressure, there was an incentive to start negotiating a way out of the conflict. The final decision stemmed from a site visit by local councilors to discuss matters. Company spokespeople told me that during this visit, the councilors and the company reached an agreement that a waste disposal licence would now indeed be issued, but that at the same time, BFES would no longer pursue the award of costs[16]. By February 1996, the company had its waste disposal licence and wastes were transported to the site for the first time.

Functioning of the plant since 1996

When I visited the plant in 1999 it was still operating, albeit below design capacity. The waste market had changed radically and it appeared that the plant was starting to become profitable after a few marginal years. BFI had withdrawn from the UK market by that time and the management of the facility had bought it in 1998, only 18 months after the plant had opened. The plant's operations were hampered by stringent requirements regarding storage of waste drums in the facility. A very limited numbers of drums had been foreseen, whereas much waste was present in drums. Consequently, an application was made to enlarge this storage capacity, a request that was approved by the Environment Agency despite misgivings by the Borough Council.

4.2.6 The role of the various institutions and discourses

The role of the private sector, experts and elected representatives

Obviously many choices in this case study were made by BFES, a private market party. Market considerations essentially led their choices, specifically the fact that they wanted to develop a facility like this in the first place and secondly the site they selected for it. After broadly establishing where they wanted to locate their new facility, the company started looking at government plans and consulting public officials. This led to the identification of an ideal site, which was chosen against the wishes of public officials and was quickly bought from the landowner. One motivation for buying the land so quickly appears to have been to overcome potential resistance as the company wanted to show the local authority it was serious about its plans through the acquisition of the land. Along similar lines, the company applied twice for planning permission and developed a certain communication strategy to win over the opinion formers in the community. The strategy largely backfired. This was not so much because of the technical quality of the proposal that was finally made, but mainly because there was social and political resistance against the proposals. In terms of technical quality, the facility proposed by BFES seems to have been well designed. Especially the fact that the facility would be fully enclosed reduced potential risks greatly.

[16] This did however not reverse the SoS's decision that such an award of costs should be granted. Litigation on this issue continued and was only resolved much later.

The ES made an attempt to defend site selection in managerial terms (several sites looked at, siting criteria applied, etc.) but this was hardly a convincing attempt to portray a rigorous site selection procedure. As was the case with Ocean's proposals, there is little opportunity to check which sites had been considered, how they scored on the criteria, and how the various criteria were weighed. In fact, the description that the ES gives of the site selection process appears to be inconsistent[17]. The impression one obtains from the ES is that siting criteria (and 'objectives' which are also suddenly introduced in the ES) were used to motivate an already taken decision (the land had already been bought by BFES). Apart from that critique, the ES seemed of good quality in managerial terms, at least that is what practically all experts attested to.

Newport's planning officers were aware of the political resistance against the proposal, but could not fault the project under the relevant legislation. This caused quite high tension between the council and planning officers, with the departure of the Director of Planning as a consequence, just like in the case of Seal Sands. The relation between the bureaucracy and council are quite interesting in this case. Note that the negative stance of the council was largely influenced by the previous experiences with the waste industry, particularly RECHEM at Pontypool, which is just kilometers away from the site that BFES selected. This negative experience implied that the council members were opposed to hazardous waste facilities per se, almost regardless of what BFES was proposing. BFES was of course not helped by the disturbing reports about its American parent company BFI. That local politicians were opposed to the proposals from day one can be gleaned from the fact that they immediately 'leaked' BFES's proposals to the press previously revealed to them in private. It can also be gleaned from the fact that local politicians played an important role in the formation of a local opposition group. The negative news about BFES's US parent company led to negative comments about BFES's proposal from a local Member of Parliament (MP) and various councilors. They established an ad interim opposition committee and organized a meeting to start an opposition group. Aware of the fact that UK planning legislation demands a neutral attitude of councilors, the councilors themselves could not lead the action group. However, the meeting had a considerable turnout and several ordinary

[17] For instance: it is describes how, out of an unknown larger subset of locations, four sites had been shortlisted, based on various (essential and non-essential criteria). One of the 'essential criteria' according to the ES was that there should be adequate wastewater treatment capacity near the site. It can thus be expected that for each of the four shortlisted sites, adequate wastewater treatment capacity was indeed available. Interestingly though, when the final selection of the Newport site is described, some of the three other shortlisted sites are suddenly discounted for not meeting this essential criterion and therefore deleted.

citizens (sometimes close relatives of politicians) were found to replace the elected officials. The action group was named Newport Against Hazardous Waste Plant, and soon became the major opponent to BFES, with the silent support of important local politicians. Ken Critchley, a member of the planning committee, described the dynamic to me. His wife Sue was an active member of the opposition group '*Sue led the protest before we could get involved in it. The action group formed before we could act, with our covert support*'[18].

The council seems to have expected that the bureaucracy would feed it with arguments to refuse planning permission as the proposal was perceived as being counter to the interest of the community. What it got instead was a report written by the Director of Planning of the borough. This report identified six main issues and discussed the question whether those issues would justify refusal of planning permission. On the environmental impact of the plant, the Director largely referred to the Loughborough report and wrote the council that no significant environmental impacts from the plant were to be expected. Regarding land use plans, the Director concluded that the site selected by BFES was designated for general industrial use. The Director suggested that the implication was that a waste treatment plant was therefore also allowed from that perspective. Although various companies had informed the Director that they would relocate from the Stephenson Street industrial area if BFES's plans were to go ahead, the Director concluded that such statements would not be acceptable evidence of economic damage at a planning inquiry as such relocations were not visible yet. The Director observed that some had described the plant as being against the interest of the community. On that he concluded that '*Central Government guidance indicates that such general grounds for refusal can only be sustained for developments which will be registered by Her Majesty's Inspector of Pollution. The proposed BFES plant would not itself be a notifiable hazardous installation. Therefore the Council would need to demonstrate why the development would be contrary to the public interest and this would be difficult in the absence of any significant environmental or health and safety consequences*'[19].

In my opinion, some of those considerations are perhaps not 100% valid. For instance, if only in the case of HMIP-regulated activities there is a requirement to take public unrest into account, then the Director could have pointed out that the facility was about to be regulated by HMIP (or the Environment Agency) under new environmental legislation that had already been adopted by the UK-Parliament. Also, Newport officials declared much later that the 'general industry' use class that applied to the site in question

[18] Interview, Newport 12 January 2000.

[19] Director's report to the Newport planning committee, p. 22.

was ambiguous and did not automatically allow hazardous waste treatment at all. This fact was not at all visible from the Director's report. There are other comments to make, but the general impression is that the bureaucracy was trying hard to make a case for granting planning permission.

In addition to the rejection of a range of possible objections to the proposal, Newport's administrators issued a warning to the council. The Chief Executive, at the request of the Director of Planning, wrote a note to the council stressing the aggregation rules that applied to the matter under the UK law. '*The proper consideration of a planning application needs the objective appraisal of all the relevant planning issues. However difficult it might be in some circumstances, a local planning authority should not let emotion, or irrelevant or non-planning issues, divert its proper consideration of a planning application*'[20]. The warning contains clear managerial language as it stresses 'objective appraisal' of issues and denounces the involvement of 'emotions, or irrelevant' considerations. Clearly, the bureaucracy felt that the council was reacting emotionally to the proposal. Associated with the warning was an implicit threat to individual councilors. Newport's officials indicated that a refusal of planning permission would create a liability for the borough. They stressed that pay-off rules in planning law prescribe that 'award of costs' is possible if applications are refused for improper reasons. Indeed, BFES had warned the officials and council that an application for award of costs would indeed be sought. Newport's officials added that councilors were potentially personally liable for the costs involved (in the order of hundreds of thousands of pounds[21]).

The result of the work of Newport's bureaucracy was that the council was void of arguments that would hold up during an appeal. In an attempt to escape from this situation, councilor Ken Critchley started a discussion of the aggregation rules under which the council was to decide: '*The unprecedented preamble to the Planning Committee clearly advises the terms of reference so to speak (. . .) I accept the need for doing that, but having said that I think we should be discussing this particular application in a far wider context recognising the seriousness and the implications of it*'[22]. Although Critchley made various other comments, the issue of the aggregation rules drew support and was picked up on by the chair of the

[20] Note of General Advice by the Chief Executive of Newport Borough Council, undated.

[21] Newport's bureaucracy had calculated the Borough's own costs for the inquiry at GBP 120,000 and assessed that BFES could have spent about GBP 300,000.

[22] Transcript of the recording of the planning committee held to discuss the BFES application for a chemical waste treatment plant, dated 22 November 1990 (p. 63). Found in the Newport Borough Council Archives.

meeting (the leader of the council). The leader concluded that '*the Planning Committee finds itself in great difficulty here today because so many of the considerations which they have mentioned here this morning, are not planning factors. And the worry that I would have is that we would arrive at a conclusion here today which could in a sense pre-empt what I believe the more important debate that should take place. (. . .) I mean a debate which is and should rove outside the question of planning criteria*'[23]. Like the councilors of Cleveland County, Newport's councilors believed that the full council would be allowed to move outside the planning system and decide on the basis of the real concerns. Indeed, when the full council unanimously refused permission, the reasons given were (1) the safety risks related to the increase in movement of hazardous goods vehicles, (2) the prematurity of the application, (3) the proximity of the site to urban communities, and (4) the perception of the local community that the proposed development was against the public interest in general and their own interest in particular. The council was keenly aware it had stepped outside the authority and scope of rules embedded in UK planning law with the fourth reason for refusal. '*Our officers helped us. They told us that we had no planning reasons to refuse. That is why we went outside the planning system*'[24]. In place of a line of reasoning within the planning system came a reference to the status of the council as the bearer of local democracy.

This line of reasoning, along with only relatively marginal technical objections to BFES's proposal, was taken to the inquiry. At the inquiry, a councilor, who spoke on behalf of the Borough, said: '*We are talking about precise planning guidelines but the world as a whole doesn't work that way*'[25]. Thus, the inquiry touched upon the authority, scope and aggregation rules that were normally widely accepted in Newport, but were now rejected. As part of its resistance to normal planning rules, the council also attacked the normally fully accepted bias in favor of development. This time Newport council suggested that BFES's plant should not go ahead unless proven safe, in the process questioning the expertise of the various experts who had told the inquiry that the plant would be safe. '*Can you show me one of those experts who has not had problems with these plants - if not they are not as expert as I would like them to be. They need to prove to me that it is safe*'[26].

[23] Ibid, p. 84.

[24] Interview with K. Critchley, Newport, 12 January 2000.

[25] 'Human error risk highlighted at BFI plant probe', South Wales Argus, 10 October 1991, p. 3. Interestingly, this point was not supported by Newport planning officers at the inquiry.

[26] Ibid.

The council's main argument against the proposal offered little opportunity for discussion at the inquiry. Instead, the council was forced to make a case there on the basis of 'substantive' reasons that should demonstrate that the proposal was against the interest of the community. The more fundamental references to local democracy hardly offered an opportunity for fruitful exchange. Instead most discussion centered on the arguments against the facility that were brought forward by a (new) expert retained by the borough. This expert argued that wrongful mixing of waste streams could occur. This could then result in emissions of untreated gasses and endanger the residents of the neighborhood and the employees at nearby factories. This scenario was quite unlikely and discarded rather soon, implying that the inquiry was over rather quickly.

The role of the community

The BFES proposal was one of the most contentious issues in Newport political history, where the population rarely ever gets very excited about local politics and voter turnout at local elections tends to remain below half of the electorate. The action group that was more or less founded by local politicians, Newport Against Hazardous Waste Plant, was one of the main reasons why the proposal received so much attention. The group was very active and organized numerous meetings, protest marches and a petition, which was eventually signed by an impressive number of 35,000 people (according to the action group). The campaign centered on the risks associated with the treatment of wastes on site and transportation of wastes towards the site. Especially on the risks associated with transportation, the group collected its own evidence. Based on newspaper archives, the group tracked down all incidents on the road towards the facility in the last couple of years in an attempt to demonstrate that trucks carrying hazardous goods were often involved in accidents. The group did indeed retrieve a variety of such incidents, mostly without gravely negative consequences however. Other arguments included the bad past record of BFI, the lack of enforcement of controls for such facilities, and the possibility of waste imports.

The group used communitarian language as a motivating tool as it portrayed the decision as a fight of the people against capital. '*They are not coming here with the sole aim of benefiting South Wales - they are coming here to make money*'[27]. '*We are taking on a multi-national company and we need people power*'[28]. The 35,000 signatures that were collected indicate a certain degree of success. On the other hand, this number is nowhere near half of the population and most signatures came from people living near the

[27] 'Waste plant is "dangerous" fear resident', Western Mail, 2 July 1990, page unknown.

[28] 'Waste war: plea to the people', Free Press, 17 May 1990, page unknown.

proposed site with the rest of the population remaining relatively unconcerned. Meetings organized by the opposition group nearby the proposed site had good attendance (100-150 people), but meetings further away had virtually no attendance. The people in one of the areas in the borough (Maesglas) had barely heard about BFES's plans and had not actively participated in the process until they heard that the residues of the treatment process could go to a landfill site in their area. At that time, long after BFES had announced its plans to build its plant, a second petition was organized and another 20,000 signatures came in protesting BFES's plans. Newport Against Hazardous Waste Plant deplored this apparent lack of unity in the population and selectiveness of participation. At the inquiry, the action group became one of the main parties and therefore faced the same dilemma that the council had faced: how to mix arguments that would surely fall in the remit of the planning system with more fundamental objections? Like the council, they decided to advance two lines of reasoning, one outside and one inside the planning system.

Outside the planning system the group used communitarian language. It argued that even if the risks caused by the plant were minor, there was still an ethical problem with the imposition of risks on the local community. Even if BFES should have proven that risks were minimal, argued the group, '*it is not its (BFES, DH) place to judge how much risk local people should be prepared to take. This risk can only be decided by local people, and they are in no position to judge unless they have the full facts before them*'[29]. The group thus advanced local democracy and a community debate on the available evidence.

Within the planning system, the group made a case that the risks associated with the operation of the facility were too high. The group had found a retired chemist with extensive experience in testing and sampling techniques at a laboratory. She argued that it would be impossible to gather representative samples of all waste transports that would arrive at the gate. The implication, it was suggested, was that the plant could not operate safely. Secondly, the group argued that the risks associated with waste transports would be unacceptable. This argument was supported by the evidence from the newspapers. The argument on road safety led to a serious discussion between the action group and BFES. BFES countered the group's arguments with two others. First of all, BFES noted that the road in question was not the road designated for use by hazardous goods vehicles by Newport. Hence, trucks bringing wastes to the site would not use this route but approach the facility from elsewhere, thereby reducing risks considerably. The action group disagreed with this argument, as they had firsthand local knowledge on the movement of trucks to the industrial estate,

[29] Newport Against Hazardous Waste Plant, Submission to the inquiry.

which suggested that hazardous goods vehicles effectively ignored the designated route. BFES countered this finding by actually driving both routes and counting gear changes. On the basis of that experiment, BFES argued it was completely unlikely that truck drivers would ignore the designated route. Secondly, BFES's experts had performed an additional analysis of national data on accidents, specifically those involving hazardous goods vehicles on the type of road in question. The conclusion was that the plant would result in a 5% increase of hazardous loads on the involved road at most, and hence imply only a small increase in risk. The experts suggested however, that the increase in risk in Newport should be compared to the reduction of risk elsewhere as wastes from Newport would no longer be exported to other jurisdictions. Although there were important gaps in the statistics brought forward by BFES (they were not specific for the road in question and also old[30]), the inspector accepted them as the only credible evidence on the matter. The data suggested that the road was not used fully to design capacity and that additional traffic would not be a problem. It seems that one managerial story line, the one of utility, was particularly helpful in convincing the inspector that increased risks should not weigh against the proposal. BFES's traffic expert noted that the BFES's facility would put an end to the export of wastes from Gwent, thereby reducing the greater risk of accidents elsewhere. In other words, by channeling the traffic to Newport, net utility of society would increase rather than decrease.

The role of the courts

The application was taken to an inquiry and after that, to the courts. Both during the inquiry and what happened afterwards, the authority and aggregation rules were an important topic of discussion. Especially the fact that the council had so clearly attempted to move outside the planning system made it vulnerable to criticism. Both the planning inspector who heard the case and the SoS for Wales emphasized the 'normal' rules of the game in decision processes like this. The inspector's advice was largely based on the rules that normally applied. The inspector concluded for instance that applicants normally did not have to justify the need for their proposal, but despite that being the case he had become convinced of the need for the facility. In arriving at this conclusion, the inspector referred to a public statement of the SoS, who had indicated that South Wales should stop its practice of exporting hazardous wastes.

[30] BFES's expert had used national accident ratio's and combined these with BFES's data on plant operations and his own assumptions. His conclusion was that the plant would result in a 5% increase of hazardous loads on the involved road at most; this would result in a small increase in the risk of an accident, but not major.

The normal rules of decision-making also helped in answering the question of risk and its acceptability. Whereas some had demanded absolute certainty that the facility would be safe, the inspector wrote that '*Clearly, there can be no absolute safeguard*'[31]. Therefore, somebody had to accept a certain risk from the facility. The inspector indicated that government standards for risk were *by definition* written to assess the acceptability of involuntary risks. In other words, the inspector rejected the argument that the facility should be refused because the community opposed it. The inspector agreed that there was ample public opposition to the plant. He noted that this opposition was related to a perceived danger and that the existence of such a perception should indeed count against the development. However, because the perception was not based on substantive evidence, it could not lead to the conclusion that permission should be refused. The inspector clearly felt that Newport had gone against national planning legislation. He concluded that costs should indeed be awarded against the borough as BFES had requested, which occurs only very rarely. The reason for that was that the borough had not followed national planning guidance, by basing its refusal on non-planning reasons, but also by delaying the discussion on risk until the inquiry.

Political influence over the inquiry? Litigation on the award of costs

After the inspector issued his advice, it took the Secretary a remarkably long time to take a decision on the appeal (one year). The decision on the award of costs came even later (18 months after the Inspector's advice). This might suggest, just like in the case of Ocean's proposals for Seal Sands, that the timing of the decision was somewhat politically influenced. However, in the presentation of the SoS's decision, the conformity of the decision with the advice of the planning inspector was stressed, implying that the political character of the decision was somewhat hidden. After the decision, the borough and one local MP started lobbying Parliament to speak out against the SoS's decision. 100 members of the House of Commons signed a petition against the decision, but this had no effect on the outcomes of the debate.

The fact that the SoS agreed with the inspector that the Newport council had acted unreasonably and should pay the inquiry costs of BFES enraged the elected members greatly. They announced an appeal to the courts against the award of costs on the basis of arguments that stress local democracy. They indicated that '*it is outrageous that the Secretary of State should conclude that the Council acted unreasonably. Is it reasonable to ignore the opinions and justifiable concerns of the people the Council is appointed to represent? What is the point of the public consultation on Planning matters*

[31] Inspector's report, p. 38.

if the response, however strongly opposed, is to be discarded?'[32]. Newport's appeal failed in the High Court, but the cost order was eventually quashed by the Court of Appeal[33]. In the verdict, Lord Justice Aldous pointed out that fear in the public of a proposal is a valid planning consideration, be it that this fear should normally not determine the decision. The judge found that Newport should have substantiated the grounds of refusal, i.e. provide reasons why the fear was just, but the fact that the authority had not done so was not unreasonable behavior. This verdict led to a certain degree of concern in the professional community of waste planners. It was feared that a 'scaremonger's charter' had been created[34].

4.2.7 Review

This case largely stands out because a proposal was rejected by explicitly referring to the interest of the community. It is interesting to see what was the evidence for this argument and secondly what the effect of the use of this story line was. To start with the evidence that the interest of the community was indeed involved: the case suggests that certainly not everyone in a relatively industrial borough such as Newport identifies with the neighborhoods that are affected by the proposal. Witness the low turnout at some of the meetings in some of the more detached areas and witness the 20,000 signatures from previously inactive citizens that were suddenly collected in an entirely different part of the borough as soon as it transpired that this part would also be affected. In this sense, the population of the borough was certainly not unified. With the absence of a clear indication that the entire population of the borough was against the proposal, I find the council's argument that the proposal was not in the community's interest somewhat suspect. The fact that local politicians themselves helped to organize the resistance only adds to this suspicion. There also is an impression that the councilors had to 'invent' the interest of the community as a reason for refusal (the reports of their meetings attest to that) after they had decided they were against the facility and after their administrators told them there were no valid planning reasons to refuse planning permission. It is thus clearly a story line that gives meaning to a course that was already chosen.

Whatever the reason for the reference to the community's interest, the fact of the matter is that the story line did not impress many of the other parties engaged in the decision process. The appeal process, including the

[32] Borough of Newport, Press Release 'Chemical waste treatment plant', undated.

[33] See: Ian Kinloch, 'Dual controls and public fears', Waste Planning, March 1999, pp. 22-23.

[34] Ibid.

inquiry, is clearly biased towards a very different line of argumentation and the interaction process changed very rapidly towards the more technical arguments. The pleas about the interest of the community were neatly listened to and noted, but were quickly labeled as considerations that could not decide the matter. In a sense this is logical, because the argument of the community's interest is more or less philosophical in nature and is hard to debate. The top flight lawyers and experts who are present at the inquiry quickly attempted to translate any such argument in the direction of arguments that can be discussed, that is risk calculations, traffic impact calculations, air dispersion models, etc. In the end, the arguments must be made in these terms. The parties to the inquiry are well aware of that and they orientate themselves towards this direction.

The outcome of the decision process was largely based on the mixture of conservative pluralist and managerial reasoning embedded in the law. To give a few examples that demonstrate this: the site selection by the firm was largely left undiscussed, and an important consideration for acceptance of transportation risks was that there was a net reduction of transportation risks in the region which should take precedence over the risk to one community (that was moderate anyhow, at least according to the relevant experts). Only one small correction was made in the end, which is that the interest of the community is an acceptable story line in the context of such debates, be it that its use must be supported by substantial reasons. The consequences of this approach are visible in the way the facility operates today. The acceptance of the facility by the community is rather limited. Although operated by a new management, both the neighbors of the plant and the council have never accepted the facility. The company is the usual suspect for various complaints about odors, which could well be caused by other installations in the area. The typical pattern is thus that if there is a foul smell from the industrial estate, neighbors phone the authorities to inform them that the waste treatment plant smells again. When I visited the plant, I was told that the action group that had opposed the plant had been right about the transport of hazardous goods to and from the plant. Trucks often do not follow the designated hazardous goods route, as the citizens had feared. In this sense, it is remarkable that this argument was so easily rejected by the inspector hearing the inquiry. Although I am not aware of any incidents with waste transport to and from the plant, it does seem that the risks of this occurring are actually higher than what was assumed during the decision process.

4.3 Rotherham: expertise from the local population

4.3.1 Introduction

The last UK case that I have studied for this chapter is a proposal to create a landfill for polluted soil in the metropolitan borough of Rotherham[35]. Rotherham, which is near Sheffield, is an industrial area that saw great decline of employment in the 1980s and 1990s because of the collapse of its major industries (e.g. eight out of nine major mine pits in the area closed in about 15 years). The pollution caused by the imploded industries is tremendous. There is a great desire to redevelop the area, but this is seriously hampered by the presence of polluted soil. The solutions to such pollution vary; in some cases the soil is concentrated and isolated on site or cleaned in situ, in other cases it is dug out and transported to landfill or soil remediation facilities.

The proposal that is discussed here was made by VHE, a construction company from the area, which had gone into the business of soil remediation and disposal as a side activity. As much polluted soil was encountered during its own construction activities, but also because there was demand more generally, the company desired landfill capacity. Because the Rotherham/Sheffield area was its home region, this is where the company started looking for a location. The main criterion, apart from the site being in the Rotherham/Sheffield area, was that the selected site should have good road access. Because of this criterion, the company restricted its search to two eclipses around the main motorway in the area, the M1. Figure 4 depicts the selected site.

The size of the envisioned landfill was 8 million tons, to be filled over a period of 15 years. Although local authorities in the area were advocating the use of existing excavations (i.e. derelict mines) for landfilling purposes, VHE selected a piece of land of 87 hectares near Brampton en le Morthen that would have to be excavated before landfilling could start. The site was locally known as the Straight Mile and the company's proposals were generally referred to as the Straight Mile landfill proposals during the decision process. The formal reason why VHE did not select existing derelict land for the site was that a newly-created landfill could be better engineered than an already existing mine which would have be adjusted for landfilling purposes. However, the fact that the company itself did not control such existing derelict land will have played a role as well. VHE was aware that one of the collieries in the area, Thurcroft, was about to be used

[35] Helpful comments to earlier versions of this case study were received from Mrs. P. Foster and Mr. R. Newman (both of the Laughton and District Conservation and Historical Society).

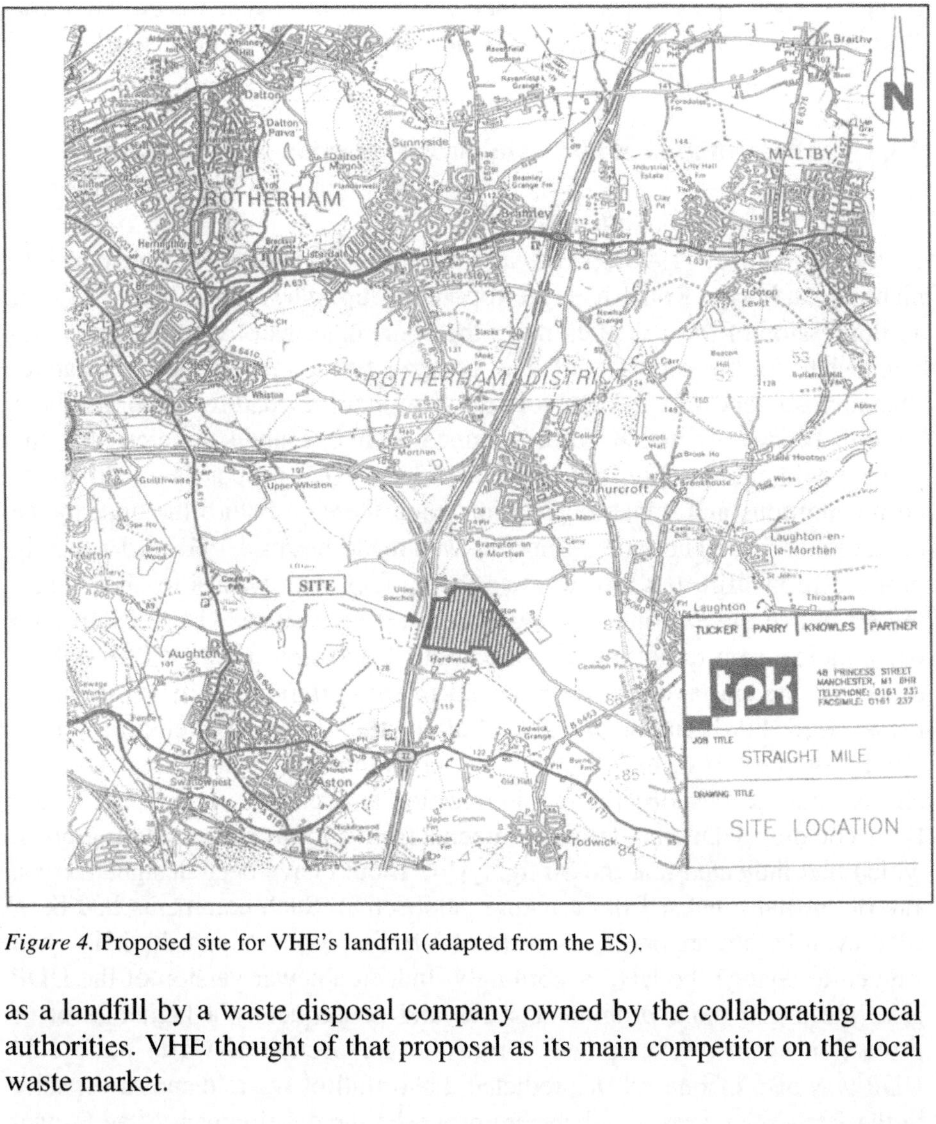

Figure 4. Proposed site for VHE's landfill (adapted from the ES).

as a landfill by a waste disposal company owned by the collaborating local authorities. VHE thought of that proposal as its main competitor on the local waste market.

Just like BFES in the previous case study, VHE went about the acquisition of the site very quickly. Without the actual users of the land (farmers) knowing of the plan, VHE reached an agreement with the landowner who was willing to sell the land at a profitable rate. According to the Environmental Statement that was made later, the site had been picked because of various reasons: it had good highway linkages, little residential development, suitable geology, and a landscape that lacked features of outstanding value. Unlike the two previous case studies, VHE met with a fair amount of criticism on its site selection. One element of the criticism was the allegation that the site selection criteria in the ES had been adjusted with the purpose of selecting the site that the company had already chosen. In reply, a company spokesperson admitted that '*To a certain extent we did have the Brampton site in mind. But I did not fully commit myself until I had considered other sites*'[36].

4.3.2 Presentation of plans and broad rejection

Contacts with public officials

By September 1993, VHE approached officials at the responsible planning authority, Rotherham Metropolitan Borough Council (MBC), to get an impression of the acceptability of the plans in terms of land use planning. Somewhat to the surprise of the company, the reaction was negative. Planning officers for the MBC told the firm that they had reservations about the site because it would be at variance with Greenbelt policies, that the plans would cause a lot of nuisance, and that VHE's proposals would create an incongruous and unnatural landform and thereby disturb the appearance of the landscape. The latter comment was made because VHE did not only propose to landfill, but also to raise the surface with wastes to about eight meters above the existing surface level, creating a slope. The relevant land use policies were to be found in the South Yorkshire Structure Plan, which described the site as a place of special interest from a nature conservation perspective. Furthermore, the first draft of the Unitary Development Plan (UDP) of the MBC had been silent on the specific spot that VHE had chosen, but did describe the site in question as a part of the national Green Belt. The draft UDP had drawn comments from various conservation groups, which had indicated that the Straight Mile should have been designated as a site of special interest from a nature perspective. Such comments had been filed even before anyone was aware of VHE's proposals and the MBC had wanted to change the UDP accordingly. Indeed a newer version of the UDP gave the site a more prominent status and consequently a higher level of protection against development. From the perspective of waste policy the UDP was also of interest. It predicted a shortfall of waste disposal capacity in the Rotherham area and therefore proposed the development of additional capacity in the borough specifically at Thurcroft. The MBC could thus rightfully claim it was contributing to the solution of the waste problem.

VHE press on

It thus seems that VHE's proposal did not meet the relevant land use plans nor the emerging waste policies of the MBC. In the eyes of the company however, the objections of the MBC were mainly intended to serve the competing proposal at Thurcroft. VHE noted that the MBC, as one of the

[36] 'Lining would ensure only a minimal risk, says engineering expert. Toxic waste leak fears', Rotherham Advertiser, 26 April 1996, p. 6-7.

parties to invest in the development of that proposal, would naturally object to a competing proposal. This interpretation of the resistance from Rotherham MBC implied that VHE felt it should ignore the negative feedback and continue to work on its proposals. The company started performing environmental assessment, for instance by constructing boreholes to test the geology and groundwater flows on the site. VHE also started consultations with various bodies (including with the National Rivers Authority (NRA)) about its proposals.

It was at this stage when the news about the company's plans became public. The groundwork on the site did not go unnoticed and VHE notified the tenants of the land in generic terms of its intended activities. Somewhat later, a story about the proposal appeared in the local newspaper that caused a certain level of unrest because it described how VHE would dispose of hazardous waste on the site and import wastes from elsewhere, even abroad. VHE's approach was not to respond to the stories in the press, mainly because it was still preparing its ES. In doing so however, the company accepted the fact that negative stories dominated the press, which in turn influenced the public perception of the proposal. The pattern continued when even more disturbing news transpired. The local paper had found out that low level radioactive waste regulations were about to be changed so that they could be landfilled at hazardous waste sites. Opponents brought this news in connection with the fact that VHE was in the process of acquiring a business involved in the decommissioning of nuclear plants. It was suggested that VHE, if granted permission, would also bring in low level radioactive wastes.

Planning application is submitted

The company finished its ES and an application for planning permission was formally submitted in January 1995 along with the ES. The application was the first public document describing the company's plans in their own words. The application described a landfill site that would mainly be used for the disposal of construction, demolition and excavation wastes (partly 'special' wastes). However, the company desired a certain level of flexibility in that respect. Because the market for said wastes was unpredictable, permission was also asked for the disposal of industrial and household wastes. Because the landfill facility would consist of several different and independent cells, each waste stream could be disposed of separately. In this sense, the proposal was for 'mono-disposal', not the 'co-disposal' of wastes. All cells of the landfill were to be connected to a central control tower, where leachate from the facility would be collected and analyzed. The landfill would have a double liner system, with a seepage detecting system underneath. One of the projected liners would consist of asphalt, the second of excavated stone.

The ES indicated the presence of a minor acquifer deep below the site, i.c. 10 meters below the secondary liner. Because the acquifer was not actively used for drinking water purposes and the strata between the liner and acquifer rock were undisturbed, there would be no chance of pollution of drinking water according to the ES. The ES described a range of preventative measures to be taken to prevent any other environmental damage. For instance, after filling up, all cells of the landfill would be capped by a clay layer to prevent entrance of precipation. In combination with the leachate management system, the capping would result in stabilization of the landfill in the long run, which would make the wastes inert. Furthermore the facility was to have a waste water treatment works (on the advice of the NRA). The facility would not only consist of a landfill. In addition, an asphalt plant for the construction of liners, recycling facilities and installations for waste treatment on site were projected. Finally, the company wanted to show its 'commitment to the local community' by adding a visitor center.

Consultations; broad rejection of the plans

Unlike the two previous cases, discussions on VHE's planning application resulted in several concerns from the statutory consultees. The Ministry of Agriculture Food and Fisheries (MAFF) questioned the reason why agricultural land should be changed into a landfill and wanted to be informed about the safety of agricultural operations after the active life of the landfill. The Highways Agency indicated that the capacity of the secondary road in the area leading to the Straight Mile was not sufficient to carry the additional traffic. The Rotherham Department of Environmental Health and Consumers Affairs questioned the predicted noise, odor and dust levels from the site and indicated they had not been calculated in the correct fashion. The National Rivers Authority (which was later absorbed into the Environment Agency) expressed concerns over the possible pollution of drinking and surface water and over the fact that the proposal would imply the transfer of polluted grounds to a relatively unpolluted catchment area. The MAFF, Highways Agency and the NRA all requested that the application be held until they had obtained reassurances on topics of concern to them: they filed a so-called holding objection. If their concerns would not be addressed, the proposal would likely not go ahead.

VHE was in touch with all these consultees and was making some progress here and there, but the statutory time limit passed quickly with a lot of concerns still open. VHE started to become frustrated by this perceived delay. In the eyes of the company, many of the problems with the application were caused by the MBC, which was stalling its application in the eyes of VHE because it wanted to assist the Thurcroft proposal. VHE's agenda now started to become both the speeding up of the consideration of its own proposal and the blocking of the competing Thurcroft proposal. Regarding the Thurcroft proposal, VHE filed an objection to the emerging UDP, specifically the part that designated Thurcroft as a suitable site. But the company also contacted the opposition group that had formed against the Straight Mile proposal and asked them to broaden their opposition to the Thurcroft proposal, which, in the words of VHE, was a far worse proposal from an environmental perspective. The action group refused however. Regarding the speeding up of the consideration of its own proposal, VHE launched an appeal against the deemed refusal of planning permission in June 1995, about two months after the statutory time limit for determination of the application had passed.

By the time of the appeal, the NRA, the MAFF, the Highways Agency, the South Yorkshire Archeology Service, English Nature and the Rotherham Department of Environmental Health had all objected to the proposal, or quickly did so in order to influence the decision process and maintain their right to be involved. Rotherham MBC indicated that it would continue to collect information on the proposal so that an informed opinion could be formed in light of the pending inquiry. Indeed the procedure was continued and about a month later the council met to discuss its opinion. Granting of planning permission would have potentially had the effect of canceling the inquiry, but the council rejected the proposals. Councilor Billingham, who told me that he thought there was a certain NIMBY-aspect to the public reaction towards the landfill, set the council apart from the general public: '*Forget about NIMBY, or however you want to call it, it was a straight policy issue*'[37]. Two basic reasons were adopted for this negative stance: the proposal was contrary to the development plan of the MBC and the proposal would cause problems for the highway system in the MBC. The MBC concluded that the proposal was contrary to the *development plan* because of five reasons. These included Green Belt policy (no intrusive landforms), damage to archeological and ecological values, borough policy that landfill should take place on derelict land, possible water pollution from the site, and finally breach of maximum noise levels.

4.3.3 Inquiry

Period leading up to the inquiry: attempts to sway opponents

The appeal would be heard at a local inquiry that was to start in April 1996. The Thurcroft proposal had by then also received opposition and the

[37] Interview, Rotherham, 19 January 2000.

UDP that designated that specific site for landfilling activity was most likely to end before an inquiry as well. Proponents of the Thurcroft site, who felt their proposal was clearly better than VHE's and who feared that approval of VHE's plans would reduce the usefulness of its own plans, wanted to delay the Straight Mile inquiry so that the two proposals could be studied at a linked inquiry. Such attempts failed however as the Planning Inspectorate concluded that VHE was too far ahead to wait for the Thurcroft proposal to reach the inquiry stage. The period leading up to the inquiry was still relatively long and VHE made good use of its time to alter its plans in an attempt to convince some consultees to withdraw their objections. The company promised to contribute to the improvement of the roads in the area for instance, but also withdrew the visitor center from its plans, which had concerned English Nature, and changed its plans somewhat to reduce the loss of valuable agricultural land. As a result of VHE's activities, the MAFF, English Nature and the Rotherham Department of Environmental Health all withdrew their objections. Attempts to sway the NRA were largely ineffective however.

VHE had informed the NRA that its plans were in conformity with NRA policy, but the NRA itself was not certain. The debate between VHE and the NRA essentially focused on the question whether NRA policy forbade landfilling in areas with sensitive water bodies per se, or that it allowed such landfilling only under the condition that leachate water would be treated. VHE argued that the wastewater from the facility would receive tertiary treatment, but the NRA expressed concern that the effluent would contain 'intractable chemical compounds' even after treatment. In response to a letter on the issue from VHE, the NRA changed its interim objections to definitive rejection of the proposals in June 1995. The company and the NRA kept talking until September 1995, but then the relationship cooled down and correspondence started looking somewhat unfriendly. In September 1995, the NRA informed VHE that it would no longer comment upon VHE's letters but would start preparing for the inquiry. This indicates the transition from consultation to a more adversarial approach.

It is my impression that VHE used the period leading up to the inquiry as an extended consultation period, which was necessary because the company had been in a great hurry to stay ahead of the Thurcroft proposal. The gamble paid off to a large extent, note the withdrawal of various objections, but also resulted in yet unforeseen problems regarding certain 'details' that had not been discussed yet. For instance, VHE had simply assumed that it could put pipelines from the facility to the wastewater treatment works along the highway. The Highway's Agency now indicated however that this element of the proposal was unacceptable under its policies for road maintenance. Also, the company realized relatively late that public footpaths in the area would have to be diverted during the active life (15 years) of the landfill. This too met with objections from the responsible officer of Rotherham MBC, who had not been consulted until then.

Actual inquiry: an extensive technical debate

The actual inquiry was planned to last for 12 days only initially, but took much longer than that in the end. To a large extent, the length of the inquiry can be attributed to the extremely detailed and extensive discussion of the technical aspects of the proposal. The SoS had outlined various topics for debate at the inquiry. These included the need for the development, the effects of the proposal on the countryside (landscape), implications for nature conservation, the question whether the proposal would be counter to planning policies (Green Belt), and finally how the proposal should be regarded from the perspective of pollution and sustainability. The number of documents submitted in this case was extremely large, and every document led to a rebuttal from the opposite party, which in turn led to a commentary on that rebuttal, and a rebuttal of the commentary on that rebuttal. The main parties to the inquiry, VHE, Rotherham MBC, and one action group left no stone unturned in the process. Other parties, such as the tenants of the proposed site and groups representing the interest of walkers and horseback riders (e.g. the Rambler's Association) played a lesser role but were also active.

The reason why the need for the development was so prominent on the agenda of the inquiry appears to be located in the fact that the MBC had based its decision partly on the waste policies contained in the emerging UDP. The difference with the two previous cases was therefore that the planning authority in this case did have recent waste projections and had used its authority under planning legislation to steer the waste-siting debate. Yet, because every application must be considered on its own merits and because the UDP had not been approved yet, the SoS wanted to have the issue of need debated. At the inquiry, the MBC presented data on local waste disposal needs and its calculations indicated that VHE had overrated its proposal for 'reasons of profitability', which is no surprise because VHE wanted to bring wastes from other authorities as well (be it in the same region). VHE countered that the MBC had an interest in a competing proposal, and secondly that there was a shortfall in disposal capacity in the region. Much of the debate focused on the status of central government guidance that seemed to prescribe that all UK regions should have about 10 years disposal capacity available at any time. The inspector agreed with Rotherham MBC on this issue. According to him, there was no legally binding requirement to have 10 years of disposal capacity available at any time. The implication was that Rotherham's disposal needs were not as critical as VHE had argued.

One interesting dynamic of this particular inquiry was that the inspector himself did much research on the issues before him. This was both the case for the countryside issues and the issue of nature conservation. On countryside issues, all parties had supplied information to the inspector. The tenants of the land had argued against the loss of valuable agricultural land, the MBC and action group had argued that the proposal would harm recreation in the area, and the action group and MBC were arguing that hedgerows on the site deserved protection under newly established jurisprudence. The inspector did not accept any of those arguments, but he did conclude that the visual attractiveness of the landscape would be negatively affected. Based on his own visits to the site, he judged that VHE's photomontages presented to the inquiry were not giving the correct impression of the landfill site. The inspector also noted that perhaps little would be visible of the site after 20 years or so, but that the large bunds that would work as screens to block the view of the interior of the landfill would for a decade be '*prominent and ugly features*'. Regarding the issue of nature conservation, the inspector concluded that the evidence supporting the special nature of the site was of poor quality: '*The scientific basis for protection rests on old records of hedgerow and use of the area by some scarce species of bird, but the information inspires little confidence that this interest survives, or is of importance to the appeal site, rather than elsewhere on Brampton Common*'. The inspector noted that the draft UDP committed the council to the protection of the county's heritage and to seek improvement of it. As the site was mainly important for bird life, the inspector considered whether there were birds at the site that deserved protection or not. The inspector noted that the evidence on this aspect was sufficient to arrive at a conclusion, even though the '*information collected by voluntary groups was in some cases so vague as to location that it was useless*' and the information '*collected by the appellant had deficiencies*'. By this he hinted at the fact that VHE had not been collecting data at the right time for bird observation. The approach he took was to combine the information from various sources and see whether it corresponded. In addition, he had done his own observations and had noted the presence of lapwings for instance. By doing so, he concluded that the site was of high importance for a number of valuable bird species. He also concluded that the landfill would be detrimental to these species as predators would be attracted by changes in the landscape or by making the landscape less attractive for vulnerable species itself. Even though the site did not enjoy special protection, the inspector concluded that it was valuable and that it fell under the remit of the UK Action Plan on Biodiversity, which sought to conserve and enhance the existence of native species.

The fact that the SoS had ordered the inspector to study the implications of the proposal in terms of pollution and sustainability is reflective of the increasing prominence of the sustainability concept in the UK environmental policy discourse. The discussion at the inquiry and consequent advice of the inspector to the SoS (who had called in the proposal) would have been interesting from the perspective of studying the UK interpretation of the concept. However, the debate stopped short of a discussion on 'sustainability' because the pollution issues were already detrimental to VHE's plans, which is something I will explain in the next section[38]. The conclusion that the proposal would cause an unacceptable risk in terms of pollution gave the inspector an escape route from the requirement to establish new aggregation rules and perhaps set a precedent. The NRA had argued that the UK sustainable development strategy had introduced the precautionary principle in the decision process. As a consequence, it was up to VHE that its proposals were safe, not up to the NRA or other government bodies to prove the proposal was unsafe, said the NRA. This issue was debated at the inquiry and it seems that the inspector agreed with VHE. He indicated that the precautionary principle did not '*reverse the normal presumption in favour of development, or requires the proposer to justify his position. Nor, in this case do I consider that it requires a special approach to the proposal*'. To the inspector's mind it was perfectly possible to assess the risk of damage to the environment in this case, despite a certain level of uncertainty. '*It is recognised that scientific knowledge may not be conclusive. Nevertheless, the likely costs of potential pollution along with other environmental costs can be balanced against the benefits in the decision to be made. Within this balance must also be a consideration of the probability of harm accruing*'. Because of this, the inspector did '*not consider that a separate consideration of the precautionary principle as it might apply to the appeal proposal is appropriate at this stage*'[39].

The matter of highway safety, finally, was the only issue on which the inspector concluded in favor of VHE. He accepted that VHE had under-estimated the traffic generated by the proposal, but generally found that the traffic flows from the facility would marginally contribute to traffic flows already present. Therefore, he found '*no clear cut reason to recommend refusal of the proposal on these grounds*'[40]. The overall conclusion of the inspector was that the proposal was not acceptable from the perspective of five of the seven discussed topics: Green Belt, risk of pollution,

[38] See 'experts versus experts' in that section.
[39] Ibid. p. 106.
[40] Excerpts from the Inspector's report to the SoS for the Environment, various pages

sustainability, nature conservation, and impact on the countryside. The inspector concluded that on the basis of each of these topics the development could have been refused, but that together they formed 'a compelling case against the proposal'. Besides that however, the need for the development was not as urgent as VHE had suggested. The inspector's recommendation was to deny the appeal and that the SoS should refuse planning permission. Just in case the SoS would decide otherwise, the inspector also wrote some recommendations on the planning conditions that had been agreed and he wrote a recommendation on the diversion of public paths. The diversions should have been allowed according to the inspector, but this recommendation would have little meaning if the landfill would not go ahead as the inspector had recommended.

As had happened in the other two cases that I have studied, the Secretary of State followed the advice of the planning inspector and the SoS decided not to allow the proposal. VHE was obviously disappointed and suggested that not all evidence submitted to the inquiry had been considered by the inspector, which seems rather absurd given the length of the inquiry (nine weeks in the end) and the very thorough report that had been written by the inspector. The company appealed the decision in court, but this appeal was withdrawn shortly after it had been filed. I have been told this was because the decision process had already cost the company a large sum of money and the chance of success of judicial appeal was considered small[41].

4.3.4 The role of the various institutions and discourses

Introduction

This case study differs in certain respects from the previous two. Most notably the freedom of the private sector to propose waste disposal facilities was not so much taken away but certainly heavily curtailed by the planning powers of the MBC. VHE was interpreting the events that occurred after it filed its application in terms of competition and to some extent, this perception was real, note the attempts of the Thurcroft proponent (a company partly owned by Rotherham MBC) to postpone the Straight Mile inquiry. To a larger degree however the actions of those participating largely conform to the legal framework, which does confer certain authority to public bodies at the expense of the private sector. In the end, the appeal implied that the decision process was taken away from Rotherham MBC and thrust into the field of the national realm. The planning inspector, himself an

[41] This information does not come from VHE itself. The main employee responsible for the Straight Mile proposal had retired and the current personnel had not been involved with the proposal. I have contacted the former VHE employee, Mr. Sidall, but he declined to comment.

expert, played a very active role in this particular inquiry, especially by adding relevant information to the equation. The debate was extremely technical, but remarkably enough, this did not stop the action group that was involved from the beginning of the inquiry. In fact, this group played a large role in determining the outcome of one of the issues of contention. In what follows below I will discuss the more salient aspects of the case study from the perspective of institutions and discourses.

Tension between the action group and the MBC

Once again, this case study involves a proponent of the facility that largely keeps the preparations for its proposal in-house. Even at the risk of the build-up of a negative local public opinion, the company refused to discuss its proposal with anyone but planning officials of the MBC. Interestingly, VHE found an ally in this approach of the decision process in the MBC-officials and members of the council. The news about VHE's plans, as I have indicated, broke when VHE started performing tests on the site to generate data for its ES. The tenants noted this, and one of them then spoke about it to Roy Newman. Newman, then a member of Anston Parish Council, was a former county councilor and an individual with great organization skills. On behalf of the tenant, Newman phoned VHE to ask for information on their plans. When he was invited to the company's office to listen to a presentation, he asked whether he could bring a friend and in fact took a journalist with him. VHE, unaware of the presence of a journalist, was very open about its plans in an attempt to win Newman over. They told Newman that the landfill would be intended for hazardous wastes and that the landfill would also receive wastes from other jurisdictions. These facts figured prominently in the story that the local newspaper consequently published and caused the first public unrest.

Newman became the major opponent to the proposal. His motivation was an emotional attachment to the landscape of the Straight Mile and an intuitive resistance to the landfilling of hazardous wastes[42]. Using the already existing Laughton and District Historical and Conservation Society as a vehicle, he started organizing the opposition by holding public meetings on the proposal, which were really more opposition rallies. The Society held its first public meeting in August 1994 in one of the villages nearby the proposed site and many more followed. Because of these public meetings, the relations between the action group and Rotherham MBC went sour. The people at the public hearings were very angry about VHE's proposals, but part of their anger was also directed at the local authority that was accused of keeping the public in the dark. Council members, who sometimes attended the meetings, were however of the opinion that the public meetings caused a

[42] Interview, Rotherham, 21 January 2000.

needless arousal of the public opinion. The council felt it was irresponsible that the public was informed of the proposal at a stage where little details were yet known about VHE's plans and about the MBC's position on the application. During one public meeting, the chair of the planning committee took the word and argued that the public had no right to obtain information on VHE's plans yet. The vision of the council was that it would have to be informed of VHE's plans first and that it was the council that decided on planning applications, not the public. The fact that the incoming application had not been made public and had not been discussed in council was defended by one of the members I spoke to later. '*I take the view that the planning department should assist development. This is in our legal system. I have no problem, even though there is a lot of activism, with the planning department giving advice to developers. I wasn't upset about planning officials talking to VHE about their proposal*'[43]. The result however, in relational terms, was a certain level of distrust between the action group and the MBC. The secrecy surrounding the proposal became an effective story line to mobilize the public. Newman told the press on various occasions that he considered the secretive approach indecent and used it as a motivation for strong resistance. '*Here we have a proposal to dump waste containing God knows what on a green field site and they can't even tell what it is all about. (. . .) As far as we are concerned the gloves are off and we shall be doing all we can to fight this*'[44].

An argumentative turn

The approach of the action group was initially quite radical and aimed for the mobilization of resistance at the grass roots. The secrecy of VHE and the idea that just about every sort of waste would be disposed at the Straight Mile were used as the main arguments to prove that the proposal was against the interest of the community. Although quite large numbers of people visited the meetings of the action group, Newman was somewhat bitter about the turnout at certain gatherings and the overall response. He considered that the public was too trusting of experts or too selfish to participate. Commenting upon a particular public meeting he said: '*65 people came out of a village of 3,000, that's disgraceful! (. . .) We tried to get people motivated about this landfill and they will say they will protest it. But if you give them a letter with a stamp, they'll use the stamp for their own mail. (. . .) The only way to motivate people is to hit them between the eyes with it. What motivates them is when their property is affected. Our society is inundated with the attitude that we should leave it to the people who know.*

[43] Both quotes from an interview, Rotherham, 19 January 2000. Ironically, the Labour Party suffered a defeat during the 2000 municipal elections and Mr. Billingham lost his seat.

[44] 'Waste firm will not meet objectors yet', Rotherham Star, 9 September 1994, p. 1.

You need groups like us, that demonstrate that it affects their purse and their children'[45].

Added to the fact that VHE was apparently not backing off, the inability to actually start a massive protest against the plans made the action group rethink its approach during remainder of the decision process. In addition, a member of the Royal Town and Planning Institute (RTPI) was invited to give a presentation to the group about the way the rest of the decision process would be set up. This member, who gave this speech as part of a public education scheme of the RTPI, said he had noticed that the action group's campaign was largely based on emotional issues. In no uncertain terms, he then informed the members of the action group of the dominant language embedded in the planning legislation. He said that '*this campaign has to be fought on a policy level not an emotional one. (. . .) The whole campaign hangs upon the strength of policy planning arguments*'[46]. The planner warned the action group not to overestimate the importance of the Greenbelt policy, as it did not inhibit waste development per se.

The warning had a certain effect upon the action group. Although their activities to stir up public opinion continued, and resulted amongst others in a petition signed by about 14,000 people, they started devoting a great deal of energy to the development of a more informed argument against the proposals. The group succeeded in attracting members with various fields of expertise, notably a former police superintendent with knowledge of traffic issues and a retired mining engineer, Mr. McIver. McIver had extensive knowledge of the soil underneath the site as he had 33 years of mining experience in the area and had been a Deputy Manager at two collieries. McIver visited both the NRA and Yorkshire Mining Authority after the ES had been submitted. He informed both agencies that the information in the ES on the suitability of the soil under the site was incorrect because the strata below the site were fractured. As a consequence, leakage from the landfill to the acquifer and then surface water was more than likely. It appears that his intervention contributed to the hesitations that were present at the NRA. Both the NRA and Mining Authority had already been in touch with VHE and it seemed that their initial objections were being met. However, the NRA renewed its opposition, partly based upon McIver's comments.

It is of interest that VHE's appeal against the deemed refusal of the application was greeted with mixed reactions. The local MP, opponent of the proposal, said that '*Local people and local planners do not want the Brampton Straight Mile site to be used for landfill. The company knows this*

45 Interview, Rotherham, 21 January 2000.

46 'Tip protestors warned they face a hard fight', Rotherham Advertiser, 11 November 1995, p. 9.

and their appeal is an attempt to ride roughshod over residents who view this development as totally unwelcome'[47]. The action group however, perhaps still concerned over the outcome if Rotherham MBC was to decide the issue, was happy with the appeal. They indicated that the strength of public opinion would be a powerful weapon at the inquiry and welcomed the public examination of proposals that an inquiry would imply.

The issue of resources

Having adjusted to the idea that the inquiry would be largely determined by a technical debate and having certain expertise in place, the action group felt confident they could beat VHE. The group realized that it needed more professional representation at the inquiry, especially experience with the planning system. They consequently hired Mr. Heseltine, the Planning Institute representative who had given them advice earlier on, as their counsel. The group also realized that the inquiry would cost them a large amount of money. To enable the group to operate effectively an attempt was made to raise GBP 20,000, for instance by selling personal belongings of UK celebrities at an auction and organizing a concert. This attempt largely succeeded, amongst others whit the cooperation of UK celebrities such as Richard Branson, the CEO of Virgin. Friends of the Earth also helped the group.

The action group, like the other groups in the UK cases described, was still at a serious disadvantage in terms of resources when compared to VHE or the Borough Council, but was determined to match the other parties. As had happened in the Seal Sands inquiry, this desire was difficult to maintain as the planning inspector did not very actively protect the action group. The amount of documentation related to the case, already enormous, kept rapidly growing as VHE was constantly supplying new information and altering its plans after the submission deadlines for the inquiry had expired. In a sense the submissions were necessary to give the parties involved an idea about the way VHE was accommodating some of the objections it had received, but they did put an enormous strain on the action group. At a certain point the group asked the planning inspector to call off the inquiry because too much new information was coming in and the changes in the proposals were considerable. The group argued that the changes were so large that there was in fact a new proposal on the table and that VHE should restart the application procedure. The planning inspector however denied this request and the inquiry went on.

[47] 'Waste tip firm set for action', Rotherham Star, 28 June 1995, p. R15.

Experts versus experts

The action group's turn towards expertise was if anything a pragmatic one and does not suggest that the action group had become a member of the managerial discourse coalition. Newman for instance, simultaneously relativizes and stresses their importance. Looking back on the decision process, he noted that scientists can disagree on many issues without reaching a firm conclusion: '*You can let scientists argue with each other and it'll be like discussing god*'. At the same time however, he was keenly aware that in order to win the debate, he needed experts. '*The inspector does listen to experts. The more money you have, the more experts, the more likely it becomes that you win. (. . .) I was a bit frightened of the technical experts, they tried to by-pass us as a little irritation. (. . .). Their scientists were good and were paid very well. But we had Heseltine and McIver and we slowly wore them down*'. He also said '*We did everything professionally, we would match them figure for figure. (. . .) One thing you have to do is give your information to the other side and other parties at the inquiry. We issued further documents and rebuttals. It was a nightmare*'[48].

Was the action group any good at the debate between the experts? The answer seems a resounding yes, as the group contributed greatly to the defeat of the proposal. Their information on the condition of the geology of the site was accepted as more credible than VHE's by the inspector. The group had made a great deal out of the potential risks that the proposal presented to ground and surface waters in the area. Their expert, Mr. McIver, indicated that there were four coal seams beneath the site at various depths (Barnsley, Thorncliffe, Clowne and Hazel) and two of those had been worked at a depth of 700 and 400meters respectively. The methods by which they had been mined varied, but some were known to lead to fissures up to the ground surface. This tendency would be strengthened by the presence of the two faults near the site; mining subsidence was known to cause fractures up to 10 meters on either side of the two faults. The maximum stresses caused by one of the nearby mines would cross three of the four phases of the landfill. Fissures could be invisible as earth and clay on top of them could very well mask them. A second mine nearby had used the total extraction method, with one pillar of coal left. Mr. McIver's experience with such pillars was that they could collapse under great forces or become less stable. McIver expected that the forces associated with pillar interaction had left a 'nest' of faulting beneath phase 1 of the landfill. McIver countered VHE's argument that faults would be sealed by rock flour; his own firsthand experience within a certain mine was that groundwater could easily wash away rock flour, creating a ready pathway. While working as a mine engineer, he had

[48] These quotes from an interview with Roy Newman, Rotherham, 21 January 2000.

encountered water originating from the Mexborough Rock at a mine nearby, at a depth of 360 meters. The water had been transmitted through fault lines. McIver also suggested that groundwater could lead to washdown of the base of the landfill, which in turn would lead to subsidence and cavities underneath. If this were to occur, monitoring systems in the second liner would not observe the passing through of leachate. This mechanism had not been studied in the risk assessment. Moreover, the Society had found evidence that permeation was not the only mechanism by which clay landfill liners (chosen by VHE for the secondary liner) could be penetrated by pollution. Diffusion could also occur and this mechanism was equally capable of transporting pollution according to an article in Environmental Science and Technology. '*Bearing in mind that well respected engineers had proposed clay as a lining material, the assurances now given by expert engineers should be treated with caution*'[49].

This argument completely undermined VHE's case, which was argued by a geologist and hydro-geologist. VHE had initially not assessed the risks of pollution to ground and surface water, essentially because their approach had been to optimize the design of the facility. Specifically for the inquiry however, a risk assessment had been performed for various scenarios, including normal operation of the facility, blockage of the seepage control system and rupture of the liners. It was calculated that under the first two scenarios, breakthrough of the liners would take 242 years and under the third scenario 16 years. After breaking through, it would take another 16 years or so for any pollution to reach the acquifer rock underneath the site. By the time pollution had arrived there it would be greatly diluted (factor 1 to 250,000), but this dilution was not taken into account in the rest of the risk assessment. Instead, the maximum level of pollution coming from the landfill (without dilution) was assessed and it was calculated what would be the effect of this on the groundwater in the rock. It was found that the water would then still be within safe drinking water limits. It would consequently take 35 to 350 years for the water to travel from the rock to the nearest boreholes indicating that the pollution would indeed affect drinking water. However, this would not be a problem because the water from the rock, with the pollutants from the landfill, would meet drinking water standards. VHE suggested that the calculated amount of pollution was unlikely anyway as the landfill would be capped and no rain could enter the facility. Because of this, a 'head' of water on the landfill liner creating pressure and perhaps damage to the liner was unlikely. All of this was however based on the assumption that faults in the strata under the site could not create ready pathways for pollution. VHE argued that even if such fissures or faults existed, rock flour would have sealed them.

[49] Ibid., p. 74.

The NRA had also presented a range of arguments against the proposal, but these arguments had not swayed the inspector. This may be because the NRA arguments were very much couched in policy terms (e.g. mono landfill was not acceptable in areas with an acquifer[50] and the safety of asphalt liners was doubted in recent guidance) or relatively generic. To give an example of a relatively generic comment: the NRA argued that VHE's risks assessment had overlooked various possible types of incidents, in particular incidents related to human failure. This comment was supported by an example. The NRA posed that a release of 1,000 cubic meters of leachate was an example of an event that could occur because of human failure and the risks should have been assessed. Further generic comments included the NRA's opinion that risk assessment was '*in its infancy*' and '*There were substantial problems to foresee the unforeseeable*'[51]. In my opinion, such statements are rather remarkable given the fact that both in the Seal Sands and Newport cases risk assessment had been carried out and accepted without any hesitation by HMIP, by then part of the same organization as the NRA (i.e. the Environment Agency). In his report to the Secretary of State, the inspector took a hard look at the arguments brought forward by the NRA. He concluded that the policies of the NRA should indeed carry much weight in the decision because they had been supported by the government's Sustainable Development Strategy. The inspector did not agree however that the NRA had valid policies in place against mono-disposal in acquifer areas, because a statement possibly indicating such a policy had not been motivated and had been tucked away in a footnote of the relevant guidance. In fact, the inspector noted that the proposal met all parts of the NRA guidance, except that part of the policy that was set against landfills above saturated zones. The degree to which the proposals were in conflict with that policy was 'amenable to analysis' however, and the inspector noted that the NRA policy had not been examined in public, thus reducing its force. Therefore, conflict with this part of the policy would not be lethal to the proposal unless actual harm to the groundwater could be demonstrated. Interestingly, this had not been done by the NRA according to the inspector: who said he '*was not convinced of the mechanism whereby the EA postulate a release of leachate into Anston Brook, and hence to the River Ryton*'.

The information provided by the Society on the other hand, was reliable. 'The evidence on effect of coal working beneath this site however was expert and based upon many years of practical experience of the behaviour of these strata under the methods of mining practised in the locality. The rock into which it is proposed to build part of the landfill has been greatly disturbed in the immediate geological past, and I consider the effects of this

[50] See the Inspector's report to the SoS for the Environment, p. 316.

[51] Inspector's report to the SoS for the Environment, p. 61.

bring into substantial doubt the risk assessment carried out on the appellants behalf'[52]. Note the use of the word expert here as a positive qualification; although the information had come from a member of the general public it was considered to be expert and hence credible. Used as such, the word denotes a reversal of who is expert and who is not as the inspector chose to believe the Society, or more exactly Mr. McIver. The inspector's conclusions closely followed the Society's statement. He concluded that the faults might not have provided a ready pathway for pollutants, but that the mining had changed this, as it must have led to fissuring. The inspector also concluded that disturbances to local groundwater had been observed and thus that there was a real risk that contaminants would spread throughout the water-bearing rock in the area. Finally, the inspector also concluded that the facility would generate leachate in some quantity forever, because it had not been demonstrated that the cap would prevent ingress of water. Therefore, it was important to look at the liners and the leachate management system. The inspector believed that they would be very effective in stopping leachate, but found that the issue of timescale had to be regarded here. He considered that even if the landfill could remain intact until the 22nd century, it was almost certain that containment would be lost at some point. In addition, the inspector did not believe that the liners would perform as well as VHE had predicted on the basis of US EPA publications which he had found himself and on the basis of the landfill floor deformation argument put forward by the action group.

4.3.5 Review of the case

This case stands out in various ways from the two others. An important difference is that permission was refused at the local level, and consequently upheld. The reasons why this was the outcome are multiple, but the case study points to a few specific factors. Important was the fact that Rotherham MBC was developing waste policies that aided the authority in making a decision and provided the MBC with reasons to refuse planning permission that were fully acceptable under the planning system. Another important factor seems to have been the fact that from the beginning various statutory consultees were also expressing concerns about the proposals, weakening VHE's case considerably. The company took too little time to consult with many of these authorities so that the appeal process was started with extensive continued consultations. Finally, the role of the action group should not be underestimated either. Although the proposal was considered unacceptable by the planning inspector for a range of reasons in the end, the

[52] These quotes from the Inspector's report to the SoS for the Environment, p. 92.

issue of pollution was an important one and the group supplied the information that was the nail in the coffin on that one.

In terms of institutions, the case largely demonstrates a similar pattern as the previous ones, although land use planning policies did effectively structure the decision process, thus taking away certain room for decision-making from VHE. Again one can not truly say that entire communities were involved in the decision process, although the population of the villages near the proposed site were probably en bloc against it. In Rotherham itself however, the proposal generated much less excitement. The political element in the decision process was easily obscured in this specific case, because all elected politicians involved, at the local and the national levels, could safely decide according to the advice of their experts. There are some indications however that politics attempted to obtain a certain level of influence on the decision process. Especially the desire to link two inquiries seems motivated by a desire to help the Thurcroft proposal, in which Rotherham MBC had a certain interest. The attempt to influence the timing of the inquiry on the Straight Mile was resisted however. In this case 'no credence to assessments of private market parties' was given (paraphrasing the Seal Sands case in Cleveland) but more was given to the assessments in land use plans, even though the figures were intensely reviewed. Another difference between this case and the previous two is that there was no conflict between planning officials and their political masters. In fact, while the action group was up in arms about the proposal, elected politicians pleaded for some trust in the planning officials. I am not certain to which degree this was because politicians knew that the proposal would not make it anyhow. There are indications against this because the withdrawal of objections by the Rotherham Department of Public Health caused no political reaction and the politicians involved were prepared to go against the opposition it seems. Despite the similarity in the positions of the action group and the MBC that resulted in the end, the relatively large degree of secrecy that was associated with this plea was unacceptable to the action group and stimulated opposition in itself. A final difference seems to be the more active role of the planning inspector during the inquiry. The inspector in this case obviously had a knack for studying the situation on the spot and it seems that this resulted in additional information or an independent check on some of the information which the parties to the inquiry were presenting. The character of the information collected in this fashion would seem to be of the 'local knowledge' type, information that the layman, if using his wits can also 'find'.

In terms of discourses, the case was dominated by the amalgam of conservative pluralist and managerial language embedded in the law. There was a bit more managerialism present than in the other two cases because of the fact that the need for the facility now had to be legitimized by data on waste arisings. Especially the excerpts in section 3.3.4 give a flavor of the highly technical nature of the debate. The fact that there is a very clear turn in the way the action group presented its opposition to the plans makes the case very interesting. In theoretical terms, a member of the 'discourse coalition' that maintains the legal system, in this case a planner, came by and made sure that the action group became versed in the relevant language. The main reasons why the action group followed the advice was that it was experiencing some difficulty in motivating the entire community to oppose VHE's plans and secondly because it realized that referring to the community's interest would not win the inquiry. The latter reason of course is a very pragmatic one, and it suggests that the interest of the group - preventing VHE's plans to go ahead - determined the selection of a certain discourse. The action group acted upon the advice in an excellent fashion and deliberately attracted the expertise that helped win the inquiry. Whether or not that expertise was crucial in the end is hard to establish, as there were various reasons why the proposal was rejected in the end. The action group was not however responsible for all of the various reasons. Still, it is very significant that 'ordinary' citizens were able to organize themselves so effectively and muster a successful campaign against the plans, despite a lack of resources.

4.4 Conclusions

4.4.1 Introduction

The cases just described offer the possibility to draw several types of conclusions, mainly some that comparing the three cases with each other and others that translate back to the theoretical framework developed in chapter two. In this final section I will pose both types of conclusions.

Various issues stand out the UK cases. The image of a rather deferential public inarising from chapter 3 was not confirmed here. Instead, in each case there was active opposition to the proposals. In terms of institutions, one can not however conclude that 'the community' played a major role in the debate. This is a consequence of the fact that all three proposals were made in industrial areas with large populations that were largely disinterested in the discussions that did not affect them. It is also a consequence of the fact that 'the community' is not a very important concept and the community's interest not a very relevant consideration under UK legislation. The decision processes largely conformed to the relevant legislation, implying that the legislation structures the debate well. The initiative for the proposed facilities came from the private sector in all three cases, and it was a private company that played the role of proponent. The cases demonstrate a differing degree of deference to the private sector, but generally speaking the issues of need for the facility and finding of a site are left in the hands of the proponents, who performed a certain degree of managerial window-dressing in all three cases. Such window-dressing, especially making the site selection look like a rigid scientific progress, is rarely probed but if probing occurs, it tends to have little consequence. On the issue of need, statements from central government, specifically the SoS, have in two cases somewhat pre-empted the debate on this issue at the inquiry as the SoS had already indicated at the outset that there was a need for the proposed facility. However, especially in the Straight Mile case, the issues of need and site selection were extensively discussed. This was possible because the relevant local authority had an up-to-date plan in place that provided guidance on these issues. In the two other cases, such guidance was not present, consequently allowing private parties to make the decisions.

It should be noted that in all three cases, the local authorities refused planning permission. It seems that the local councils did share the concerns of the action groups that sprung up in all three cases. Especially concern about health effects, the risks associated with incidents and a negative impact on development were reasons to refuse planning permission. The influence of previous experiences with waste facilities seems great, especially in Newport and to a lesser degree in the Seal Sands case. In two out of three cases (Seal Sands, Newport) the elected representatives rejected the advice of their own experts (planning officers and other administrators) which caused a certain level of stress between them. The different outlook of planning officers and elected representatives somewhat weakens the position of the elected representatives. If their planning officers tell them to approve an application and they want to deny approval, then the elected representatives must 'fabricate' the reasons for rejection themselves. They tend to do so with respect to the appropriate terms under the planning system. Only in the case of Newport was there a degree of defiance, but this remained inconsequential as the proposed facility there went ahead anyhow. That particular case was relevant for future decision processes because a community-based motivation for refusing planning permission was in the end accepted by the nation's highest court, at least to some degree.

The courts are a little-used avenue in the UK, which is partly related to the fact that the inquiry already provides a quasi-judicial arena for opponents to cross swords. The desire to go to court after the lengthy hearings (up to seven months!) and very detailed discussions at the inquiry appears to be minimal. In most cases, the parties to the inquiry accepted the advice of the inspector. The inquiries do seem to succeed well in creating 'political control with the benefits of judicial authority'. Most parties tend to accept the terms of reference for the inquiry, even though these are often geared towards a managerial discussion of issues at a very detailed level and do not call for the achievement of community consensus (to mention an alternative possible purpose of an inquiry). Parties tend to be allowed to go beyond the terms of reference, but will be informed that their submissions will not be considered in the advice of the planning inspector.

In all three cases, the planning inspectors succeeded in conveying the image of an independent third party that will treat the competing parties fairly. However in two cases there were complaints about the fact that the inspector let the submission of documents continue well beyond deadlines, thereby harming in particular the groups with the least resources, i.e. the action groups. These groups are at a severe disadvantage anyhow because they lack important resources such as money, experience and scientific information. In all three cases, opposition groups undertook efforts to collect such resources. The success of such efforts depended in part on their organizing skills. The expertise that the action groups mustered was of varying quality, but ranged from good to excellent. However, the information presented by the action groups failed to affect the outcomes of the decision process in two out of three cases (Seal Sands and Newport), largely because the information was not 'scientific' enough (e.g. the health study in Cleveland). The lengthy duration of the inquiries and their adversarial nature offer a more than extensive opportunity to debate all relevant issues (within the terms of reference then) and when the parties do not bring forward relevant information, the inspector can and does supply additional relevant information. In all three cases, the parties in the inquiry were satisfied with the opportunity for debate that was offered to them. What was less acceptable for most parties was that the SoS did not always completely follow the advice of his inspector. This then created a certain level of bitterness among the parties at the inquiry.

4.4.2 Cross case comparison

Introduction

There is a large degree of similarity between the three cases studied in terms of institutions and discourses that drove the decision process. The similarities derive from the fact that the legal framework succeeded quite well in structuring the debate. The differences are mainly located in the fact that Rotherham MBC had a better prepared planning framework for discussing the proposals of the private sector and could therefore advance arguments that were related to need and site selection. This in contradistinction to the two other authorities that were somewhat caught by surprise by the proposals of the private sector. Furthermore the Newport case is very distinctive because the argument of the community's interest figured so prominently. The story line was largely ineffective, but did work to clarify the importance of community resistance in future cases. The Seal Sands inquiry was very interesting because it was a linked inquiry.

The role of the private sector

When looking at the role of the private sector, it is clear that private companies largely drive the decision process. The impact of 'the market' for waste treatment is very large. So, even though the various ES's portray site selection as a rigid scientific process wherein environmental factors are very prominent, this is not the way site selection actually occurs. In all three cases, the proponent foresaw, largely as a consequence of developing government policies, a certain market for waste disposal in a certain part of the country. He then targeted a certain region and selected a site that offered good operating characteristics, specifically good road access. In order not to alert competitors, proponents tend to postpone publicity until their proposals are well advanced and are practically complete. The role of government in the site selection process is limited, although all proponents acknowledged the importance of waste disposal plans (these partly determine their market) and land use plans (these determine where their facilities are acceptable). It is interesting that within the local authorities (Newport, Cleveland), the first point of contact tends to be the economic department that has an interest in increasing employment and economic activity in the jurisdiction and that tends to offer a certain site. Later contacts are then with planning officers who tend to take a more hesitant outlook. It does seem that the proponents are greatly helped by the fact that hazardous waste facilities are 'industrial use' under national UK planning guidance (an incinerator in heavy industrial use). The implication is that land use plans are really not that restrictive when it comes to siting because there tends to be ample land designated for industrial or heavy industrial use. The designation 'industrial use' however does not tend to be given with waste facilities in mind. As a consequence, the local authorities and neighboring companies are somewhat taken by surprise by proposals for waste facilities. In the case of Newport, the neighboring companies objected to their new neighbor. The important role of the private sector is underlined by the fact that, even though the authorities may approve of a plan, the facility may in the end not actually be constructed because of market considerations (Seal Sands). However, the planning permission issued for that site is itself commercially valuable and is kept for future plans.

The role of experts

When looking at the role of experts, one can observe that their role is large, if not crucial, but that they sometimes compete with elected representatives for the most important place at the decision table. The major role of experts is demonstrated by the fact that they are the ones who design the facility and tend to write the ES that is crucial for determining the acceptability of the proposal. Experts hired by the proponents tend to consult with experts employed by the authorities, and their debates are largely technical, usually structured by certain guidelines that are prepared nationally and which exemplify 'good practice', or by more or less informal consensus among the experts in the field. It is clear that the proposals by Ocean Environmental and BFES conformed to the standards of good practice and therefore most experts agreed that they were acceptable, or even 'eminently acceptable'. Both proposals went ahead in the end. In the case of VHE's proposals there was disagreement between the experts on the technical design of the facility (and its location) and remarkably, this is the proposal that did not make it. It is remarkable that, in that specific case, the most trusted expertise was found outside the sphere of consultants employed by the proponent and government regulators but with an action group. In this sense, the fact that an inquiry was started and an action group admitted as a main party to the inquiry was a very beneficial action.

In all three cases, the inquiries were a clash between experts. As Mr. Newman remarked in the Straight Mile case, such a clash is somewhat detrimental to the status of the experts, as their discussions become a bit 'like discussing god'. It is remarkable that in all three cases, experts could be found to argue for and against the proposal, which could mean various things. It is my impression that to some extent such discussions point to a certain bias among experts towards the actors that employ them. The way the site selection processes in all three cases were described by experts, giving an essentially market-based decision a managerial flavor, points to this. On the other hand, there is also evidence that there are all sorts and kinds of experts and that there are boundaries to what the experts can provide. To start with the argument that there are various kinds of experts: it does seem that the perception of the proposal by various experts is affected by their discipline, training and working experience. In the case of Newport for instance, a retired chemist who severely criticized BFES's proposals for sampling incoming waste streams, assisted the action group. The criticism was largely based on her own experience working in a laboratory, where strict protocols applied for sampling. Such strict protocols are however not the norm for the waste industry, mainly because waste acceptance would become way too time consuming. Such differences of opinion are then largely resolved by referring to some standard that reflects acceptable practice in the field (e.g. contained in the Waste Management Papers of the DoE). If such a standard is lacking, or the information on the question whether the standard will be met is lacking, then the credibility of the experts in question becomes important. The credibility of Mr. McIver in the Straight Mile inquiry is a case in point. Mr. McIver's knowledge of the underground at the proposed site was much more detailed and based on firsthand experience. Furthermore, he held his own valiantly when cross-examined by VHE experts. Combined, this gave him a level of credibility that surpassed the credibility of the well-paid experts hired by VHE in the eyes of the inspector.

The role of elected representatives

When I say that experts sometimes have to compete with politicians, I immediately introduce the topic of political control. In a sense, the decision process under the UK legal system is a bit like a 'blind date', where two parties interact without really knowing eachothers' history. For instance, what did the planners at Newport know about BFES? Very little as they had never had any dealings with the firm before. This implies that deciding the company's proposal in terms of the planning system and on the basis of descriptions on paper requires a certain level of faith in the company's intentions and reputation. Certainly, plans need to be outlined and motivated and they are critically evaluated, but whether or not the proponent will actually do what he says is a different matter, that receives no consideration under the law. Politicians on the other hand clearly let past experiences with 'the waste industry' (which really does not exist as a unity) and images play an important role. In Newport, there were various negative experiences with waste disposal plants. In Cleveland, council members felt they were dealing too often with proposals for waste disposal facilities and feared for the image of their county.

Given the fact that such considerations have no place in the relevant legislation, the tension between administrators, who advance on the basis of the legislative framework, and elected representatives, who pride themselves in representing the local community, is predictable. In both the cases of Newport and Cleveland, the council felt it should refuse the application, whereas the administrators actively tried to get council to consider the issue 'on the appropriate terms', i.e. not emotional or rational, and on the basis of valid planning considerations. My impression is that here and there (but especially in the Newport case), the administrators let their opinion on the proposal influence the advice they gave to the council. In the Newport case, factors in favor of the proposal were stressed, whereas the factors about which there could be doubt were de-emphasized.

The tension between planners and elected politicians is detrimental to both the planners (they make themselves impossible) and politicians (they have difficulty in making a case against the proposal). In all three cases, politicians were aware that their resistance could not really be framed in terms of the community's interest or trust. This implies for instance that they are hesitant to openly and enthusiastically embrace action groups. They must avoid this to remain to be seen as neutral actors. It also implies that they must frame their objections in terms that are relevant under the planning system. This happened in both cases (Newport, Seal Sands) where planners and elected politicians were at odds with each other. The role of the leader of the council was major in both cases, as it was the leader who drafted the council's motion to reject the application. In both cases a remarkable shift took place from consideration of the application in the planning committee to consideration by the full council. This shift was based on the idea that the full council would have greater freedom to consider the broader issues (i.e. trust and the interest of the community), whereas the planning committee was thought to be required to remain within the narrow confines of the planning system. This line of reasoning is clearly incorrect but indicative of the way councilors saw the proposal. I find it remarkable that in all cases, the council was unanimous in rejecting the application, as many parts of the area would not be affected by the proposal. It seems that leadership in the council may have played an important role here, as is the fact that little political gains can be made from support for a waste disposal facility.

It appears there is no tension between elected representatives (i.c. the SoS) and experts (planning inspectors) at the national level. In cases where the SoS (it can be doubted he actually takes the decisions himself - but nevertheless) deviates from the advice of the planning inspector - this seems to be the case mainly where inspectors advise imposing some conditions that seem absolutely crucial for local acceptance (if still possible) of a proposed facility - the SoS acts mainly as an agent of the system. What I mean by that, is that the SoS seems little concerned with opposition to his decision (which can develop to reasonably large dimensions - see the 100 MP's who signed a petition in the case of Newport) and remove planning conditions that are not acceptable under the legislation (i.e. the condition of public monitoring in the case of Seal Sands). The fact that such deviations are so widely resented, even more it seems than the actual decision to overturn local resistance, indicates both the belief in and the need for determination of the conflict by a non-political figure (the inspector) in the UK. Or does it indicate a sense of unfairness that somebody who was not involved in the inquiry (the SoS) should look at the outcomes again and then delete some hard-fought gains for the opposition? It seems that both factors do carry weight to some extent.

The role of the courts

The courts are largely absent from the decision process. It does seem that the extensive discussion at the inquiry contributes to that to some extent. I have asked various parties why they did not challenge SoS decisions they did not agree with, and they tended to reply that the inquiry is the event where a case must be made. It was seen as a 'one-off event', a 'fair fight' of which the outcomes had to be accepted. The inquiry thus functions as a substitute for a court case, but differs from it in various respects. One difference is that the terms of reference are not set by the parties to the inquiry, but by the SoS. Another is that the planning inspector is far from a passive judge, listening to the presentations and then indicating that a certain party was right. Instead, the inspector himself undertakes some fact finding here and there, plus he must always translate what is being said at the inquiry (there are few limitations on that) to the terms of reference and to the relevant guidance. In a sense the inspector also functions as a buffer for the SoS. Much anger and frustration is vented during the inquiries and there is quite a bit of argumentation going on that is not really relevant to the outcomes as it is not relevant under the legislation. Whereas parties are told that their contributions will not affect the outcomes if they start to rant and rave, the fact that the parties were allowed to express themselves volubly helps to vent some anger. The inspectors who heard the three inquiries of interest have been scrupulous in avoiding any indication of partiality. I have been told that the inspectors refused to mix with the parties outside the hearings and refused even the slightest offers such as a drink from any party. Where there was an attempt to mix with the inspector from one of the parties, for instance during a site visit, inspectors refused.

The inquiries all clearly took place on the terms identified as relevant by the legislation. The discussion was largely technical, whereas an exchange on issues such as trust and the community's interest hardly took place. In fact, to raise such arguments makes the local council vulnerable to an award of costs as it could easily be regarded as wasting money by raising irrelevant considerations (see Newport, but also Cleveland). The possibility of an award of costs thereby functions as a very effective 'agent' that enhances the structuration and institutionalization of the discourses embedded in the legislation. The fact that the inquiry is heard by another expert (the planning inspector, in one case assisted by a technical assessor) helps this structuration.

The role of the community

The role of the community was rather limited in the three cases. It seems that 'the community', if existent at all, does not correspond to the political boundaries of the three authorities involved. The pattern that repeats in all three cases is that the most active members of the community, the ones who form the action groups, live relatively close to the proposed site. Other people in the locality tend to care less about the proposal, unless they later get the impression that it will affect them as well (the Newport case is a good example). But even in the areas relatively close to the proposed site, activism is not common. This pattern of motivation expresses itself through the support that the action groups receive; especially the relatively 'easy' way of supporting the action group as the signing of a petition is popular. Each of the three action groups succeeded in collecting thousands of signatures, which implies however in none of the cases that a majority of the population signed. When it comes to demanding forms of participation, i.e. becoming a member of an action group, the numbers of people willing to participate decline rapidly, to dozens of people rather than thousands. And even fewer people actually go to an inquiry, write submissions, and lobby politicians.

The use of the various languages in the three UK cases

It is clear that decision-making in the UK is dominated by the combination of conservative pluralism and managerialism embedded in the law. For traces of conservative pluralism one only needs to look at the case of Seal Sands. There the issue of need became hotly contested with parties throwing statistics and results of questionnaires at each other. In the end, the planning inspectors who heard the linked inquiries accepted the projections of the private sector, essentially because they came from the private sector. This may sound somewhat tautological but it does make sense because the private sector will invest in the facility and therefore stands to gain or lose a considerable sum of money. Other people, who also make projections, do this without running any such risk and they can therefore ponder about waste problems freely, but will not have to bear the consequences if they are wrong. The presence of the managerial discourse in the debate seems somewhat more prominent though, especially in the sense that the debates are largely about the technical quality of the various proposals. And occasionally there is mention of the fact (Newport) that the risk of one community helps reduce the risks in a range of other communities, thus increasing net utility for society.

The finding of a prominent presence of the managerial discourse is somewhat surprising as the reasons why the proposals were refused in all three cases have much more to do with a lack of trust in the private sector (typically a progressive pluralist argument) and the perception that the proposals are against the interest of the community (a communitarian argument). However even those who are outside the managerial discourse coalition (constituted by practically everyone with experience in the planning system and pollution control system, including members of the local planning committees) are forced to argue on the basis of that discourse. This is especially so during the inquiries, which are effectively constrained to consider technical issues only, and where the main strategy seems to be to throw experts at each other. Especially the disagreement between experts that was visible at every inquiry and the fact that inquiries regularly cross the boundaries of the world that has been studied by scientists pose questions regarding the managerial emphasis. I must say that through their openness and litigious character the inquiries do seem to bring out the best technical information available and create an atmosphere where scientific findings are rigorously debated. I also note that the inquiry is remarkably effective in reaching consensus or in bindingly deciding what should happen. It does seem here that the presence of a supposedly unbiased third party does wonders here. Also, the set-up of the inquiry is so that parties collide and collaborate at the same time. During every inquiry, the parties are forced to work on the conditions they think should apply if planning permission were to be granted. In some cases (Newport) they are forced to work together on risk assessment. These requirements bring out the problem-solving tendencies in most of the experts involved. In the end, it often transpired that experts opposing each other started agreeing on the proper basis for the decisions after a certain amount of discussion (take Newport again). The outcome of such a process is not so much a rigid scientific answer to the problems at hand, but much more akin to an inter-subjective consensus: the best bet on how it will be. The willingness to, in the end, accept the Inspector's report seems to hinge largely on his capability to capture this consensus in his report or to judge on the matters left undecided in an unbiased fashion, but perhaps more importantly on the fact that the inquiry provided the opportunity for a tough fight, where everyone could make his strongest possible case. In this sense, the pluralist character of the inquiry aids its acceptance, which is only disturbed when the SoS consequently modifies some of the outcomes. Whereas inquiries have been described as a means to force unwanted facilities upon localities (see chapter 3), this does not seem to be the exact truth. Actually, it is true that the local interest does not weigh very heavily in the outcome, but this is an intrinsic element of the managerial discourse that structures the debate. One should not forget however that the technical quality of the proposal is a second element of the managerial discourse, and certainly the inquiries I have studied focused very strongly on that topic. The importance of this element is supported by the outcomes of the Straight Mile inquiry, where the technical quality of the proposal was not sufficient in the end. So even though there is a willingness to force facilities on local communities (see Newport, see Seal Sands), this is not done if the proposal fails to meet technical quality standards (see Rotherham).

4.4.3 Theoretical feedback

Introduction

In chapters one and two I have indicated that the case studies should shed some light not only on the theoretical framework itself, but also on some of the empirically testable assumptions embedded within the discourses. I think the three case studies provide some insights on both issues, but mainly on some of the assumptions embedded within the various discourses.

Interests and language

As noted above, there is practically no use of the communitarian language in the three case studies of interest here, with the occasional exception of some actor referring to the community interest. When actors do use this language, the goal tends to be to create a certain level of opposition, take the 'we need people power' statement from Cleveland County Council. Progressive pluralist language was also largely absent, with the exception of the statement in Cleveland and Newport that they did not want to become the dustbin of their regions. The story line of distributive justice was not very useful in either case as both localities were net exporters of waste and could thus not rightfully claim the title of dustbin. In fact, some other jurisdiction was being their dustbin without them noticing it.

There are various indications of the relation between interests and language in the cases I have studied. First, it is clear that there is quite a strong discourse coalition that considers it important to adhere to the legislation. The SoS, the planning inspectors, the statutory consultees, and the local planners have all demonstrated the importance they attach to adherence to the law. Whereas this strategy is relatively risk free for the majority of these actors, it is not always so for local planners and for the SoS. These actors could make themselves infinitely more popular if they gave in to local councils and reject certain proposals, even if they are acceptable in technical terms. The pressure does seem to influence the SoS as the timing of decisions upon appeal tend to be taken at politically opportune moments. On the other hand, even a petition of 100 MP's could not sway the SoS for Wales to change his decision on BFES's proposal. The local planners clearly had an interest in maintaining their jobs, but even in the face of tensions with the council, they chose to argue in favor of the proposals (in Cleveland and Newport) and to do otherwise does not seem to have crossed their minds. What this implies is that the relatively managerial nature of the decision process has its share of true believers, people who adhere to it despite their own political or personal interest. Interestingly, the planners in Cleveland who had initially suggested the application for planning permission should be approved, did argue against the application at the inquiry. This however, made them an easy target for the opponents because the planners themselves were on record saying that the proposal should be allowed. This mistake was not made, as far as I could see, in the case of Newport.

So there is evidence that interests do not always drive the use of a certain language. On the other hand, there is ample evidence of the opposite too. For most action groups, the managerial language was not their 'natural' way of speaking about the proposals, but they were still forced to use the language in order to advance their interest and they adapted. The action group in the case of Rotherham provides the best example, because there the story line that the proposals were against the community's interest was replaced with managerial language after the advice of a planning expert. The group was made aware that if it wanted to win the debate, the group should start speaking a different language. Politicians, especially local councilors who have received positive advice from their bureaucracy on the proposals but want to reject them, have a similar problem. Here too does one see complete (Cleveland) or partial (Newport) managerial framing of objections, even though the 'real' objections were originally not in the technical sphere. It is indeed clear that in order to be effective, managerial language must be used. Not to do so would result in a very brief inquiry (Newport), 'punishment' in the form of cost awards (Newport again), and almost certain loss of the argument. The briefness of the inquiry in Newport, like the court cases on the cost award, does already indicate how difficult it is to talk about the proposals in communitarian terms. The Newport case resulted in the end in a clarification that the community interest was an acceptable story line in hazardous waste siting, but only to the extent that the claim could be substantiated with real arguments. In this sense, the litigation on the award of costs has not resulted in a watershed outcome.

Feedback into some of the assumptions

In chapter 2, I have identified some assumptions that are embedded in the three discourses (or a criticism of them) that would potentially be empirically testable. These assumptions have been formulated in the form of negative and positive expectations about the way certain groups of actors would behave in hazardous waste siting. In this final section of this chapter, I will present some indications about the various assumptions made.

(1) The role that experts play

The expectations formulated about experts in chapter 2 were, on the positive side, that they are objective and neutral and will separate values from facts. On the negative side, the expectations were that experts would not be able to separate values from facts, and secondly that they will deliver the results that the persons paying for them want.

There is little evidence in the three case studies on the neutrality of experts. The little evidence there is is somewhat conflicting. Take the experts employed by the local authorities involved. In two cases there were very clear signals that their political masters wanted to reject the proposals, yet the planning officers suggested to approve of them and actively tried to influence the council. The language used by these experts is clearly based on a separation between facts and values as the councilors are asked to decide the application in a rational and unemotional fashion. So there is a certain degree of evidence that experts are sometimes as neutral (or at least independent) as they are portrayed to be under the managerial discourse and do separate facts and values (although perhaps the influencing of council could be considered the mingling with values and although one should not forget the bigger picture of bias in the planning system).

There are on the other hand some different signals. One could, for instance, very well have the impression that it is almost too coincidental that ES's, documents prepared by experts, in all three cases suggest that the proposals are not detrimental to the environment, that the proponent has selected exactly the right site and that projections of waste accumulations that were made by the proponent are correct. What this suggests, and I think this can most obviously be recognized in the site selection process, is that the experts making the ES are somewhat biased towards the person who pays them. At the stage of the inquiry this impression applies to practically all experts. All parties tend to be able to find an expert who will argue their case. The implication is that, during an inquiry, expert A vehemently argues a certain issue and expert B adamantly denies the very same point. To some extent this can be explained by the fact that different experts have different outlooks and that scientific knowledge is rather inconclusive in a great many areas and that assumptions must be made that can be skewed in a certain direction. The inquiry largely functions as a method to discover the assumptions underlying the judgements of various experts and to move towards a common understanding of the issues before the inquiry. It seems that in some cases (Newport) experts are willing to cooperate and can develop a common understanding.

(2) Market parties

The assumptions formulated on market parties were on the one hand that they would know their own interest best and will defend it if need be in court or in politics (pluralism). Also, it was assumed that they define their own interest so that it includes the interests of others (pluralism - the harm principle). These positive assumptions, largely stemming from the conservative pluralist discourse, have their counterparts in the assumptions that market parties will have difficulty in defining their interest in complex issues and will not think of anybody but themselves.

In the cases in this chapter, market parties are mainly landowners and the proponents are well aware of their interest and certainly defend it in court or in politics if need be. Landowners hardly played any role in the three cases studied here because in all three cases they were willing to sell their land to the proponent. This should be no surprise because the costs of land acquisition tend to be small when compared to the facility that is constructed. The proponents (waste companies) tend to be driven by the idea that there is a certain market for the facility they are about to construct. If one comment can be made about the knowledge of their own interest, then it probably is that this market is in reality very unpredictable, note especially the Seal Sands case. In this sense, it is hard for market parties to know their interest. The expectation that market parties have difficulty in defining their own interest in complex issues is thus true to some extent. It does seem however that when the proponents have decided they want to actually construct a facility, they tend to be very adamant about achieving their goal. Given the investment needed (often millions of pounds) these companies tend to have relatively deep pockets, and they tend to use that money during the appeal process to an extent that is rather intimidating to other parties.

There is hardly any evidence about the proponents being very concerned about the interest of other parties. Although of course the companies do want to build a safe facility (both BFES's and Ocean Environmental's facilities were generally considered extremely advanced) and do want to adhere to the regulations, they clearly do not propose the facility for anything else than profit making. The only proposal where 'the community' was mentioned at all, was VHE's proposal, where a visitor's center was planned for the community. Whether or not the community really wanted this center was however never asked. Certainly I would not say that the case studies have shown that the harm principle - private parties defining their interest to include the interest of others - has a very strong presence.

(3) Politicians

The assumptions about politicians that were formulated in chapter 2 stem from the pluralist and managerial discourse. The first assumption was that politicians will be accountable to the people because they need their support. There are however two negative assumptions as well. The first is that politicians will be accountable only to 'special interests'. The second is that politicians will 'misuse' hazardous waste decisions for political gain, even if it is strictly a technical issue (managerialism).

In the three cases studied here there are some similarities and some differences in the way politicians have responded to the proposals. In two cases (Seal Sands and Newport) politicians broke the news of the proposals to the public as soon as they heard about it. In Rotherham, councilors chose to let planning officers do their job. If openness is an indication of accountability, then it might perhaps be argued that Cleveland County Council and Newport Borough Council were behaving more accountably than their colleagues in Rotherham. Another question is however to whom are councilors accountable, and whether they are perhaps accountable to special interests only? The question is very hard, if not impossible, to answer. Very often accountability to special interests in environmental issues is associated with responsiveness to polluting industries. If we look beyond the fact that waste disposal firms are in the business of reducing pollution, we could probably view them as polluters too, at least to the extent that their facilities have environmental effects. I am quite certain that the politicians in the described cases were not especially accountable to the waste industry. In fact, in the case of Newport, local councilors and the MP from the area could hardly wait to fight the proposal. In fact they covertly did oppose the plans even before they had decided on the proposal formally.

Only in the case of Rotherham MBC was there a suggestion that councilors had an interest in a competing proposal and were thus accountable to a competitor, but in that very same case study the council was also seen as acting in concord with VHE. Another issue is whether councilors were perhaps accountable to a certain segment of the population at the expense of others. One could make the case that local councilors and MPs do jump on the bandwagon of resistance relatively quickly. Indeed, in the cases of Newport and Cleveland it was already clear what the council thought long before their bureaucrats had advised them. It should be remembered that in each case, a minority of the population was to be negatively affected by the proposals, whereas other parties stood to gain (employment, responsible waste treatment). The latter interest was however nowhere formulated and not picked up on by any politician. If the goal of the council is to be accountable to the population at large, it could perhaps be expected that this should have occurred. Did the councilors then trample over the local interest? This would probably be beside the point as the rest of the population probably sympathized to some degree with the action groups. One should also consider the fact that the proposals were among the most controversial issues in local politics in the three localities for years. Local politics is mostly ignored by most members of the population and therefore a petition with 12,000 signatures is pretty impressive and does perhaps indicate a bit more than just a small minority in disagreement.

The case of Rotherham MBC, if anything, demonstrates a certain paternalism towards the public. Council members were actually annoyed that the action group was campaigning against the proposal, even before VHE's proposals had received adequate consideration from the planning department. The councilors were not overly happy either about the fact that VHE's plans had become public. During an interview, one of the former members of the planning committee told me that it was his responsibility to go against the public sometimes: '*we are fortunate because we are Labour consolidated and we don't run the risk of losing the public*'. He also indicated that the council had wanted to stick to the ordinary approach to planning applications, which involved a major role for the borough's planners: '*I take the view that the planning department should assist development. This is in our legal system. I have no problem, even though there is a lot of activism, with the planning department giving advice to developers. I wasn't upset about planning officials talking to VHE about their proposal*'[53]. In a sense, the attitude of Rotherham MBC implies a less biased approach than the one taken by the other two councils. In that specific sense too, the council was perhaps less responsive to special interest than their colleagues.

(4) Voters

The assumptions formulated in chapter 2 differ in their perception of the voter. The more positive assumptions about voters stem from the pluralist discourse, the negative ones from the criticism of pluralism and managerialism. The first assumption was that voters would hold politicians accountable if they do not deliver the right decisions. The second assumption was that they would form interest groups to influence decisions in the right direction. The three remaining assumptions were logically the opposite. The third assumption was that voters would not have an interest in decision-making and would not be sufficiently educated to form an opinion. The fourth assumption was that voters would behave selfishly, favor economic development over the environment and will express 'absolute values'. The last assumption was that voters would form interest groups to further their own particular interest and prevent achievement of the public interest.

In none of the cases was the issue a matter that influenced local elections, although politicians did fear an effect (see e.g. the SoS's timing of decisions). Part of the reason why the issue would have not influenced local elections is that most politicians quickly spoke out against the proposals, thereby pleasing the most passionate part of the electorate (the opponents)

[53] Both quotes from an interview with Mr. Billingham, Rotherham, 19 January 2000. Sadly, the Labour Party suffered a defeat during the 2000 municipal elections and Mr. Billingham lost his seat.

and not really harming the other parts. Many of the assumptions made in chapter 2 are about the motives and capabilities of voters, or more broadly speaking, the general population. It does seem that the interest in the decision process is largely confined to the area quite close to the proposed site and is close to zero farther removed from the site. In this sense, the motivation for participating in the decision process is somewhat selfish. This selfishness should however not be interpreted in materialistic terms, but rather in terms of health. Health is the dominant concern of most of the people who participate. Also what figures prominently is a concern for the landscape, for a concentration of bad development in the area, and harm to economic activities nearby. It does seem that people who organize into action groups have a tendency to formulate absolute values, meaning that they argue against the proposal without actually identifying alternative locations. The argument that there is no need for waste treatment is often heard as well, but one cannot escape the impression that this argument fits nicely to an already determined opinion. To say however that the action groups prevented the attainment of the public interest would go too far, although there does seem to be a certain tension between planning officers, who claim to bear the standard of the public interest, and the action groups. The organization of an action group is quite a demanding activity in itself. Certainly when the proposal goes to an inquiry and the action group wants to continue in a prominent role, this involves a certain level of activity that can hardly be combined with an ordinary daytime job. However, in all three cases action groups did succeed in organizing and in presenting a coherent case to the inquiry. In the case of the Straight Mile inquiry this organization was even extremely effective as the action group succeeded in attracting a fair bit of money and the required expertise.

(5) Parties to a court case

Especially under the pluralist discourse, parties to a court case are expected to improve decisions by making the strongest possible case for themselves and thereby support truth finding. A criticism of pluralism is however that parties in a court case make the strongest possible case for themselves and thereby *hinder* truth finding. A third assumption derived in chapter 2 was that parties in a court case will not rest until the last legal resort has been tried, prolonging the decision process.

Having already indicated that the cases studied for this chapter have not ended before the courts, and having already touched upon some elements of the inquiry, I can be relatively brief about the assumptions formulated about such parties. I conclude that truth finding was supported by the behavior of the various parties at the inquiry. Had there not been an inquiry, the action group in the case of the Straight Mile proposal could probably not have mustered their strong attack on VHE's proposals. Given the widely differing opinions represented at the inquiry and the force with which they were represented, combined with the fact that planning inspectors sometimes added to the discussion, I do think that most proposals were effectively improved in managerial terms (although Ocean Environmental's proposal made it through the inquiry practically unscathed). In all three cases, the parties largely accepted the outcomes of the inquiries, thereby falsifying the assumption that they will use every possible route for judicial review. An important reason seems to be the perceived fairness and comprehensiveness of the local inquiries.

(6) Members of a community

The communitarian discourse contains very positive and optimistic ideas about decision processes that take place within communities. One assumption from this discourse predicts that community members will be engaged in coercion free relationships and are in dialogue with each other. Another states that members of the community have local knowledge and that such knowledge is crucial for the decision process. Far less positive assumptions can be derived from the criticism of communitarianism, and from the managerial and pluralist discourses. The third assumption, therefore, is that community members will be parochial in their outlook and not look beyond their own community. The fourth and fifth assumptions are that community members do not have scientific knowledge, which is crucial for the decision process and that they do not want to participate in the decision process because of a lack of motivation.

In the cases reported here, I have not detected the extensive dialogue among community members that the communitarian discourse envisions. The main pattern seems to be rather that a small group of individuals actively opposes the plans, with many others looking on. It is true that there was little coercion between the various members of the community, but in a sense that was a problem from the perspective of the various action groups as they could not force everyone to protest.

I do find that members of the community brought forward local knowledge. Specifically the Newport case, where the population had insight in the movement of transportation trucks, demonstrates that such information is easily ignored yet can be proven correct after a while. To say that this information was crucial for the decision process is somewhat beside the point. The truth is that the information was contradicted by information from experts, and what these brought forward was considered more reliable. Other knowledge that ordinary members of the community contributed is difficult to qualify as local knowledge. In the case of Seal Sands, but also in other cases, community members brought forward information they had obtained through their work (e.g. as an operator of a private incinerator) or by their training (e.g. Mr. McIver). Certainly, there was not a broad comprehension of the technical details of the proposals among the population, but equally certain is the fact that in all three cases some people had a remarkably high amount of relevant information and a respectable number of people were highly educated and were able to form an opinion about the proposals. Especially in the case of the Straight Mile proposal, the knowledge of the action group actually surpassed the knowledge of the hired experts. In that sense, that information was rather crucial to the decision.

Were the members of the community parochial in their outlook or were they very passive? To some extent both questions can be answered with a yes. The reason why both questions are answered with a yes will be clear from the above. The parochial character of the participation has already been pointed out when it was noted that only people nearby the proposed site tended to participate, whereas they usually did not provide alternative proposals for the one being made by the private developer. Also, a majority of the population, even those who lived somewhat nearer to the proposed sites, remained largely passive.

Chapter 5

The Netherlands: participation through the court system

The three Dutch cases described here demonstrate certain peculiar features of hazardous waste decisions in the Netherlands. One of these features is the abundant use of the courts as an avenue for decision-making, hence the name of this chapter. Because of this, the cases give an excellent opportunity for looking at some of the supposed advantages and disadvantages of decision-making through the courts. I conclude that the courts are a relatively slow institution that does not really solve conflicts. Another feature is the practice of 'gedogen', the silent approval of activities without formal licences. One implication of 'gedogen' and the consequent omission of formal procedures is that citizens or environmental groups cannot participate in the decision process as they should. It seems that decisions are largely made in the constellation of private firms, administrators, political heads of the bureaucracy, the Commission for EIA and the courts. The conservative pluralist and managerial languages embedded in the law structure the debate quite well and force all participants to speak in managerial terms. Yet the influence of the political heads of the bureaucracy in the decision process is large. This influence is largely concealed however by constant reference to expert decision-making. Interestingly, in the one case where an attempt was made to explicitly structure the decision process on managerial terms and take away the right to propose locations from the private sector, the courts protected the right of a private market party. The community played a minor role. Only individual members of the community participated, along with environmental groups that were consistently present in the debate. Remarkably all these actors use managerial language to structure their contributions. Only in one case did they really affect the consequent decisions.

Figure 1. Cases in the Netherlands

5.1 North Refinery: an example of the art of 'gedogen'

5.1.1 Introduction

The first Dutch case I shall describe is a very protracted one, involving several proposals, authorities and procedures over a long period. The case demonstrates various points. The first is that citizen participation can be made practically irrelevant by a phenomenon called 'gedogen' in Dutch. Gedogen implies that a government knows about a certain illegal situation and yet does not take measures to counter it. The second is that the courts (besides the licensing authorities) play a large role in hazardous waste decision-making in the Netherlands and that there are problems associated with that. This case has been very contentious, at least judging by the amount of litigation over it. Yet, it would be difficult to point out situations where courts were very clear about the merits of the pressing issues put before them.

A remarkable feature of this case is the difference in reaction to the waste plant by ordinary people and environmental organizations. For much of the period that the decision-making process lasted, 'ordinary citizens' were inactive, whereas one environmental group played a leading role in what happened. The case study period encompasses the period from 1983 until 1999, which is extremely long and prohibitive of a description in detail. I will therefore only give limited factual background concentrated in the early part of the case history.

5.1.2 Various proposals

In this case, it is hard to draw a line in time and identify the beginning and end to the decision process. One complication is that developments on the workshop floor and the formal decision-making process were never in sync, implying that reality and legal problems continued to catch up with each other. Several different proposals have been put forward to the authorities, which could perhaps also be studied in isolation. However, exactly because of the continuous flow of events such isolation would be somewhat artificial. The table below gives the details of the various proposals that have been made.

Table 1. The North Refinery case. Proposals made in time.

Period	Proposal details and licensing authority
1983-1987	Distillation and refinery plant to separate usable fractions (toluene, gasoline, diesel oil, furnace oil, etc) from oily wastes with tank park and waste water treatment capacity. Changed to include heavier oil types and crude oil slops in 1985, changed plans for waste water treatment plant in 1986. Proposed to the province of Gelderland and the municipality of Beuningen, and later to the province of Groningen and the municipality of Delfzijl. Involvement of the Department of the Environment and the Department of Transport and Public Works.
1988-1996	Additional distillation capacity, hyper-filtration unit, new furnace and scrubber. Together composing the Central Treatment Unit for Waste Oils (CBE). Later proposed regardless of CBE-decision. Proposed to the province of Groningen and the municipality of Delfzijl. Involvement of the Department of the Environment.
1997	Same plant. Increase of capacity for hazardous waste treatment to 100,000 tons annually. Same authorities involved. Plan abandoned quite soon because of quashing of licences.
1997-1999	Recycling and Utilities North (RUN). Dewatering, gassification, cutting pyrolisis, and smelting. Responsible authority: province of Groningen and Waste Consultation Organ (AOO).

5.1.3 The art of 'gedogen'

Introduction

The history of the case starts somewhere else other than where the plant was eventually built, namely in the province of Gelderland. A relatively marginal yet successful company there, Holding Bos, was active in distributing oils and as a correlate of that, started taking in used oils and became involved in the treatment of such oils. Business apparently went well, because the Ministry of Transport and Public Works, concerned with oil discharges to public waters, was willing to subsidize a new facility that the company wanted to build. An impressive looking 'surroundings analysis' ('omgevingsanalyse') was made by an environmental consultancy firm and a site was picked on the basis of that report. The proposals failed in the end because Holding Bos became a suspect company in the eyes of the authorities in Gelderland. The company was found to have illegally stored waste oils and suspicions arose over the actual level of treatment that the company provided to the oils it collected. Holding Bos quickly acted once its bad record emerged. Instead of fighting off the opposition, the company decided it would relocate and start from scratch. One month after receiving critical comments to its surroundings analysis from the province of Gelderland, the company succeeded in convincing the municipality of Delfzijl in the far North of the country that the treatment plant was a good idea. The company was attractive to Delfzijl because it offered 20 to 30 much needed jobs to the economically depressed area. The town was attractive to the company because there were few land use planning restrictions and there was a friendly climate for chemical industry in the area. National subsidies to stimulate economic investments in the North were also attractive for Holding Bos. Figure 2 shows the site of choice.

Licences for Holding Bos

The plans in Delfzijl required various licences from the authorities, including one under the Nuisance Act (later under the Environment management act), under the Pollution of Surfacewaters Act (PSA), and one under the Chemical Waste Act (WCA, later incorporated with the Environment management act). The land use planning regime posed virtually no limits as the relevant local zoning ordinance dated back to 1965 and deemed practically any type of industrial development in the area acceptable.

The plans were discussed with the municipality first and received with enthuse. The municipality was responsible for the required licence under the Nuisance Act and set out to act as quickly as possible. The municipal haste was so great that it desired to have a licence written before the application was even formally submitted[1].

[1] Hand-written report by the attending inspector of the pre-application meeting between Inspectorate, Holding Bos and Delfzijl Municipality, 3 November 1983.

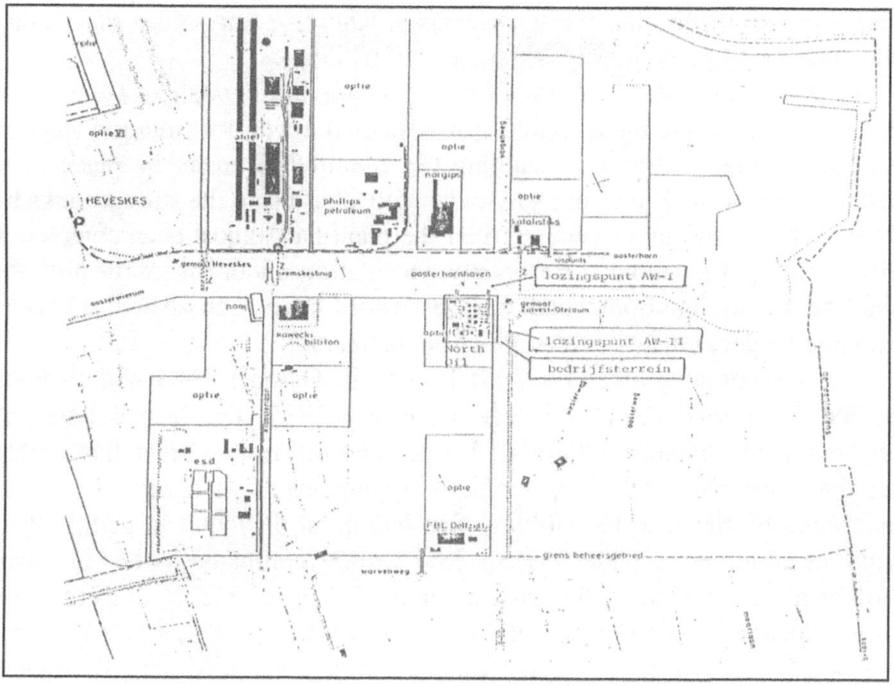

Figure 2. Location of the facility on an industrial area (Adapted from a licence application).

This despite the fact that Holding Bos refused to be very open about its plans and evaded critical questions asked by the authorities[2]. Then the Regional Branch of National Inspectorate for Public Health (from here on 'the Inspectorate') became involved as a statutory consultee to the municipality. The Regional Branch of the Inspectorate was very suspicious of the company because its colleagues from Gelderland had warned that Holding Bos was not trustworthy. The municipality wanted to issue a licence as quickly as possible and turned to the Inspectorate for advice on the appropriate conditions as the municipality lacked expertise in this area and did not want to hire a consultancy firm for advice. The Inspectorate however, was not minded to assist the municipality in the intended way. It advised the municipality that it should hire external expertise and advised the municipality not to accept the initial licence application that was on the table as it was too vague in the eyes of the Inspectorate. The Inspectorate wanted to proceed strictly by the book and be sure that certainty that the licence would meet all existing guidelines. Although the municipality accepted this advice and a new application was required, the municipality doubted some

[2] Indicative of this attitude is the fact the existing 'surroundings analysis' was not handed to Delfzijl and the Regional Branch of National Inspectorate for Public Health (RIMH), which was a statutory consultee to the procedure.

of the requirements that the Inspectorate wanted to impose on the facility and continued to plea for quick issuance of the licence.

Because decisions were not forthcoming rapidly enough in the eyes of the municipality, the municipality slowly drifted towards a situation wherein start-up would be allowed without the required licences in place. The municipality allowed the company to actually start using the storage tanks by 29 June 1984[3] and by December 1984 the plant had almost been completed. Municipal spokespeople were openly speculating about the start-up of the plant before the environmental licences would have been issued[4]. Pressure from the Inspectorate prevented this from happening.

The municipality issued a draft licence in October 1984 and strongly pushed for a final decision before 1 January 1985. The Inspectorate had substantial comments on the draft licence and did not consider the desired deadline feasible[5]. The Inspectorate's suspicions of Holding Bos were reinforced by the fact that Holding Bos had great difficulty in getting their tanks certified as suitable. Out of four testing companies, only one was willing to attest to the safety of the tanks. The municipality did not share these concerns, but a second version of the draft licence was prepared that was somewhat different from the first one (e.g. it included the requirement of floating roofs in the storage tanks). The Inspectorate still did not approve this draft because not all of its comments had been heeded and the two authorities seemed to be headed for gridlock. The municipality contested the Inspectorate's position on the basis of various reports, including one from the local fire department. Furthermore the tanks had by then been certified, a result that was not trusted by the Inspectorate however.

Bankruptcy?

By April 1985, the company declared that it would need to start operating in May: otherwise it would go bankrupt. In a meeting about the situation, the municipality pressured the Inspectorate to approve. Working without licences was standard practice suggested the municipality and besides, no societal interests would be served by closure of the plant[6]. The Inspectorate strongly disagreed and reminded the municipality of its commitment to a proper environmental licence. Concerned about what was to happening, the

3 Letter Municipality of Delfzijl to Holding Bos BV, 29 June 1984, found in the Province of Groningen Archives.

4 'Fabriek bijna klaar, nog geen hinderwetvergunning', Winschoter Courant, 12 December 1984, page unknown.

5 Minutes of the meeting on 9 October 1984, found in the Inspectorate's Offices in Groningen.

6 Minutes of a meeting between Inspectorate and municipality by one of the inspectors present, 1 May 1985. Found in the archives of the District Office of the Inspectorate for Public Health, Groningen.

Inspectorate reviewed their legal options and decided to involve the District Attorney. After an explanation of the situation, the District Attorney indicated his willingness to stop the plant from opening using police force. This fact was communicated to the municipality during two meetings on the topic. The municipal response was defiant and the mayor said that attempts to block the opening would be regarded as '*an attack on the municipality*'[7]. With approval from the municipality, the plant started up quickly after the second meeting and neither the District Attorney Office nor the Inspectorate intervened. The Inspectorate had not received support for its drastic stance from the Waste Branch of the Ministry of the Environment, whose dominant concern was an increase in waste treatment capacity. Holding Bos fitted the bill for the Waste Branch. Judging by subsequent events, the Branch seemed not at all concerned with the indications it was receiving from the Inspectorate. In fact, the Branch issued a draft licence under the Chemical Waste Substances Act on 10 June 1985, which was significantly more lenient than the draft licence under the Nuisance Act[8]. Issuance of the draft licence is remarkable because of two reasons. First, it was known within the Ministry that the criminal investigation of the company's plant in Beuningen had found evidence of malversations that could lead to withdrawal of the licence at a later stage. Second, the law required that other licences should be in place before a licence under the Chemical Waste Substances Act could be issued. Despite this being the case, the plant could now start treating hazardous wastes.

Not at all satisfied with the way things were progressing, the Inspectorate started stringent enforcement of the plant's operations and found that the plant did not meet licence conditions by far. The Waste Branch at the Ministry now apparently got the message and retracted. One month after issuing a draft licence, the Branch withdrew it[9]. The company, strictly operating illegally, responded with a remarkable move and filed an appeal to the State Council against the Ministry to ensure that both the facilities in Beuningen and Delfzijl could remain open. At the appeal, the Ministry did not argue their case on the basis of the Inspectorate's findings, but on the basis of police reports on the company's wrongdoings in Gelderland. The judge did not admit the police reports and consequently, the Ministry had

7 Telex from the regional branch of the Inspectorate to the headquarters, 6 May 1985. Found in the archives of the District Office of the Inspectorate for Public Health.

8 For example, whereas the Ministry would allow PCBs in the feedstock to the order of 50 ppm (parts per million), the municipality had allowed 'only' 1 ppm.

9 'Bos in Delfzijl gaat definitief dicht', Nieuwsblad van het Noorden, 2 August 1985, page unknown; 'Nederlandse bedrijven dumpen massaal chemisch afval in Anwerpen', ibid.

virtually no case[10]. The judge ordered that the plant be reopened and displayed a certain sensitivity to the illegality of the operations by requiring that the Ministry should decide the licence application within ten days. Shortly after the judge's verdict, Holding Bos announced new construction plans for the site: a refinery unit would be added, which would make gasoline products from raw oil and generate 60 new jobs[11]. Frantic discussions between company, municipality, Inspectorate and the Ministry followed. The company made various promises during these meetings on improvements it would make to the facility and its operations, but the Ministry would not give in. On 9 August, the Minister decided that the existing licence of the plant in Beuningen should be repealed and that the application for a licence for the Delfzijl plant was refused. Holding Bos appealed this decision to the State Council. The hearing was largely a repeat of the previous one. Even though disturbing evidence from the Inspectorate was available on the functioning of the Delfzijl facility, the Ministry decided to once again stress the illegal operations of the company in Beuningen. As the Chair of the State Council had already refused to admit such arguments, the logical outcome was once again to grant the appeal[12]. The Chair of the State Council ruled that the Delfzijl branch should at least be opened again until the full State Council would decide the issue[13]. To prevent the company from going bankrupt, he even decided to allow treatment of phenol-containing wastes that had been stored on site. Such wastes had been brought to the site despite the fact that the draft licence under the Chemical Waste Substances Act would not allow treatment of such wastes. The decision of the Chair to allow treatment of these wastes thus bypassed the explicit will of the administration and implicitly implies the condoning of illegal activity. What may have raised the confidence of the judge in the company somewhat is the stepping down of the two directors of Holding Bos, in return for a reliable looking replacement. A former Associate Director-General at the Ministry of Economic Affairs had become the director.

Emergence of North Refinery

The problems associated with Holding Bos in Delfzijl continued as the firm did not comply with its licences. This issue became far less relevant

[10] See: 'Ministerie legt afvalbedrijf Bos lam', Volkskrant, 2 August 1985, page unknown; see also 'Uitspraak Raad van State, Afdeling Geschillen voor Bestuur, dd. 2 Augustus 1985, no. G05.85.0737s'.

[11] 'Bos Delfzijl wil ook olieraffinaderij bouwen', Nieuwsblad van het Noorden, 3 August 1985, page unknown.

[12] 'Staatraad wil bedrijf Holding Bos niet brandmerken als milieumisdadiger', NRC-Handelsblad, 21 August 1985, page unknown.

[13] 'Uitspraak Raad van State, Afdeling Geschillen van Bestuur, no. G05.85.0786S en no. G05.85.0787S'.

however when the company went bankrupt in October 1985. Negotiations started between the official receiver and various candidates for a take-over of the plant. The leading candidate was a new firm to be founded by the banks which had held shares in Holding Bos and an oil trading firm, Forsythe International. The former Associate Director-General of the Minister of Economic Affairs led the new firm, which was initially called North Oil, but was quickly renamed North Refinery. Because of the take-over, the issue of licences became relevant again. The licence under the Nuisance Act that Holding Bos had held (this licence was still under appeal and in this sense far from definitive) was transferable, but licences under the Chemical Waste Substances Act and the Pollution of Surfacewaters Act were also required. The Minister of the Environment decided to hire expert consultants and these warned the minister that only under a great number of new conditions should he approve of the re-opening of the facility[14]. The conditions were never fully met by the company and it therefore received no licence. However, despite the fact that the company could not treat chemical wastes, it did obtain permission to start treating other wastes and North Oil started importing certain quantities of crude oil slops from the UK. These crude oil slops caused quite a shock at the Ministry because analysis of samples demonstrated that the oils actually classified as hazardous wastes under Dutch law. By then they had been treated. This fact had the potential to ignite a debate about the Dutch policy for classifying hazardous wastes, as the norms were apparently very different from those of other countries. Furthermore, treatment of the oils at North Refinery had thus been illegal, which was also embarrassing for the Ministry. In order to prevent possible public embarrassment, the Waste Branch and Inspectorate Headquarters agreed it would be best to licence North Oil. The 'policy factor' in the decisions of the Ministry increased even more when it started entertaining the idea that North Refinery could potentially play a future role as a central treatment unit for waste oils ('CBE'), an important part of the Ministerial agenda[15].

On 9 July 1986 the Ministry sent a letter indicating the Minister '*was not minded to use his powers under the Chemical Waste Substances Act*' to prevent the company from treating hazardous wastes[16]. In writing this, the Minister ignored reports from the Regional Branch of the Inspectorate that

[14] Telex Minister of the Environment to North Oil, Delfzijl and Groningen, 19 December 1985. Found in the archives of the District Office of the Inspectorate for Public Health, Groningen.

[15] The earliest trace of this line of thinking I could find was a letter by North Refinery to the Waste Branch of the Ministry, dated 29 July 1986, which refers to a meeting 'one month ago' (hence before the decision to allow the plant to operate without licence). Letter DGMH 3176502.

[16] Letter of the Ministry to North Refinery (DGMH/AST no. 0376522), 9 July 1986.

the facility was not functioning properly. Furthermore, the decision was taken outside the public domain. The good news for North Refinery kept piling up when the State Council rejected various appeals against the licence under the Nuisance Act and thereby conclusively forced approval of the licence. Somewhat later, in April 1987, the province issued a licence under the Pollution of Surface waters Act (PSA), and on top of that, the Minister finally went public with his endorsement of North Refinery by publishing a draft licence under the Chemical Waste Act (CWA) in January 1988. All these licences were objected to or appealed however.

Vulnerability

Both the decisions under the PSA and CWA seemed somewhat vulnerable. The PSA decision appeared vulnerable because emissions with very high Chemical Oxygen Demand were allowed that contained various black list substances for which the standstill principle had been introduced. In issuing a CWA licence, the Minister of the Environment had ignored strong advice from both his Inspectorate and from the National Accountant's Office ('Rijks Accountants Dienst'). This office had written a damning report on the financial soundness of North Refinery and this would have normally implied refusal of the licence, as financial soundness of waste operators was a criterion for licence decisions under the CWA. However, the Ministry wanted to have North Refinery around as long as the Ministry had not taken a definitive decision on the Central Treatment Unit ('CBE') for used oils it was planning. Indicative of the Ministry's purposes was the fact that the licence was only to be valid for two years (deadline of the Central Treatment Unit decision).

The two licences were appealed to the State Council, but the appeal to the licence under the CWA became irrelevant because it was not decided within the two years during which the licence was valid, and the Ministry did not renew the licence after it expired. This decision was again related to policy considerations. North Refinery would not become the Central Treatment Unit. Furthermore, the amount of hazardous wastes treated at North Refinery was very limited as the firm was mainly treating waste oil (non-polluted leftovers of the large refineries)[17]. Whereas the CWA decision was a bit of a setback, the State Council's decision on an appeal against the PSA-licence was an outright nightmare for North Refinery and the province. The verdict in the matter was reached on April 1991[18] and was a major surprise to most parties. The State Council found that the province had wrongfully allowed a fundamental change of the application after the public inspection period had

17 Decision Minister of the Environment, 16 February 1990 (DGM/A nr 3089513/19).

18 'Afdeling Geschillen van Bestuur Raad van State, Uitspraak dd. 29 April 1991, G05.87.0516 [AB 1991/561])'.

started. Indeed, North Refinery had found a simpler way of treating the wastewater after applying for a licence for a certain type of treatment plant and had notified the province, which had subsequently accepted the new facility under the old application. The proper response from the province, judged the State Council, would have been to restart the procedure or to decide on the original application. Because the province had failed to do either, the decision was judged to be in violation of the law and was quashed.

'Gedoog' licence after 'gedoog' licence

The pattern of interaction that now ensued would be repeated until 1997 or so. Space prohibits me going into the events in detail, but I will summarize them very briefly. The province consulted with the company on the costs of transporting the wastewater so as to avoid the discharge. The costs, in the eyes of the company, were prohibitive and based on this assertion, the province issued a 'gedoog licence', implying that it would allow the discharges during the decision procedure that would lead up to a new licence. The province assured the public through the media that the 'gedoog licence' would not be any more lenient than the licence that had been in place. Also the province justified the action on the basis that a procedural error had been made and no real problem existed around the plant[19]. Embarrassingly the province now had to reveal that North Refinery had not been meeting its licence conditions, especially COD levels which were much higher than allowed[20]. The province had extensively studied the issue, even with the help of the National Institute for Inland Water Management and Wastewater Treatment (RIZA) and various measures had been tried, but to no avail. Countering the public justification for the 'gedoog licence', the COD limits were made less stringent by the province. North Refinery in its turn used the quashing of the licence to present expansion plans, that included a re-entry to the hazardous waste market.

I have said that this pattern was repeated more often in time. By that I mean that the authorities involved continued to stumble, North Refinery continued to change its plans and not meet some of its licence conditions, the courts continued to quash licences, and the authorities continued to 'gedoog' the situation. The authorities stumbled in the sense that the province now

19 'North Refinery mag toch blijven lozen', Nieuwsblad van het Noorden, 12 June 1991, page 10.

20 Referring to the 'many consultations' it had had with officials of the provincial water authority, the company pleaded for a loosening of the standards from 500 mg/liter to 1,500 mg/liter of effluent. Letter North Refinery to the Province, 7 June 1991. An advice by the RIZA concluded that dioxane and dioxol derivates were most likely to be the cause of these high COD values. Dioxane was suspected of carcinogenic properties and is poorly biodegradable.

became the coordinating organ for the licence applications. In that capacity the province had declared the application for a CWA-licence valid, even though the Ministry did not yet approve. This resulted in loss of time (at least six months) as all procedures were restarted. The authorities then decided not to require environmental assessment. It appears that such assessment was indeed not required under the then current Dutch regulations, but there were already signs at that point that these regulations were a too narrow interpretation of European regulations. The Inspectorate for instance warned the province, municipality and Ministry to that effect, but this advice was ignored[21]. Consequently three licences were issued in 1992. Once again, the company had long started construction of the new facility at the time the licences were issued, even despite a stop work order from the municipality[22]. Especially the province underwent great embarrassment when the State Council quashed the 'gedoog licence' under the PSA that North Refinery had been using because it did not meet newly developed national guidelines for 'gedoog' situations[23]. Yet another gedoog licence was issued, with the provincial executive claiming that the State Council had quashed the gedoog licence for 'futile reasons'[24]. The new gedoog licence was appealed but not quashed and the Chair of the State Council did not stay the three licences that were consequently issued either[25].

The decision not to perform environmental assessment however then came back to haunt the decision-makers when in 1997 the full State Council did quash the three licences because it found that the applications should have been denied by the various authorities because no assessment report had been prepared[26]. In response, the province and the Ministry (the municipality was no longer involved) issued yet again a pair of 'gedoog licences'[27]. The provincial licence was salient because once again the licence conditions were made less strict, despite denials to that effect from the

[21] Minutes of a meeting between province, municipality, North Refinery, Inspectorate and the CWA-Bureau, 27 January 1992; letter North Refinery to the province, 25 May 1992.

[22] Letter Municipality of Delfzijl to Forsythe International (parent company of North Refinery), 30 September 1992.

[23] 'Raad van State. Afdeling geschillen van Bestuur. Uitspraak dd. 26 Maart 1993. No. G05.92.0706'.

[24] 'Delfzijlster bedrijf moet ophouden met gevaarlijke lozing', Grongings Dagblad, 14 April 1993, page unknown.

[25] 'Uitspraak Raad van State, Afdeling Geschillen van Bestuur, dd. 15 Juni 1993, nos. B05.92.1439/1442/1443'.

[26] The verdict hinged on the judgement that North Refinery's proposal had constituted a new project, and not an expansion of an existing facility. Because of that, and because the proposal was for a refinery (a schedule I project), the capacity of the 'new' refinery was irrelevant and assessment obligatory.

[27] See for instance the press message from the province (no. 1997/136), 25 August 1997.

province[28]. The quashing of licences came at a time when the company was already considering its future (mainly in the form of an expansion of hazardous waste treatment capacity) and the company now started changing its plans even more radically. The result was the Recycling and Utilities North Project (RUN), which was a cooperation with various other companies and which would apply the most modern waste treatment techniques and render electricity to the net[29]. The gist of the plan survived environmental assessment and the advice of the Commission for EIA. In the assessment report life cycle assessments had to be made for every waste stream entering the facility. The latter requirement indicated the increasing sophistication of the Dutch regulatory system, with the Waste Consultation Organ and the Ministry of the Environment guarding against uses of wastes that do not conform to the waste hierarchy. The facility was approved in 1999 and the licences did stand up to judicial review. In my analysis of this remarkable case I will occasionally refer to the RUN-project.

5.1.4 The role of various institutions and discourse in this case

The role of market parties, elected politicians and experts

It is clear that Holding Bos, later North Refinery, was driving the decision process through their site selection, design of the plant, and development of new plans with the new facility. However, their plans had to be approved by various authorities with various different interests. Within each of these authorities there is a certain division of labor between experts and elected representatives, with the experts, the first ones to engage in the debate with the company, in need of political sanctioning for their actions. Involvement of elected bodies was minimal throughout the entire case study

[28] The considerations in the licence suggested that the maximum acceptable quantity of wastewater be 110 cubic meter per 24 hours. As far as I could check, this in itself was an expansion of 10% in comparison with the previous licence. The licence conditions further state that the maximum acceptable quantity of wastewater is 200 cubic meters a day, should production levels go over 120,000 ton per annum. Again supposing that the previous licence accepted 100 cubic meters per 24 hours for a production level of 400,000 tons per annum, the allowed increase seems considerable!

[29] The new plans, partly developed in association with a company specialized in energy production units (Gybros PEC), were supported by North Refinery's new parent company, Thermo Electron Corporation. In the new facility, various waste streams (solid and liquid; hazardous and non-hazardous) would be used to create usable products and generate power for the power grid. Waste streams would be treated in six steps, at increasing temperatures (dewatering of wastes or gassification, cutting of solid wastes, drying of sludges, pyrolisis (500 centigrades), smelting (1450 centigrades), gassification of flammable liquids. As a result of the process, most of the wastes would become usable products and little rest products would remain. The total input of wastes would between 500,000 and 600,000 tons per annum.

period, although the Committee for the Environment in the Provincial Legislature of Gelderland in the end decided that plans would not be realized there. In Groningen, the issuance of licences was a task delegated to the provincial executive, and there was little discussion in the Legislature about the case. Only when the provincial decisions continued to be quashed by the State Council was there some concern. But the provincial executive strongly countered any interest by telling the one Member who had asked critical questions that he should not 'wrongfully accuse North Refinery' and that the State Council had only condemned the procedures that the province had followed[30]. Apart from the member of the provincial executive for environmental affairs, only the Minister of the Environment, The Queen's Commissioner for Groningen and the mayor of Delfzijl were personally involved in the matter, all in their capacity as head of the bureaucracy. The latter three shared party affiliation and this gave the mayor and Commissioner easy access to the Minister. They used this on various occasions to plead the case of North Refinery, most notably around the time of the Central Treatment Unit decision. The agenda of the local politicians was very much to advance the creation of jobs in the area and this, to some extent, explains the way things progressed. The Minister of the Environment was of course less concerned with this employment, for him treatment capacity and the Central Treatment Unit were major concerns.

There was a certain level of disagreement between the various experts involved in the process, with the strongest disagreements being between the Inspectorate and the municipality. My impression is that the disagreement between the experts in this case was much more a fight over the definition of the appropriate guidelines than a real substantive discussion. Take the issue of the storage tanks. Three consecutive certification firms refused to testify that the tanks met the relevant standards. The fourth firm did not contest this, but argued that the tanks were old and thus fell under different standards. The official agency government overseeing certification agreed to this interpretation and the issue was settled. The Inspectorate however continued to demand that these tanks should have floating roofs (these are required to channel pressure upwards in case of an explosion). Again the debate was whether guidelines require such roofs or not. In this matter the Inspectorate 'won' but it was a rather meaningless battle, because the tanks had long been placed on site when this decision was reached. Shining through all these debates is the issue of trust. The Inspectorate obviously did not trust the company because it had been warned of its wrongdoings elsewhere. However, none of the demands that the Inspectorate made was framed in

[30] Minutes of the meeting of the Environment and Water Management Committee of the Provincial Legislature ('Provinciale Staten') on 20 October 1997.

terms of trust. They were all framed in managerial terms, referring to technical standards. The debate between the parties had a somewhat artificial character because of this, because the real issues were not played out. In fairness it must be said that the Inspectorate was probably (and rightfully) convinced that the province and municipality would not entertain a discussion about trust at all. The Waste Branch of the Ministry did play out the trust issue after refusing to issue a CWA-licence. They did so before the State Council, which was by no means the right place to play this card.

The company, the province, the municipality and the courts often requested expert advice on the issues of contention[31]. It would be difficult to go into every issue but two issues seem remarkable. The first issue involves the ability of scientific advisors to actually sort out certain issues. Experts could not productively solve some issues. Especially the emission of certain substances to surface water is a case in point. Even though North Refinery was a unique case in the Netherlands, the RIZA experts set out to develop general emission norms for this type of plant. The RIZA experts failed in determining appropriate norms. Part of the problem was that, because of the presence of black list substances, Best Available Technology should be applied. RIZA could not determine which technology would constitute BAT because all technologies essentially failed to solve the problem so that the 'B' in BAT was hard to find. On the other hand, experts were apparently able to agree on questions of best available technology in the context of the environmental assessment report on the RUN-initiative. The Dutch government is in the business of regulating hazardous waste mainly in order to achieve a properly functioning waste market. The translation of that goal is managerial to an extreme extent. If a waste company (such as North Refinery) wants to introduce new treatment capacity, it must study existing treatment capacity, the demand for capacity, the technologies in use and demonstrate that the proposed plans meet demand and are more environmentally friendly than what is currently being used. Although the requirement leads to excessively large environmental assessment documents, with the results strongly influenced by the various assumptions being made, it must be said that the experts involved in this debate, especially from the side of the Waste Consultation Organ, were very critical and made the approach work. On the other hand: the added value of the approach seems somewhat limited as very little waste streams that the company wanted to treat were refused in the end. The second issue concerns the tendency of the Dutch courts (in this case just the State Council) to strongly rely on experts. This tendency is confirmed in this case, where experts have practically

[31] See e.g. the advice from the RIZA to the State Council, dd. 19 February 1993 (no RABB/PB/1542) and the advice from the Advisor Environmental Appeals, dd. 26 April 1993 (no. BABM/27134/S).

determined the outcomes of court cases provided the State Council did not find a procedural error. In chapter two I have noted that experts often make assumptions and that going along with the experts thus implies acceptance of these assumptions. In this particular case, it can be observed that especially in the appeal against the first Nuisance licence, such assumptions were made, but that they were not very realistic. The appeal was a somewhat strange affair, as the licence under discussion did not accurately reflect the plans that North Refinery had at the time, but reflected the plans of Holding Bos (that were different). An official appointed by the Minister of the Environment advised the State Council on the issue. The expert concluded that many of the conditions that the original licence had contained were no longer necessary because North Refinery wanted to build a far simpler plant than Holding Bos. Provided that the licence would be changed, the expert saw no need to quash it[32], a somewhat contradictory advice in itself. During the hearing before the State Council, the municipality indeed indicated its desire to bring the licensing conditions in line with the existing situation. The State Council consequently advised the Minister to deny the appeal and on 2 December 1987, and the Minister followed this advice[33]. However, instead of fully adopting the licence to the existing situation, the municipality subsequently repealed only the floating roof condition, thus weakening the licence. The expert's assumption of a tailored licence was thus incorrect and the licence remained way 'too encompassing' for the firm.

The role of the courts

From the start, the case has been before the courts. Especially the involvement of the State Council, the highest administrative court in the country, is remarkable. All issued licences, but also all refusals of licences and quasi-licences ('gedoog' licences) ended before the court. The outcomes of this large amount of activity are rather unimpressive. The verdicts of the court in this case were in no sense decisions on the substance of the matter, although indirectly they had the effect of allowing the company to continue its operations. Despite requesting expert advice quite often, the court itself seems to have looked for, and found procedural mistakes practically every time, except for the licences to the RUN project. Through its verdicts, the State Council did clarify certain legal aspects such as the requirement of environmental assessment but achieved little in closing the issues. With the province and the municipality withstanding much of the pressure resulting from the damning verdicts of the court, the actual effect of the court was rather limited.

[32] Letter from the Minister of the Environment to the State Council, 30 March 1987.

[33] 'Koninklijk Besluit nummer. 1987/60, 2 December 1987'.

The role of the community

Members of the ordinary public were largely absent from the decision process in this case. Only starting with some of the procedures that were implemented in 1992, two or three farmers and private citizens began expressing concern over industrial development in the area. The concerns were relatively generic, informative, and added virtually nothing to the largely technical debate although it was noted that certain facts that had been assumed in the environmental assessment report were not correct. A small concentration of houses was much closer to North Refinery than the makers of the assessment report had assumed. The RUN project did generate somewhat more participation, but then largely from Germany and not by ordinary citizens from there but from local and regional authorities. These authorities had been triggered by the intention to start pyrolysis of waste. A professor at the University of Braunschweig had said that this technology was not reliable and would lead to dioxin emissions above acceptable concentration levels. Because Delfzijl is close to the German border, this caused concern with these authorities. The approach of the province was to try and reassure the neighboring authorities, which could only be achieved by touring them at the plant, providing German translations of relevant documentation and talks about the technical aspects of the matter. Yet some of the authorities did object to the licences and asked De Werkgroep to also appeal the licence to the State Council on their behalf (remember everyone has this right). The appeal failed because of formal reasons.

My impression is that the lack of participation from ordinary citizens from Delfzijl must be explained by the fact that the inhabitants of Delfzijl had no problem whatsoever with Holding Bos or North Refinery. The population largely shared the economic development agenda of the administration. I have spoken to one of the citizens who participated in the procedures, Mrs. Remmerswaal. She told me that she was relatively new to the area and had assumed that she had moved to a house by the sea in a nice, rural setting[34]. This turned out not to be the case as the industrial complex of Delfzijl was having a relatively large impact on the surroundings. According to Mrs. Remmerswaal, the industrial complex provides for most of the employment in a town starved for jobs and its presence is therefore little discussed. She told me that she herself has received negative reactions from other community members about her participation in the decision-making process[35]. She also added that she had planned to appeal the Recycling and Utilities North project to the State Council but was stopped by financial reasons; for her the Dfl. 500.00 required for such an action was prohibitive.

[34] Interview with Mrs. W. Remmerswaal, Delfzijl, 20 April 2000.

[35] Ibid.

Given the fact that the Chemical Waste Act - and the subsequent chemical waste regulations under the Environment Management Act - are concerned with regulating the hazardous waste market, it should be no real surprise that North Refinery's competitors also participated in the licensing procedures. Tanker Cleaning Amsterdam and Tankcleaning Rotterdam were among those competitors to argue against the expansion of treatment capacity for waste oils as it would undermine profitability of other companies operating in the field (notably they themselves). The impact of such representations was minimal. Certainly in this case of the RUN project, the life cycle assessments made provided sufficient ammunition to resist competitors' claims. Furthermore, the Ministry could refer to the fact that North Refinery was the only company of its kind in the North of the country and thus had an important regional function. The competitors did not take the issues further, although they hired a lawyer at some point. The reason may be related to the fact that one of the companies was closed down after criminal investigations and the directors sentenced to jail.

Real participation

The 'real' participation in this case came from the Werkgroep Eemsmond ('Working Group Eems-mouth'), a local branch of a nationally operating environmental organization. The group is concerned with the protection of the Wadden Sea (the part of the North Sea between the Friesian Islands and the mainland). As an outgrowth of its concern, it also plays a role in industrial development on the land near the Wadden Sea, including Delfzijl. Participation of the group was not made easy by the authorities involved. This can already be gleaned from the simple refusal of Delfzijl throughout the process to send documentation relating to North Refinery to the Werkgroep. The Werkgroep had to rely on notices in the State Paper ('Staatscourant') and a local paper for the Delfzijl area that is not widely circulated in the rest of the province. The group in fact missed the publicity surrounding the first licence application from Holding Bos and this would have been lethal to their opportunities to participate any further if the Inspectorate had not demanded a whole new licence application and had the group not noticed the second advertisement by chance. Participation in the decision was not enhanced either by the fact that legal procedures were so often bypassed and replaced with 'gedoog licences'. The Werkgroep had run into the 'gedoog' habits of the municipality and province on various occasions and decided to take the issue to the national level. The behavior of the Minister of the Environment in the North Refinery case was presented to friendly members of Parliament, which led to Parliamentary questions[36], and when these were not answered either, to the Ombudsman. The Ombudsman indeed took up the case and in the subsequent report, the Minister's behavior was severely criticized. The Ombudsman concluded that the Minister should have seen to the coordination of the procedures, that the minister should have made public the National Accountant Service and Inspectorate reports on the company (be it in modified forms), and that the Minister should have more actively enforced the licence. The Ombudsman used the case for a far-reaching consideration of the 'gedoog' phenomenon. While acknowledging the 'fact' that in some cases it would be necessary to allow a company to keep operating when the licences were not in order, the Ombudsman suggested that these situations should be exceptions. The Ombudsman pressed for explicit 'gedoog' decisions (open to appeal), involving conditions to protect the environment. Absolutely necessary for any 'gedoog' decision in the eyes of the Ombudsman were amongst others an exceptional situation, a valid licence application, and positive advice from the statutory consultees. It was clear that the Ministry in the case of North Refinery had not met several of these conditions. The Ombudsman report and associated publicity from the Werkgroep led to national attention and a governmental policy to end the phenomenon in the long run[37]. The government policy largely appropriated the Ombudsman's suggestions for good practice in existing 'gedoog' situations, and the State Council later used the Ombudsman-criteria to quash one of North Refinery's 'gedoog licences'.

The Werkgroep was the appellant to all licences that were issued and was the main initiator of the litigation that occurred. It was a frustrating experience however, as the verdicts of the State Council had so little real impact and came so late. When the last 'gedoog licence' that North Refinery held withstood review by the State Council, the Werkgroep expressed this frustration. '*The verdict implies that the province may allow an illegal discharge. This way the pollution control regulations are undermined. (. . .) The legal possibilities to fight this type of practice are (. . .) an empty shell*'[38]. The Werkgroep was one of the actors to warn the authorities that environmental assessment for North Refinery's plans in the early 1990s was

36 Letter Werkgroep Eemsmond to the Ministry of the Environment, Waste Branch, 21 November 1986; 'Tweede Kamer der Staten Generaal, Vergaderjaar 1986-1987, Aaanhangsel van de Handelingen no. 617, p. 1227. Vragen van het lid Veldhoen'.

37 The Second Chamber of Parliament met to discuss the issue on 6 March 1990. The government prepared a 'gedoog letter' outlining its policy on 28 May 1990 (DGM/B/JB nr. 0640055). The idea was that 'gedogen' as a phenomenon would end in 5-10 years. There are few signs that this is realistic even as of today, even though much has improved.

38 'North refinery mag blijven lozen op Waddenzee', Nieuwsblad van het Noorden, 14 September 1993, page unknown.

necessary. This was to no avail either, even though the European Commission confirmed the requirement in reply to questions that one MEP asked at the request of the Werkgroep[39].

The communitarian language was not present

Important for my analysis is that the Werkgroep - or any other actors involved - never referred to communitarian language throughout the entire decision process. Instead, the line of argumentation that the Werkgroep advanced was pretty much managerial. The group lacked the technical qualifications to go into the details of waste treatment, but used a common sense approach, its knowledge of the legal procedures, and some political connections (see the Parliamentary questions). To give some examples of the line of reasoning applied by the Werkgroep: their comments to the application for a licence under the Nuisance Act included the observation that process descriptions and a list of incoming substances were yet missing from the application. The group also noted that the application listed the allowed pollutants in the oils but not their maximum concentrations. Furthermore the Werkgroep doubted that the wastewater treatment plant would be effective in removing various water-soluble contaminants, they suggested the effluent would contain black list substances, and noticed a lack of safety measures at the plant. Finally the group observed - at that point already - that the plant would require various other licences and that the various procedures should be coordinated. Later, the group called attention to the possible violation of the 'proximity principle' (treatment of waste near their point of generation) if North Refinery was to expand, external risks created by such an expansion, and the appropriateness of the site in terms of national planning policy[40].

The effectiveness of the group's argumentation was close to zero, even though in various respects it was right. What the Werkgroep would have liked was a debate about the question whether Delfzijl was a good location for the facility, but the leniency of local and national plans prohibited this (the national plan grants special protection to the Wadden Sea but not the area near Delfzijl). The Werkgroep saw the environmental assessment process as possibly a good occasion to re-open the debate about the location of the plant but the Commission for Environmental Impact Assessment did not agree. Instead, the Commission recommended the province to require of North Refinery that the focus be put on the proposed treatment techniques and their suitability for various waste streams. In terms of siting, the company was to explain why Delfzijl was a good location but was not required to study alternative locations and the question received little

39 'Schriftelijke vraag nr. 488/93 van de heer Muntingh (S) aan de Commissie'.

40 See letter Werkgroep to the province, 23 June 1997.

attention. During further hearings (e.g. on the assessment report itself), the Werkgroep continued to push its agenda. But as Mr. Van Dijk, the then member of the provincial executive for the environment told me: '*At a hearing, much has been decided already. At the stage of consultations, positions have been taken*'[41]. Both Mr. Van Dijk and Mr. Op den Kamp, who served as an advisor to North Refinery for many years, were keen to point out that the Werkgroep was different from the 'ordinary citizens' who wanted to participate[42]. In their opinion, this difference was mainly related to the fact that the Werkgroep had the resources (a certain degree of technical knowledge and time) to gain insight into the proposals and to develop an argument concerning these plans whereas ordinary citizens had not.

There was some initiative to go beyond the law from the side of the proponent. North Refinery (not so much Holding Bos) on various occasions tried to approach the Werkgroep. The purpose was usually to stress the fact that good waste treatment, certainly of waste oils, was in the interest of the environment and that the plant would be safe. Obviously, the firm did not at all like the uncertainty that flowed from the constant legal quibbling. In April 1988, North Refinery approached the Werkgroep on the basis of this motivation. '*To avoid the problems of the past we would appreciate an opportunity to consult with you at an early stage, instead of having discussions afterwards (be it at the State Council or not) which are very time consuming and perhaps unnecessary*'[43]. In the case of the RUN proposals, the company even went further and promised the Werkgroep to keep waste shipments over sea at a constant level, despite the increase in capacity of the plant[44]. This did not have the desired result because the Werkgroep still went to the State Council to fight the licence, also on behalf of the various German parties. The motivation was far less great this time however. The Court warned the Werkgroep that formal proof was needed that the Werkgroep was allowed to represent the German parties. This proof was never produced (a letter with a signature would have sufficed) and because of this, the entire appeal was declared invalid[45]. One member of the group informed me that resources and motivation had played a role here. The Werkgroep's technical knowledge needed to fight the proposal was limited; secondly the Werkgroep's prime area of interest was the Wadden Sea, not hazardous waste treatment and therefore other priorities had come to the

41 Interview, Groningen, 9 June 2000.

42 Interviews in Den Haag, 16 May 2000 and Groningen, 9 June 2000.

43 Letter North Refinery to the Werkgroep Eemsmond, 12 April 1988.

44 E-mail message Werkgroep Eemsmond to North Refinery, 19 May 1999 and letter North Refinery to Werkgroep Eemsmond, 1 June 1999.

45 See 'Uitspraak Raad van State. Voorzitter afdeling Bestuursrechtspraak, no. 199901983/P11'.

forefront[46]. The licence was in the end only appealed by one competitor, who appealed about only a minor part of the licence. This appeal had not been decided at the time of writing this book.

5.2 Kaliwaal: how to take rational decisions and still end in Court

5.2.1 Introduction

The second case I have selected for study is yet another example of a long and protracted decision-making process that involved much litigation. Again it is difficult to single out a specific period and call that 'the' case study period. Like in the case of North Refinery, there have been discussions about various different proposals throughout a period of at least 15 years. The case is interesting for various reasons. One is that there was a certain level of competition between 'the' market and 'the' state for the greater part of the case study period, with a private party offering a certain location for waste disposal and the authorities refusing it. The language in use with the responsible authorities was rather managerial, but the scientific approach to siting that followed from it was wholly unsuccessful, and led to acceptance of private sector initiative in the end. The private sector, in order to make the proposal more attractive to the surrounding population, added certain 'nature development aspects'. This worked to convince some of the actors involved but did not lead to less resistance to the plan from a small but very determined group. The courts played an important role in the decisions that were made. Their decisions, like in the case of North Refinery, reflect certain weaknesses, specifically a delaying effect and an inability to actually solve the conflict.

The waste involved in this case is dredging sludge from the soils of the large rivers, specifically the river Waal, which is the main arm of the Rhine in the Netherlands. In the Netherlands, such sludges used to be landfilled or used for landraising, but since the early 1970s it became clear that the chemical characteristics of these wastes made them unsuitable for these purposes. Since then, there has been a discussion about what to do with these wastes, which in a decade or two evolved towards acceptance of landfilling of the waste in the surface water system. Part of the evolution was the development of a separate regulatory regime for polluted water soils and underwater waste disposal sites that is different from the regime that applies on land. The land waste regime is for instance far more hesitant with the landfilling of hazardous wastes. But are the sludges actually hazardous

[46] Interview with D. Stoppelenburg at the Werkgroep Eemsmond, Groningen, 20 April 2000.

wastes? One element in the discussions that I describe below is the question of whether this is indeed the case. The outgrowth of a separate regime for dredging sludges has implied that the definitions of the 'normal' waste regime hardly apply. The classification of dredging sludges is based for instance on four classes, with class I being the relatively clean sludges and class IV heavily polluted sludges. The word 'hazardous waste' is not at all involved in this classification. But it is important to note that class IV is a 'rest category' for which only a threshold amount of pollution is defined. In other words, there is a minimum value above which sludges are named class IV sludges, but beyond that value there is little discrimination. The proposals that I describe here involve class III and IV sludges. The sludges for which disposal is sought contain a pollutant, arsenic, in such concentrations that, under the land waste regime, they would be hazardous wastes. Combined with the fact that I could not identify proposals for hazardous waste landfill that met my case selection criteria (see appendix 1) outside the sphere of sludge disposal, I have considered it justifiable to include a case of sludge disposal in a study on hazardous waste siting. The specific case involves a pond near the river the Waal. This pond, named the Kaliwaal, was the result of sand excavation. The company that had created the pond, Delgromij, was under legal requirement to fill the pond up again, but by its sheer size (60 hectares) this was very difficult. The pond, initially a rather unattractive part of the landscape, was transformed to an important habitat for birdlife in the years after the excavation work had stopped and received protection under the land use planning system as such. Because of this, it became extremely difficult to use the pond for sludge disposal once Delgromij decided to pursue that idea. Figure 3 shows the site.

5.2.2 The various proposals made

Introduction

The table 2 provides an overview of the three proposals that were made since 1978 in this particular case. As with North Refinery, every proposal could have been studied separately, but the effect would have been somewhat artificial because of the connections between the various proposals. Such connections are made in various ways, amongst others through the various procedures that have continued almost throughout the entire period.

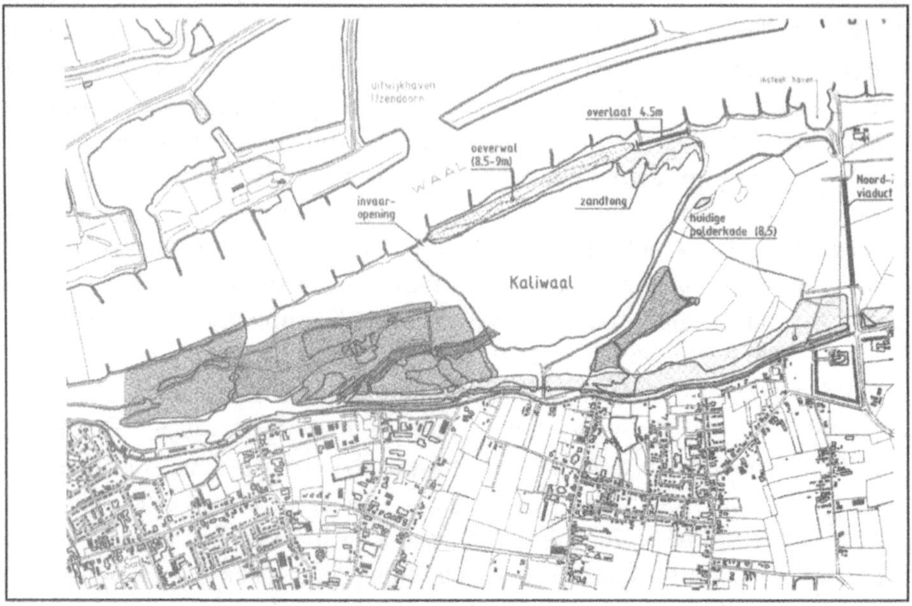

Figure 3. Location of the Kaliwaal, adjacent to the river Waal (adapted from the EIA-report).

Table 2. The Kaliwaal case. Proposals made in time.

Period	Proposal details and licensing authority
1978-1984	Disposal of dredging sludges from Rotterdam harbor. To be licensed by Rijkswaterstaat, Province of Gelderland and municipality of Druten.
1989-1994	Disposal of dredging sludges from the province of Gelderland. Exact size modified in order not to undergo environmental assessment, then modified again. To be licensed by Rijkswaterstaat, Province of Gelderland and municipality of Druten.
1995-2001	Fan of Trenches[47] project: sludge disposal and nature development. To be licensed by Rijkswaterstaat, Province of Gelderland and municipality of Druten.

The first proposal: Grontmij as a policy broker

The problem of contaminated sludges transpired first in Rotterdam harbor. The harbor was the source of sludges that had been used to do landraising in a polder that was used for agricultural purposes. When traces of pollutants beyond acceptable levels were found in the produce, the sludges were the prime suspect. Slowly more news about the issues transpired. Already in 1975 the responsible authorities had decided to study

[47] 'Waaier van Geulen' in Dutch.

alternatives for more responsibly disposing of the sludges. A Steering Group for the Disposal of Dredging Sludge was established in 1975 and this Steering Group asked Grontmij, a technical consulting company, to perform an environmental assessment (which was still an experimental instrument at the time) on the various options available. A draft assessment was completed by July 1978; one of its key recommendations was that cleaning the sludges was not a realistic option because of the costs involved. The report (which was heavily criticized by a supervisory committee) indicated that dumping in ponds or at sea was the most preferred option for these sludges. Only the most severely contaminated sludges would have to be disposed of on land. The assessment report mentioned a few possible sites on land and in addition a few ponds for the disposal of lesser polluted sludges. Among these sites were some in the province of Gelderland, which is the area where Grontmij's daughter company, Delgromij, had been active in excavation activities.

The company was apparently so inspired by its own report, that by 1977 it had concluded negotiations with the Ministry of Transport and Public Works and the city of Rotterdam to take dredging sludge for the purpose of filling up sand excavation ponds in Gelderland that Delgromij controlled. The contract was conditional upon the granting of all appropriate licences. Delgromij approached the province of Gelderland in 1977[48] and applied to the Ministry of Transport and Public Works for a licence under the Pollution of Surface waters Act 1971 (PSA) and a licence under the Rivers Act on 30 January 1978. The applications were accompanied by a report called 'Disposal of dredging sludges in a sand extraction pond near the Waal', which was akin to an environmental assessment document. In reply, Rijkswaterstaat (The Public Works Agency) informed Delgromij it would take one to two years before the application could be decided. Land use planning permission was required from both the municipality of Druten and the province of Gelderland, in its supervisory role of the municipality. The province was attracted to the idea of sludge disposal in ponds but was hesitant about the possible environmental effects and wanted to experiment on a small scale first. Landfilling of the Kaliwaal, which was to contain 10 million cubic meters of sludge according to the Delgromij plan, was too large to qualify as an experiment because the results would be irreversible. The municipality of Druten quickly requested expert advice on the proposals, which came in the form of a small research note, 'The usefulness of dredging sludge from Rotterdam for landfilling of the Kaliwaal', written by a small consultancy bureau from the area. The report was well-balanced and indicated that from a nature perspective much could be gained by filling the pond, on the condition that large parts of the water surface would remain.

[48] Letter Grontmij to Gedeputeerde Staten of Gelderland, 10 May 1977. Found in the municipal archives, Druten.

However, the report argued that if the pond was to be filled, then this should not be done with sludges from the Rotterdam area because there were too many uncertainties regarding the behavior of contaminants in the sludge under water. In response, the municipality took the decision to prepare a new zoning ordinance - that would forbid filling up of the pond - on 31 October 1978. The province approved the new zoning ordinance.

The province itself had become a member of the Steering Group for the Disposal of Dredging Sludge. In that Steering Group, the conclusion was that disposal of sludges from Rotterdam should take place there and not in Gelderland. However, just in case, the province of Gelderland was asked to identify suitable sites along the rivers in the province. The province followed this request and started studying the feasibility of 24 potential sites. The final results of the provincial study were not published until June 1983. The final version of the provincial study identified eight sites that were 'not immediately unacceptable', but the Kaliwaal-site was not among them. This was because of its exceptional ecological status - especially for waterfowl -, which had been confirmed by the National Institute for Nature Management ('Rijks Instituut voor Natuurbeheer'). According to the Institute, the Kaliwaal hosted more than 1% of the North European population of certain waterfowl; hence the pond met the requirements of the Ramsar Convention[49]. Despite quite a bit of a discussion between province, Delgromij, Rijkswaterstaat, the municipalities involved and various environmental groups, the provincial executive maintained that the Kaliwaal was not an appropriate site for sludge disposal. The licence applications to Rijkswaterstaat meanwhile had still not been decided. The Minister of Transport and Public Works had publicly confirmed his commitment to consultation with the province before granting the licences on 18 October 1983 in a session of Parliament. As the province was not minded to grant approval, it was unlikely that the Minister would authorize the licences.

Despite the continued negative feedback from the local authorities and environmental groups, Delgromij continued to forge its way ahead. It applied for a licence under the Rivers Act to excavate the natural dam between the pond and the river. The Minister of Transport and Public Works issued this licence (without consultation with the province or municipality). The company also applied for a planning licence from the municipality to start its landfilling activities on 24 October 1983. The municipality was quick to reject this application as non-valid because it was incomplete. Although normally an opportunity would be provided to restore any deficiencies in applications, the council felt that such restoration could not

[49] Convention on Wetlands of International Importance especially as Waterfowl Habitat, 1971.

be expected in time in this particular case and that the application could therefore not be decided within the legal time limit (one month). The apparent consequence was that the municipality would not take the application into further consideration. The line of reasoning is remarkable to say the least, but it was clear that the municipality would not cooperate. Delgromij's plans were then undermined by the Council for Water Management ('Raad van de Waterstaat'). This body advised the Minister that the costs related to transportation of sludges to the Kaliwaal (estimated at Dfl. 70-100 million) made the option 'not realistic'. Delgromij had tried to influence the council by pointing out that transportation could be done at lower costs, but this had no effect on the Council's position. Consequently, the whole idea collapsed, although Delgromij started litigation in the courts to keep its options open.

The second proposal

Pressure on the province to develop a policy regarding sludge disposal sites waned. This changed however as it slowly transpired that contaminated water soils were not just a problem in Rotterdam, but also in the province of Gelderland itself. This insight stemmed from the decision by the Regional Directorate of Rijkswaterstaat to make an inventory of the sludge problematique in the area. An alarming report appeared by July 1988. The report identified no less than 111 locations in the province where the river soil was polluted (class III) or heavily polluted (IV). Thus Gelderland now had a 'domestic' ('gebiedseigen') sludge problem. The province and Rijkswaterstaat decided that they would cooperatively develop a policy to solve this problem. Rijkswaterstaat seemed to have preferred to start using the Kaliwaal as a disposal site, but provincial planning policies at the time did not allow sludge disposal anywhere in the province, except for the three sites that had resulted from the time of the previous proposal. These had by then been embedded in a draft policy for the river foreland (Concept-Beleidsplan Uiterwaarden). Rijkswaterstaat subsequently went along with the province in its desire to develop a very systematic approach to the problems[50]. Environmental assessment should be a key part of the approach agreed upon by the two authorities, because the instrument would imply ample opportunity for public participation, which would in turn create the public acceptance that had been lacking until then.

The cooperating authorities wrote various environmental assessment documents. When they started out in 1990, there was a certain pretence that the environmental assessment process would partly be at the strategic level

[50] 'Oplossingsrichtingen berging baggerspecie', Notitie van een ambtelijke werkgroep met vertegenwoordigers van de provincie Gelderland, de Inspectie Volksgezondheid en de Rijkswaterstaat directie Gelderland, Arnhem, Februari 1989.

(studying the various disposal options) but the Start Note nonetheless indicated that options other than underwater disposal would be too costly for sludge disposal. The approach in the end boiled down to a comparative assessment of ten sites for underwater disposal[51]. The Kaliwaal was not on the list. The results of the assessment process were published in 1992[52] and met with sharp criticism from the Commission for EIA. At the outset, the Commission had pressed the province to drop its preference for the ten underwater sites and add land sites instead. The Commission had also suggested that the province should study ways of reducing the amount of sludge offered for disposal. Both suggestions had been rejected by the province and declared 'out of order' in its decision on the Terms of Reference for the assessment work. However, this decision came back to haunt the province, because the Commission would not approve its assessment work. The decision was then taken to make yet another environment assessment with the simple purpose of comparing the ten sites. The assessment report stemming from this exercise was completed in May 1995 but by then it had become clear that the sites stemming from it were by far too small to solve the problems in Gelderland. The ten sites combined would not suffice to solve the problem. Added to that was the fact that in each of the most preferred sites, strong local opposition had arisen, with as a result that several municipalities even took the province to court, based purely on the fact that locations within their territory scored high in the outcomes of the assessment document[53]. The provincial executive essentially gave up in 1995 and decided not to take any initiative regarding the actual development of sludge disposal sites, but to wait for initiatives from the private sector instead.

The third proposal

Somewhat in the background of the assessment process, much activity had taken place meanwhile. In 1989, Delgromij had announced its intention to file yet another application for a licence for sludge disposal[54], this time for regional sludges. Like in the case of North Refinery, the legal environment for this decision was changing. On the one hand court verdicts seemed to suggest that a waste disposal licence would be required for sludge disposal, which had not been the case previously. On the other hand, the Minister of the Environment had expressed his intention to require environmental

[51] Remarkably, eight of the ten possible locations that the province put forward in 1989 had not been studied by the province in 1984

[52] 'MER baggerspecieberging Glederland', Rijkswaterstaat and Gelderland Province, March 1992.

[53] 'Beroep tegen rangorde locaties storten baggerslib niet mogelijk', De Gelderlander, 7 April 1999.

[54] 'Delgromij: slib in de Kaliwaal', De Gelderlander, 5 August 1989.

assessment for all individual proposed sludge disposal facilities by a ministerial decree[55]. In December 1989, Delgromij broke off consultations with the authorities and quickly finalized its application in what seemed an attempt to pre-empt an obligation to perform environmental assessment.

The application had the potential to undermine the cooperation between the province and Rijkswaterstaat, especially because Rijkswaterstaat was apparently minded to actually decide upon the application. Rijkswaterstaat requested the advice from the National Institute for Inland Water Management and Wastewater Treatment (RIZA) about the application. The RIZA agreed with Delgromij that the filling of the pond would have beneficial consequences for natural life in it and that extra landfill capacity was needed. The institute also confirmed that the groundwater at the site was already polluted and that landfill was likely to reduce spreading of this contamination because seepage from the pond would be less. However, the RIZA also noted that the pond was on the list of 'wetlands of international importance' that had been published in the official government paper (Staatscourant)[56]. The advice of the RIZA was that these interests could best be weighed against each other by environmental assessment. The licence decision should therefore be postponed until assessment had been completed. Rijkswaterstaat ignored the advice and announced its intention to issue a temporary licence for 3-5 years. The reason why the authority was planning on doing this is to be found in the fact that national sludge disposal policies were starting to de-emphasize treatment of sludges and emphasize landfilling, that is underwater storage. However, problems emerged between Delgromij and Rijkswaterstaat regarding the desire of Delgromij to excavate a certain quantity of sand from underneath the sludge layer already in the pond. Rijkswaterstaat would not approve of this[57] as this layer was functioning as a liner and sand excavation could possibly fracture the layer. The disagreement caused delay and the statutory time limit for determination of the application (seven months) was not met. In response, the company took Rijkswaterstaat to court. The State Council proved sensitive to Delgromij's arguments and in March 1991 Rijkswaterstaat's fictive refusal

[55] Letter from the Minister of the Environment to all water quality managers, 26 September 1989. Found in the Druten municipal archives. The requirement was conditional upon water quality managers reporting proposals to the Minister. Because the province and Rijkswaterstaat did not report, the minister was asked by Druten municipality to decree that environmental assessment was required. The Minister refused, and Druten went to the State Council to have this decision quashed. This appeal failed.

[56] Parliamentary questions were asked about the status of the Kaliwaal around that time and the minister of Agriculture, Nature Management and Fisheries had replied that he was considering assignment of the area as special bird protection area; see: 'Braks: Kaliwaal Druten beschermen', De Gelderlander, 29 May 1990, page unknown.

[57] 'Rijkswaterstaat en Kaliwaal', De Gelderlander, 24 April 1991, page unknown.

of the licence application was quashed[58]. The Court held that the plans for sand excavation were not a significant alteration of the initial disposal plans or warranted a new application, as Rijkswaterstaat had argued. Rijkswaterstaat now announced its intention to issue a licence. By then however, it had become clear that underwater sludge disposal legally constituted waste disposal and thus required not only a licence under the Pollution of Surface waters Act (PSA) but also under the Waste substances act (Wsa). Grontmij was to apply for a licence under this act with the province. Because environmental assessment for sludge disposal sites was shortly to be legislated by a Statutory Instrument ('Algemene Maatregel van Bestuur') (the draft was public already), the province required that assessment indeed took place. However, the Statutory Instrument contained a threshold value, which made assessment not necessary for disposal sites below 500,000 tons in capacity. Grontmij subsequently altered its plans (and licence applications) to a disposal site of 475,000 tons. The province obviously felt that the facility would be expanded later (as the licence application under the PSA for 10 million tons suggested) and that environmental assessment of an already functional facility would be too late. The draft Statutory Instrument instructed authorities to take future expansions into account, and on that basis the province of Gelderland rejected the application in November 1992. Delgromij went to the State Council to appeal this decision. The State Council later quashed the decision made by the province, on the basis of a rather limited argumentation. '*The situation that appellant intends to expand the disposal capacity above the boundary where environmental assessment is mandatory can, now that this greater amount has not be made explicit in the licence application, not lead to the judgement that (. . .) a requirement exists to present an environmental assessment report with the application*'[59]. In reaching this decision, the State Council observed that incoming regulations that addressed the downsizing of proposals to duck the requirements had no legal force as yet. This decision did not have the intended consequences for Delgromij however. The province still refused to make a decision on the application on the basis of the argument that the application was counter to planning policies of both the province and the municipality. Deciding the application would thus be futile according to the province[60]. This decision was also appealed to the State Council as being quashed as an act against the law, which required expeditious decision-making. Delgromij's case seemed strengthened by a report from Grontmij (Delgromij's parent company). This report broadly

[58] Afdeling Geschillen van Bestuur Raad van State, 26 March 1991, no. G05.90.1008.

[59] 'Uitspraak, Afdeling Geschillen van Bestuur of the State Council, 8 March 1993, no. G05.92.2277.P090'.

[60] Letter of the Province to Delgromij, 26 May 1993.

applied the provincial environmental assessment method to the Kaliwaal and concluded that the site, if it had been included, would have ended in the top three.

Consultations between provincial officials and the municipality of Druten followed and resulted in an agreement that the planning system, specifically Druten's zoning ordinance, would provide better ground for the fight. With provincial support the municipality set out to strengthen the ordinance by inserting a prohibition of any activities that would reduce the depth of the pond[61]. The formal motivation was that the pond had an important function for birds during wintertime. By reducing depth, the pond would freeze sooner and this would harm birdlife. Despite objections from Delgromij, the provincial executive approved the new ordinance on 29 August 1991. This decision too was appealed to the Crown. Based on an advice from the Advisor Environmental Appeals, which suggested that the crucial function of the pond for birdlife in the winter was unproven, the State Council and consequently the Crown considered that the proposed ordinance was not proper for the fulfilment of its stated function. If the municipality wanted to protect the birds in the pond, it should use different means, argued the Crown. Specifically mentioned was the provincial authority under the Waste Substances Act (the State Council had not yet decided on the provincial refusal of a licence under this act at that time)[62]. Despite this defeat, the local executive was victorious, mainly because it felt that the Crown had implicitly indicated that the province should refuse the Waste Disposal Licence.

However, the province started to feel insecure about the wisdom of refusing Delgromij's plans. Also, Delgromij started to put far more emphasis on a pledge it had made long ago. This pledge implied the creation of an attractive landscape at the Kaliwaal and surrounding area. Previously the idea had not been made specific, but starting in 1991, Delgromij picked up on a plan from various nature organizations (Plan Stork, 'Plan Ooievaar' in Dutch) and the increasing recognition of the river landscape by the national government. Delgromij consulted with various organizations such as the province, the water board, the bird conservation society ('Vogel-bescherming') on the plan and had made a strategic alliance with the Dutch branch of the World Wildlife Fund (WWF). The alliance seemed to be founded on a network of high placed Grontmij personnel. The Chief Executive Officer of Grontmij was secretary of the WWF, and Grontmij hired various consultants in landscape design for the Kaliwaal project who had also prepared WWF's vision 'Living Rivers'. The Delgromij plan that was produced for the Kaliwaal was called 'Fan of Trenches', because the

[61] Ironically, the depth of the pond would decrease anyhow because of natural sedimentation.
[62] KB 93.006247, 30 July 1993.

landscape was to be dominated by a set of trenches in which riverwater could overflowe in case of high river levels. The trenches were expected to provide a good habitat for birds, but also for a range of other species attracted by a semi-wet environment.

The official 'opening' of the first trench in June 1994 was witnessed by the rank and file of provincial politics and received much positive press publicity. The new plans, along with the bankruptcy of the provincial approach, did not miss their effect. A letter from the province, written under the responsibility of a new executive, indicated that the province '*no longer intended to oppose the integral plans beforehand*'. If Delgromij were willing to replace its old licence application with one that reflected the nature development strategy, the province would take steps to adjust its planning policies, be it only in consultation with Druten municipality[63]. Delgromij responded positively and since that time, the company has worked its way through the various required procedures. The company started environmental assessment for the sludge disposal part of the Fan of Trenches project. The stated intention was to accept twelve million tons of class II and III sludges, but '*study will be made of the consequences of disposing class IV as well*'[64]. The assessment was not to focus on alternative locations, but only on the Kaliwaal. Assessment was completed in July 1996 and from then on licence procedures started. Licences were needed under a range of acts, including the River Act (effect of the disposal site on river tides), PSA (emission of pollutants to surface water), the Environment Management Act (waste disposal licence), the Soil protection act (as strange as it sounds: permission to not clean up the existing amount of polluted sludges in the pond), Soil excavation act (permission to dig trenches), and planning legislation (Land use planning act, Construction act).

The pattern of interaction was that Rijkswaterstaat and the province now backed Delgromij, as did certain environmental groups (including the WWF). The two local authorities involved (Druten and West Maas en Waal[65]) fought the proposals tooth and nail, as did several other environmental groups, various private citizens, and a group founded by such citizens, the 'Interest Group No Poison in the Kaliwaal'[66]. Each and every licence application went through every possible procedure, starting with hearings, representations to the responsible authorities, judicial appeal to the

63 Letter by Gelderland to Delgromij, 13 October 1994. Found in the Druten municipal archives.

64 Start Note, p. 13.

65 Initially the municipality was called Wamel. West Maas en Waal is a name given to the municipality after it had merged with other municpalities.

66 This group later changed to its name to the 'Society for the Conservation of Environment and Nature in Maas en Waal'. I will use the old name throughout this description.

Administrative Court ('Rechtbank') and to the State Council[67]. The parts of the procedures that were not related to land use planning were finally resolved by two State Council decisions in June 2001[68]. The State Council essentially left the licences that had been issued by then in tact. In the realm of land use planning, Delgromij was forced by the municipality of Druten to apply for separate planning permission and/or construction licence for each and every component of the facility, including certain components where such a requirement is highly unusual (e.g. a mobile weighing bridge). The municipality tried to pre-empt the procedures by formally deciding that a new zoning ordinance would be prepared[69]. This decision in itself has a 'freezing effect' on the situation, but then only for one year. Within this one year no new zoning ordinance was prepared, which then led to another 'preparation decision'. The municipality experienced several legal setbacks in the course of its activities. The municipality had requested expert advice from a private consultancy firm on the effects of the disposal site on birdlife. This firm argued that the facility would be detrimental to the bird population. Delgromij took Druten to court and the court threw out the municipality's arguments, amongst others because the existing zoning ordinance had prescribed a procedure for decision-making on proposals for activities in the pond and the municipality had deviated from this procedure by requesting advice from a private consultancy. Counter experts paid by Delgromij and statements from the Ministry of Agriculture on the natural status of the pond consequently blighted the expertise of the consultancy firm. Delgromij won the appeal[70]. It also won another court case on the question whether the municipality was allowed to look at the entire project (which would fill up the pond 8 meters beyond what the zoning ordinance allowed) or just at the first phase (which would remain under the maximum level contained in the ordinance). The Chair of the Court decided that Druten could not decide the planning application for the first phase on the basis of the full proposal, but should just look at the concrete proposal that was made at that time. The court verdicts largely took away the municipal arguments against the facility,

67 See e.g. 'Raad van State. Afdeling Bestuursrechtspraak, Uitspraak van de voorzitter, etc., 16 Juli 1999, no. F03.99.0172'. Or 'Raad van State. Voorzitter van de Afdeling Bestuursrechtspraak. Uitspraak inzake Kaliwaal/Nulonderzoek/Verzoek om voorlopige voorziening. No. 199901485/1/G6I, 9 November 1999'.

68 'Uitspraken Raad van State dd 20 Juni 2001, E03.98.0236/E03.98.0352/E01.98.0132-1 and H01.98.1772'.

69 Decision of the Council, 30 October 1997.

70 The judge concluded that Druten's decision had not been diligently prepared, had not reasonably balanced all interests involved, had not adequately checked the advice it had received, and had not adequately motivated its decision. He quashed the decision. See 'Arrondissementrechtbank Arnhem, sector bestuursrecht. Uitspraak op het verzoek om voorlopige voorziening van Delgromij, 3 November 1999, no. 99/1636 WRO 019 08'.

but on the other hand made it questionable that the second phase of the project could go ahead. The Chair in this case went so far as to temporarily grant permission to dispose of sludges in 1999. The court requested the advice of the Association for Advice on Environmental Appeal on the environmental consequences of the first phase of the project, and this advice confirmed the impression of the court that little harm would be done[71]. The full Court then confirmed the verdict of the Chair and ordered Druten to quickly decide the application now[72]. This decision was appealed to the State Council by the municipality. The State Council had by then seen more than enough of the matter and one judge publicly exclaimed: '*This type of pond is fun for lawyers*'[73]. The case was far less fun for judges because on various occasions opponents challenged judges, necessitating sessions of the full court to consider whether their colleague might have been biased. I was told that this move from opponents was taken in order to be able to challenge Dutch court verdicts to the European Courts.

5.2.3 The role of various institutions and discourses

Struggle between authorities and market party

The case just described can be viewed partly as a struggle between a private market party (Delgromij) and public authorities to establish control over sludge disposal decisions. Delgromij was and is a formidable proponent in the sense that (partly through its parent company) it has been present in the sludge disposal debate since the very early start, has excellent societal connections (such as with WWF), and was able to jump over all the legal hurdles that were put in the way by challenging them in court time after time. The company also had the financial resources to buy up most of the land surrounding the Kaliwaal so that it had at least some property rights over the land. Not all public authorities opposed sludge disposal in the Kaliwaal. Especially Rijkswaterstaat was essentially in favor. The province and the municipality however opposed the plans and particularly the province tried to take away the responsibility for siting from the private sector and develop a rational overall siting policy. Rijkswaterstaat essentially agreed to this approach for a while, but kept looking more favorably on Delgromij's proposals than the province. The municipality of Druten

[71] 'Stichting Advisering Bestuursrechtspraak, rapport StAB/34381/R, 12 Januari 2000', p. 11.

[72] 'Arrondissementrechtbank Arnhem, Meervoudige kamer bestuursrecht. Uitspraak in het geding tussen Delgromij, Gemeente Druten en de Stichting Behoud Leefmilieu en Natuur Maas en Waal, 28 April 2000, reg. no. 99/1635'.

[73] Ibid.

cooperated with the province in so far that it attempted to pre-empt Delgromij from choosing the Kaliwaal as a location.

Managerialism from the province

The provincial approach, if I may call it that, during the first two phases of the case study period was defended in managerial terms. The province rejected the first proposal of Delgromij by referring to the uncertainty surrounding the safety of underwater disposal. The province wanted to cooperate, but then only by performing some small scale experiments, which would then supply information about the potential risks. The province rejected the Kaliwaal as a site and had therefore no difficulty in approving the Druten zoning ordinance that was meant to prevent sludge disposal in the Kaliwaal. This zoning ordinance itself was based on an expert's report, which reflected the uncertainties that indeed surrounded sludge disposal at the time. Remarkably, a majority in the Provincial Legislature did not support the cautious approach by the provincial executive and this forced the executive to become very explicit about its approach. In a debate about the first proposal, some members of the Legislature questioned the sincerity of the Provincial Executive. '*On the one hand it is said that we want to cooperate, on the other hand the impression arises that everything is being done to make actual realization impossible*'[74]. The majority pressed for greater weight to economic considerations in the decision i.e. the financial gain that Delgromij could make. The member of the executive cautioned the members that the substances involved were certainly not innocent and suggested that Delgromij had not agreed to environmental protection measures proposed by the province. Perhaps reflective of its experiences with waste disposal on land, the province had amongst others demanded liners below the facility, pre-treatment to lower water content, and a water treatment system for leachate from the sludge. Delgromij had indicated that such precautionary measures would be economically unfeasible and were unrealistic. When confronted with the majority's opinion that supported Delgromij, the member of the executive denounced the conservative pluralist element in the equation. '*Certainly in the environment committee, economic arguments may not decide the issue*'[75].

At the time of the second proposal, the provincial position changed somewhat in the sense that it now recognized that polluted sludges were also a problem in its own area. This however did not lead to an acceptance of a role for Delgromij in the provision of a solution. Instead, they made a very ambitious effort to identify various sites that could accommodate the need

[74] Report of the provincial environment committee meeting on 30 September 1983, p.8. Found in the Druten municipal archives.

[75] Report of the provincial environment committee meeting on 30 September 1983, p.4.

for disposal capacity within the province. Provincial officials convinced Rijkswaterstaat that the various applications that were being made at the time involved the serious risk that the suitability of various locations would not be compared, that no assessment of other options than disposal would take place, and that a large number of disposal sites would spring up with a potential for pollution in each case. The officials indicated that the then current approach, which implied that the authorities remained passive and followed proposals from the market sector, had serious downsides such as a limited level of control over what would happen, and was therefore probably a non-optimal solution.

The courts and the Commission for EIA make the managerial approach impossible

Based on this managerial line of thinking the province tried to keep Delgromij outside the door, but was essentially hampered by the fact that the Dutch legislation is not at all based on this managerial idea of 'positive planning'. Delgromij was effectively able to challenge the actions of the province by simply filing licence applications and demanding that they be decided. The courts, specifically the State Council, supported the company in this approach simply because the law requires that decisions be taken on individual applications and the State Council wanted to uphold this rule. The ideal of managerial decision-making was invalidated in the end by the fact that it was not producing sufficient results. The problem of contaminated sludges grew and grew during the case period, and the sites that the province was considering would not be sufficient to dispose of them. Contributory factors to the demise of the managerial approach were the critical comments from the Commission for EIA, which took the managerial pretense quite seriously and continued to require that the EIA would also be about strategic issues, as the province and Rijkswaterstaat had wanted to do in the beginning. However, particularly the limited budget available nationally for sludge disposal and relatively positive experiences with certain facilities implied a practically complete dependence on underwater storage for sludges and essentially pre-empted a debate about alternatives such as land-based disposal and treatment. It seems that a certain expert consensus emerged on underwater sludge disposal during the case study period. Whereas many experts had initially expressed doubts about the practice per se, the discussion moved towards acceptable ways of doing it. An expression of this development is formed by the emergence of the so-called IBC-criteria ('Isolatie, Beheer en Controle', Isolation Management and Control) specifically for underwater sludge disposal sites. The failure of the managerial approach towards siting can be read from the fate of the three environmental assessment reports that the province and Rijkswaterstaat produced on sludge disposal. The Commission for EIA did not accept the first two and the third was essentially useless because of the sheer size of the problem. Some of the credibility of the reports was undermined by Delgromij's assessment of the Kaliwaal by provincial criteria.

Retreat from managerialism: return to normality

The province in the end retreated from its commitment to a managerial approach to siting. It could save face however by various gestures. The first was the insertion of the (essentially illegal) condition that sludge disposal would be linked with nature development. This condition implied a certain attention to distributive equity (the people around the Kaliwaal should receive something in return for the nuisance from sludge disposal). A second gesture was the public commitment of the province to consultations with the local executive from Druten. This gesture is in line with the third, which was the expression of a provincial commitment to stringently supervise the proposal and see that everything would go according to the letter of the law. A member of the provincial executive said during a hearing '*I can assure you that at the provincial offices we are working in a conscious and honest fashion, with knowledge of the most recent technologies and scientific insights in this field*'[76]. The commitment to consultation with the local authorities was not really influential in what transpired later[77], but the commitment to sound procedures and outcomes was real. The dominant mixture of the managerial and conservative pluralist discourse embedded in Dutch public law was thus re-established, largely by the inability to make the alternative work. The slight progressive pluralist accent of compensation in the form of nature creation was a concession that Delgromij decided to make at the suggestion of the province. The gesture allowed the province to save face. Reflective of its extra-legal character however, Delgromij threatened on various occasions (especially during negotiations on licence conditions) to withdraw the nature development plans if it did not get its way[78].

Because the political will to approve of the proposal was now present, the discussion could now return to 'normality', which implied a great role for various kinds of experts. The discussion about the proposals took place in the context of both environmental assessment and licensing procedures. The

[76] Report of the hearing on 16 September 1997, p. 2.

[77] This is in part due to the fact that when the province requested its advice, the local executive did not completely oppose the licence, but suggested that a municipal opinion was dependent on various factors that were as yet unknown. See the letter municipality of Druten to the province of Gelderland, 30 July 1997.

[78] E.g. in the sphere of the PSA-licence discussions, Delgromij pressed for a longer operating season than Rijkswaterstaat wanted to allow. Delgromij threatened to abandon the nature development plans if the 'disposal season' would not be lengthened. Letter Delgromij to the province and Rijkswaterstaat, 18 September 1997.

most important topic was the question whether (and later how) the facility could meet the IBC-standards. Other relevant topics included the washing away of pollutants from the pond (which was in contact with the river at high tide) and the effects on the bird population, Regarding the IBC-criteria, the Kaliwaal is above a major groundwater acquifer, which was a factor pleading against the location. However, there already was a certain quantity of sludge in the pond, which had compressed into a clay-like substance. Delgromij's argument throughout the assessment process was that the existing sludge layer would prevent emissions of pollutants, certainly when combined with clay liners on the side of the pond. Remarkably, there was already a certain level of groundwater contamination on the site, which was caused by the fact that the existing sludge layer was contaminated with pollutants that had been set free by the compression process[79].

A largely technical discussion by experts

The ensuing discussion largely focused on the effect of seepage from the existing layer because of the additional weight from the to-be-disposed of sludges. Discussion on this issue initially involved mainly Rijkswaterstaat and provincial officials with representatives of Grontmij, but later the calculations had to be presented to the Commission for EIA and the drinking water company as well. It appears that Delgromij needed a certain amount of prodding before the environmental assessment report was to the liking of the authorities. Even before the Commission for EIA looked at a report, the province and Rijkswaterstaat were expressing their dissatisfaction with the way Delgromij interpreted the terms of reference[80]. After much additional work the report was presented at a public hearing, and sent to the Commission for EIA along with the comments made during the hearing. The Commission was quite critical of Delgromij's work and threatened not to approve the report: '*the assessment report in general is too one-sided and optimistic about uncertainties that are connected with the (positive and negative) environmental effects which are related to the activity*'[81]. The homework given by the Commission in the end caused a delay of six months for Delgromij. Mid 1997, the Commission released its opinion that information in the assessment report was then sufficient for the responsible authorities to take a decision. However, the Commission's report reveals a certain level of dissatisfaction with the report as the authorities were advised to study the suitability of the Kaliwaal versus other sites and to require more

[79] Traces had been found of TCB (tri-chlorine-benzene), which is one of the more mobile micro-organic pollutants.

[80] See e.g. the letter by the province of Gelderland to Grontmij, 16 February 1996. This letter expresses dissatisfaction with a.o. the study of alternatives for Delgromij's own proposal.

[81] Draft advice on the environmental assessment report, 15 January 1997.

studies of the geo-hydrological situation on the spot. The authorities, confirming the defeat of the managerial approach, largely ignored the first recommendation. '*Because the proponent only has one location to offer, it can not be reasonably asked of him to compare his locations against other possible locations. For the province, the relevant question is whether the Kaliwaal is suitable for sludge disposal (and thus not whether the facility is best suitable), because the Kaliwaal is currently the only facility that has been taken into procedure*'[82]. The second recommendation was followed up on however. In the end, concern about the geo-hydrology of the site led to a requirement for additional tests by Delgromij and the hiring of counter expertise by the province to check Delgromij's interpretation of the findings[83]. This resulted in the conclusion that the northern part of the pond was not covered with a sludge layer. A licence condition was inserted to put a clay liner on the soil of the pond in the northern area and another condition to install a pollution absorbing sludge layer on top of the existing sludge layer throughout the pond. An important factor in the acceptation of this construction for the province was the fact that the drinking water company that had a pumping station at some distance from the Kaliwaal did not object to Delgromij's plans.

Regarding the issue of pollutants washing away, the RIZA had earlier indicated that the ideal depth of the pond was around six meters below water surface level. Because Delgromij wanted to surpass this height considerably, a dike was proposed around the facility, which would remain in place until the facility had been completed and capped. These proposals were accepted. On the issue of bird disturbance, Delgromij's parent company had expertise in house. Their expert suggested that no irreversible damage would be done to the bird population because the birds could migrate to the trenches or to other parts of the river foreland. Interestingly, the Ministry of Agriculture testified on various occasions that the value of the habitat for birdlife was marginal and did not warrant protection under the Habitat Directive[84]. Later the State Council established that the Ministry should have applied the Habitat Directive in the Kaliwaal, but agreed that the area that deserved Habitat Directive protection was far larger than the Kaliwaal, implying that birds could find refuge elsewhere in the Special Protection Area. The decision of the State Council to finally reject the appeals against the various licences was also largely based on expert opinion. The Association Advice on Environmental Appeals ('Stichting Advisering Bestuursrechtspsraak') was hired by the State Council to study the issue. Their report, based on the study of the file, various interviews with those involved, and independent

[82] Draft licence under the Environment Management Act, p. 15.

[83] 'Kaliwaal contra-expertise onderzoek', GeoCom, 25 September 1998.

[84] Letter to the provincial executive, 25 April 1995.

fact finding, was published in 18 October 1999. The report largely supported the issuance of licences[85], and the State Council refers to this report continuously throughout its verdicts of June 2001.

The role of the community

The Kaliwaal is located in the municipality of Druten, but nowhere near the population concentrations in that municipality. Instead, the site is much closer to small population concentrations that are located in the municipality of West Maas en Waal. The somewhat odd consequence is that the relevant population has to address a different municipality than the one they live in. It seems fair to say that only a small minority of the population of either municipality is particularly concerned about the proposals or is active in the decision process. Despite that being the case, ordinary citizens, some groups consisting of such citizens, and environmental organizations have been extremely active participants in the decision process. The first proposals by Delgromij already drew protest from West Maas en Waal. Citizens from that municipality united in the group 'No poison in the river bed' (Geen gif(t) in de Uiterwaarden). The group held various protest meetings and organized a petition in the various surrounding communities (Druten, Boven-Leeuwen, and Beneden-Leeuwen) which appeared to be effective in convincing the municipality that it should oppose the plans. The second proposal seems to have generated little response from ordinary citizens, which is logical in a sense because the province was not minded to cooperate with Delgromij, but the third led to responses from various individual citizens who later united in the group 'No Poison in the Kaliwaal'. Additional opposition came from various environmental groups, including the Society for the Protection of Birds ('Nederlandse Vereniging tot Bescherming van Vogels'), Reinwater, and the provincial environmental federation ('Gelderse Milieu Federatie'). Finally, opposition came from the farmers who used the land near the Kaliwaal and agricultural interest group GLTO ('Gewestelijke Land en Tuinbouw Organisatie')[86].

What connects all these groups is their acceptance of the managerial/pluralist approach to decision-making embedded in the law. The persons and groups involved never referred to the communitarian language and only to a moderate degree did they use progressive pluralist language. To give some, relatively scarce, examples of use of progressive pluralist arguments: the group 'No Poison in the River Bed' argued it was unfair that

[85] The report did mention the fact that in case of failure of safety measures, large groundwater extractions will be necessary at high costs. The report indicates that apparently the province is willing to take this financial risk.

[86] On the Environment management act licence, some 150 questions from 20 groups or individuals were asked.

excavation activities should be followed by waste disposal. The group felt that the region was being treated as 'a colony' by the rest of the country[87]. Years later, the group 'No Poison in the Kaliwaal' questioned the reason why sludge disposal should be the responsibility of a private company and not the government. Somewhat more managerially sounding, they also asked why the environmental assessment report had been written by Delgromij's parent company. Such a report '*needs to be written by independent experts. This Report however has been written by Grontmij, Delgromij's parent company*'[88]. Apart from these scarce outbursts, the action groups largely framed their actions in the language prescribed by the legislation. Take the first protest group. It presented its arguments against the facility in a research note, largely quoting scientific information. The group's arguments included the suggestion that contamination would leak from the pond through seepage and via the water that Delgromij was planning to leave on top of the sludge. The group argued that the Kaliwaal, although initially an eyesore on the surrounding landscape, had now developed into an ecologically important reserve. Surveys at the time of the proposal had indicated the presence of 149 sorts of birds, including certain red list species. The Kaliwaal was important especially for waterfowl and an important feeding and resting place for many other birds. The emphasis on expertise from the side of the opposition groups implied that the public hearings and even the court cases to some degree became fora for scientific discussion. In a comment on one of the public hearings, the local paper wrote that '*once again, as with the excavations for more than a decade now, experts stood against experts. What is absolutely safe for the one, is playing with fire for the other*'[89].

The reactions from the environmental groups throughout time do seem to mirror the development of expert consensus on sludge disposal in the Netherlands as initially there was much resistance to underwater disposal per se, whereas the emphasis was later put on the degree to which the standards would be met. Especially the group 'No Poison in the Kaliwaal' had certain members who had relevant expertise in this field. The chair of the group, Mr. Van Heck, was a forestry engineer who was engaged in several policy debates related to water pollution. Another member, Mr. Banken, was an agricultural engineer with specific knowledge about sludges. Together these people wrote an 'Alternatives Memorandum', which criticized the lack of independent contra-expertise in the environmental assessment process and presented various alternatives for sludge disposal, trying to create a discussion on the limited government budget for treatment of sludges. To this end, they invited representatives of a treatment firm that had the

87 'Slibinformatie', Geen gif(t) in de uiterwaarden, March 1979.
88 Letter to the editor by Mr. Van Heck, de Waalkanter, 12 December 1996, p. 4.
89 'Risico's zijn weggemoffeld', De Gelderlander, 19 December 1996.

technology to treat sludges but which was receiving little business from government as treatment was more expensive than landfill. The group played an important role in identifying certain weaknesses in the environmental assessment report, specifically in relation to the existing soil in the pond. Mr. Banken, for instance, noted that the assessment report described the results of various drilling tests (22). Most of them had been towards the West of the pond, where it could be expected that such a layer was indeed present. Mr. Banken however commented that only one drilling test had been done in the East Side of the pond, and that there no sludge layer had been found. Although I cannot be certain the Commission for EIA would have not noticed this omission without the input from the group, I do think it was the group that noted this first. The continuing attention to these findings, also in court cases, led to additional test drilling which revealed that indeed the existing sludge layer was not sufficient everywhere. Delgromij was forced to take additional precautionary measures including the installment of a lining in a part of the pond. The plan according to which this would be done consequently also became a bone of contention between the parties.

It seems that the group 'No Poison in the Kaliwaal' was so preoccupied with the technical debate that a certain level of mobilization of the population was considered less important. Members of the group did write various letters to the editor in the local paper in which they spoke of the project in a very alarmist tone, but these letters had little effect it seems. Perhaps this was also caused by the fact that the local paper wrote only sparingly about Delgromij's plans. Also, the fact that the proposals were couched in nature development terms and supported by the WWF gave Delgromij a headstart in the public opinion. Geerts (1999) found that the nature development part had a certain tempering effect on the perceived risks from the facility by residents of the area and on negative opinions about Delgromij's plans[90]. The letters in the local papers by the group were largely spent on attempts to break down this image. Instead of a nature development project, said the group, Delgromij's plans should be seen as a 'nature destruction project'. The group continued to stress the fact that the proposals involved hazardous wastes (even 'pure poison'), something that the responsible authorities did not agree about[91]. The only occasion where one could see an outburst of citizen concern however was quite late in the

[90] He also found however that the majority of the residents still had a negative opinion of the plans, in part because they expected that groundwater contamination by the project was unavoidable. In general, Geerts' conclusion is that nature development is not sufficient to change the positive feelings of the people in the area towards a positive one (1999:49).

[91] The authorities indicated that the wastes were not hazardous, and that only arsenic concentrations in some sludges were so high that they could legally be called hazardous wastes. But the norms regarding that issue were about to be changed said the authorities, plus natural concentrations of arsenic in the river ecosystem were already very high.

decision process, in 1998, after much of the licensing procedures had been (almost) completed. At that time, members of the disbanded group 'No poison in the river foreland' reunited and started mobilizing the villages specifically near the Kaliwaal (e.g. Beneden-Leeuwen). Their first target was the local executive of West Maas en Waal, which was lamented for not informing the public, despite the fact that all public hearings had been announced through the established channels. The executive was told that '*We have always thought that things would not be so bad. You should have better informed us*'[92]. The mayor also continued a story line that this municipality had been using since the early 1980s. He suggested the municipality had little expertise in this area and '*If all sorts of experts tell us there is no danger, then we must believe this*'[93]. This reply caused a huge outcry and was not believed: '*Let us not stick our heads in the sand: the Kaliwaal is on Druten territory but we will suffer from the consequences*'[94]. Pressured by the sudden resistance, the local executive of West Maas en Waal promised opponents an additional meeting on the Kaliwaal plan. Other authorities, which were invited to present their opinion, had great reservations about its usefulness, but still came. The meeting had an attendance of 500-750, and was a media magnet. Those (few) members of the audience who got to speak used the meeting as a forum to express their frustration and distrust in government: '*We should have been informed earlier. All parties in the council are against the plan. Why has no appeal been made yet to the State Council?*' and '*We will fight for our community!*'[95]. One member of the public, a representative of a small environmental group, lashed out at the other environmental groups: '*The WWF very cheekily betrays its members. Through their acts, the integrity of all nature organizations in the Netherlands is harmed*'[96].

The only visible outcome of the meeting was a list of questions asked by those present (about 100) and the fact that the municipality of West Maas en Waal joined Druten which had already appealed the licences to the State Council. The local executive said: '*That is our obligation to our citizens, although I want to refrain from causing unrest. You will never hear me say*

92 'Dorp wakker geschrokken', De Gelderlander, 15 January 1998, page unknown.

93 Ibid. The municipal council of West Maas en Waal (then called Wamel) discussed Delgromij's application for licences in the 1970s. One member of the council said: '*We should try to trust the experts somewhat more*', a statement that was to indicate that council could rely on Rijkswaterstaat to assess the safety of the proposals. See: 'Nu eindelijk advies' in: De Gelderlander, 19 November 1980, page unknown.

94 Ibid.

95 'Veehouders Boven Leeuwen worden gewoon uiterwaarden uitgedreven', Agrarisch Dagblad, 28 February 1998, age unknown.

96 'Mediacircus rond Kaliwaalprotest', De Gelderlander, 27 February 1998.

that we will be totally poisoned here. I still have that much faith in higher authorities that they will not make our area unlivable'[97]. Only some of the citizens' resolve to fight Delgromij carried over into the formal decision-making process. Because they had entered the decision arena so late, the group 'No Poison in the River Foreland' were not admitted to the legal procedures that followed, but this was solved by linking with the other action group. Both groups attempted to get 'busloads' of citizens to court hearings, but only at one time did a bus with 40 people actually attend.

5.2.4 Some reflections

This case is very different from the North Refinery case, yet it is similar in the sense that there is quite an involvement of the courts in the decision-making process. As with North Refinery, the courts have obvious drawbacks as an avenue for decision making, largely because they are rather slow and because they do not really solve conflicts. This case is an extreme example because not a single possibility for litigation has been missed, which implies that a certain cloud of uncertainty will continue to hang over the outcomes of the decision process until the European Court decides the issues. I personally doubt however that the action groups in this case will in the end get what they wanted from the court system.

Again the case demonstrates successful structuration of the debate by the relevant legislation as the conflict is largely fought out in managerial and conservative pluralist terms. The province of Gelderland attempted to overtake the initiative for site selection from the private sector but ran into large difficulties that are partly explained by the fact that the province could not impose a moratorium on proposals from the private sector. Also, the task of finding suitable sites by the scientific method was hobbled by the fact that national policies regarding sludge disposal were still being developed. This process was driven by new scientific insights into the art of underwater disposal and the lack of financial resources on part of central government. I find the development very remarkable because the issue of sludge disposal was initially approached by the use of scientific knowledge of disposal on land, but slowly moved to a separate system. This system can be called much more pragmatic e.g. because the landfilling of hazardous wastes is allowed (not so, or only in exceptional circumstances on land), because reuse or recycling is not very much at the forefront, and because eternal containment of wastes in the landfill is not a goal with underwater disposal (on land it is). The only argument given for such relatively arbitrary differences is the availability of money for disposal. Initially there was a concern with policymakers that if dredging sludges could be landfilled under water, that

[97] 'Onrust over baggerslib in Drutense vogelplas', Trouw, 26 February 1998, page unknown.

hazardous waste generators now paying high prices for e.g. incineration, would attempt to use underwater disposal instead. Such concerns are reflected in licence conditions and more or less informal agreements with Delgromij (they have promised not to take this road), but my impression is that this could become a real problem at some point. Returning to the managerial approach of the province; whereas national policies moved towards landfill instead of treatment, the environmental assessment process by the province had been started on the assumption that alternative disposal methods should also be studied and failed because of this assumption. Suppose however that this problem could have been overcome, then the province would have run into various other legal problems, amongst others related to its obligation to actually decide Delgromij's applications, and probably also problems related to planning policies and ownership of land at the 'most suitable locations' according to the assessment report. In a sense the province was lucky that a very powerful proponent such as Delgromij was more than willing to play the role of proponent in the end. This because I doubt the province would have had the resources and the resolve to go through the procedures that had to be run for the Kaliwaal, but then for four to five alternative locations.

After the failure of the provincial managerial approach, the decision process largely continued the way the law prescribes it. One relevant distinction is that the sludge disposal plan was reframed to a nature development plan, which normally would not have been required from the proponent. As a way to reduce resistance from the province this reframing worked quite well, and it also worked to some extent to lessen public resistance. The lack of concern from the neighboring villages may be explained in part by this fact. However, the strategy also caused a certain level of suspicion with the public, certainly later in the process when they discovered the 'real' purpose was actually to dispose of the sludges. The public discussion on the relevant licences (quite a few to be sure) took place mainly in a mixture of conservative pluralist and managerial discourse, with Delgromij often stressing the costs of certain measures including the construction of a liner, or the benefits of some options, e.g. removal of a sand layer under the pond. The responsible authorities meanwhile served as critical reviewers of proposals from the company.

My impression is that certain members of the public could actually follow the debate quite critically and could effectively participate. This is not through their claims for distributive justice, which are essentially just noted and have little effect on the licensing authorities or the courts. I think the added value of this participation is especially clear from the attention to the condition of the soil in the Kaliwaal. Citizen arguments about this soil have proven to be correct, which makes one wonder what would have happened if the disposal was started in the 1980s. In this managerial sense the decision has thus improved partly because of their participation. Their expertise, combined with the fact that the municipalities involved also gave technical arguments against Delgromij's plans, also implied that the courts have taken their arguments very seriously. Quite uniquely from an international perspective, the courts consequently hired 'counter expertise' to advise them. The consequence looks a bit like what happened in Canada with intervenor funding: the discussion becomes even more a technical debate.

5.3 Dordrecht: national policy under stress

5.3.1 Introduction

The third case involves the relocation of a plant to a new site in the same region. In contrast with the other two cases, this proposal generated little attention except from competitors, one individual and an environmental group. Their comments pertained mainly to the way the proposal would fit in with national policies regarding hazardous waste. The proposal did not seem to fit into the intentions of the Ministry of the Environment to stimulate responsible treatment of these waste streams.

The company proposing a waste treatment facility in this case, Ecoservice, is a company that had German origins and that initially was little more than a transport firm for its German parent. After changing hands to a different German company in 1990, and being renamed from GMU to Ecoservice, the company developed plans to expand the treatment activities it had assumed in the course of ten years or so (dewatering and physical/chemical treatment, partly by mobile facilities). The company, supported by its new German parent, wanted to start construction of a facility consisting of five halls, where all the activities were to take place (e.g. dewatering, sorting, neutralization, and filtration). The proposed plant would have a maximum treatment capacity of 130,000 tons per annum, but was envisioned to actually treat just 43,000 tons per year. The halls would be fully enclosed and equipped with a central air treatment system. Process wastewater would be treated before discharge to the municipal sewer system. The sewer system on the plant would keep the various wastewater streams separate so that they could receive a different level of treatment. The floor in the areas where waste-handling was to take place was to be covered with either impermeable concrete or a mix of steel and concrete that would detect and contain any seepage. The plant was a copy of an already operating facility in Germany and the plant's current safety record and the standards by which it was designed surpassed Dutch standards, which made the plant relatively expensive[98].

The company wanted to keep operating in the province of Zuid-Holland, but because of lack of space in its hometown, the expansion meant that the company would have to relocate. The company went looking for proper locations and was offered a site by the municipality of Dordrecht. This municipality owned some land on a peninsula intended for industrial use, which it was eager to sell as the municipality saw waste treatment as a growth business that would provide stable employment. Ecoservice liked the site because it offered sufficient land for the proposed expansion and because it was within the same general area within which the company was already operating. Also, the site offered good road connections, as it was less than a kilometer from a highway exit. Later a certain level of soil pollution was found on the site (which had partly been created by the disposal of sludges), but the municipality paid the bill for the clean up.

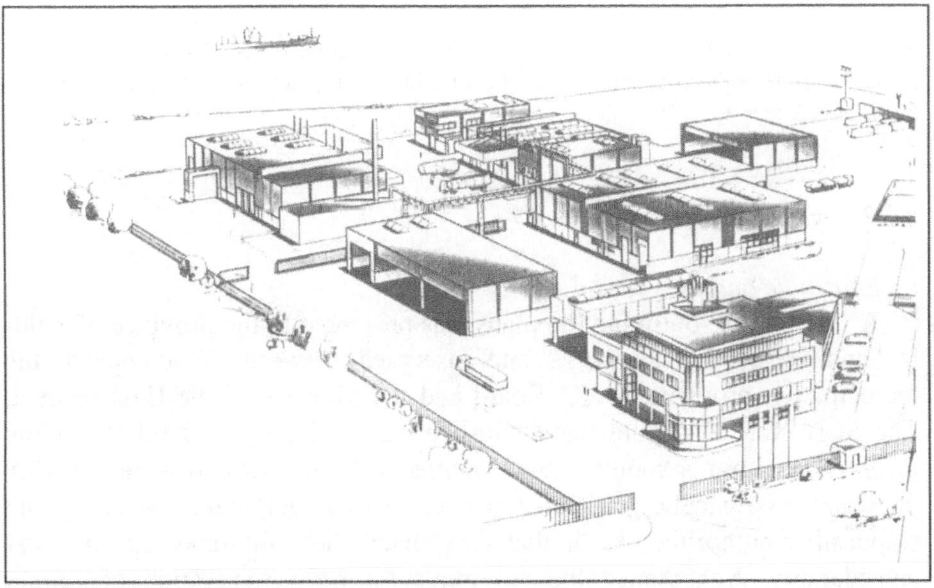

Figure 4. Artist's impression of the treatment plant. Adapted from the environmental impact assessment-report.

[98] Interview with L. van Veen, Dordrecht, 17 May 2000.

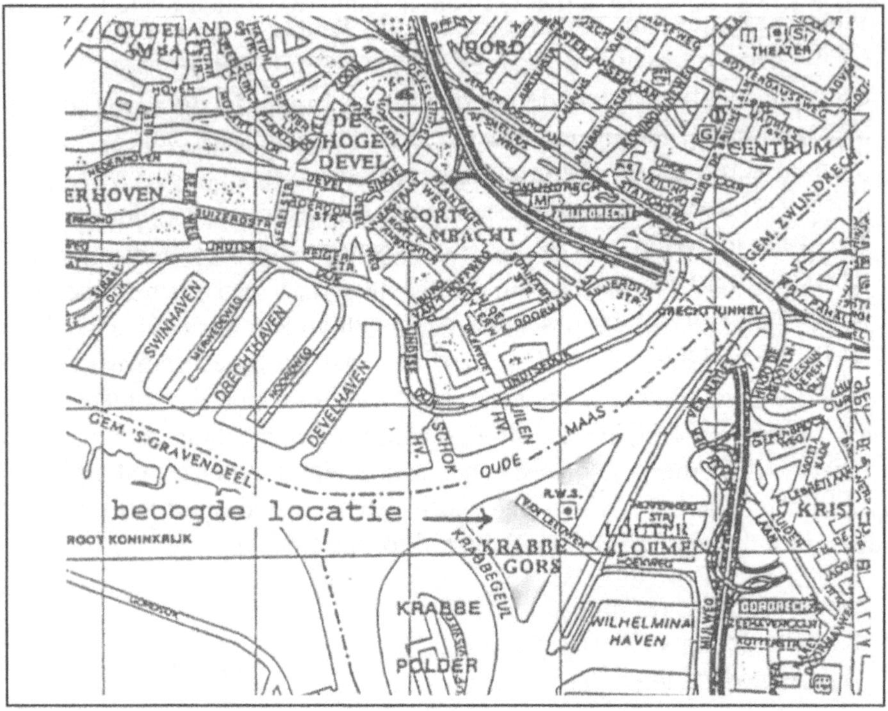

Figure 5. Location ('Krabbegors') of Ecoservice's treatment plant. The map shows Dordrecht to the East and Zwijndrecht to the North of the facility and is pointed out by the arrow. Taken from the EIA-report.

5.3.2 Formal procedures

Start note and first consultations

A 'Start Note' outlining the plans was presented to the province of Zuid-Holland on 28 January 1992 and discussed between the company, the municipality, province, Water Board and the Ministry of the Environment. The Start Note was publicized mid-February 1992 and put up for public inspection within a month. Two months later, the Commission for EIA published assessment guidelines, which were largely taken over by the responsible authorities, be it that they made them far more specific and considerably strengthened the requirement for the motivation of certain choices. The company was required to report on the various locations it had considered in the site selection process and on how these locations had compared to the chosen location in terms of planning and environmental policies. The province also required Ecoservice to indicate what was the state of treatment technology for the various waste streams it was targeting and how its own proposals compared with other possible technologies. For the wastes and products from the facility, their destination was to be outlined and in case of use as secondary fuel in cement ovens (Ecoservice wanted to blend certain types of wastes to make Secondary Fuel), the environmental consequences would have to be outlined. The latter requirement reflected the Ministry's negative stance towards waste export for incineration in cement ovens. The facility would fall under the post-Seveso guidelines and therefore an assessment of the external risks caused by the facility would be necessary. An advice from the RIZA at the request of the province concluded that discharges of black list substances from the plant were likely and careful attention should be paid to wastewater treatment. The company was therefore required to outline the wastewater treatment methods that would be used. Alternatives for Ecoservice's plans were to be studied by developing different combinations, such as different ways to implement certain parts of the facility (e.g. air treatment equipment, presses, and different treatment capacity) or adding additional safety and treatment measures. The alternatives (including a 'most environmentally friendly' one) were to be compared on various environmental issues (air, soil, groundwater, surface water, several nuisance forms, safety, ecology, visual aspects, etc.).

Environmental assessment report

The company set to work quickly and already by July 1992 there was a first draft of large parts of the assessment report, which were discussed at a meeting between the company and the authorities involved. There were many comments about the text but the authorities involved were essentially positive. The drafts kept being refined until 20 November 1992, when a final draft was presented, which was soon followed by a draft licence application. Some of the highlights of the assessment and applications include the fact that the idea to produce secondary fuel was presented as a good idea, amongst others because export would allow Dutch incinerators to take a more stable and reliable feedstock. The site selection was, despite the requirement to do so, hardly motivated by the company. The assessment report just stated that apart from the desired location, no alternatives had been investigated at all. A comparison of the existing location with the future location (also required) was hobbled by the fact that 'no data were available' on the emissions and risks at the current site, which is perhaps better interpreted as that such data had not been collected by the company. I also found it remarkable that both in the assessment report and risk assessment documents the operating capacity of the plant was put at 22,000 tons/annum, whereas the maximum capacity of the plant had earlier been estimated at 130,000. This capacity was attainable, but then only by introducing various shifts and the company suggested this was not economically feasible. Whereas this is perfectly acceptable, one would have perhaps liked to see a sensitivity analysis for the outcomes of the risk assessment, especially because the report on the external risks from the facility from (semi-government agency) TNO seemed to imply that there was a certain risk contour. The TNO-report contained a quantitative assessment of both group and individual risk, which suggested that the $1x10^{-6}$ contour (maximum acceptable individual risk) was at 250 to 350 meters from the plant, whereas the $1x10^{-8}$ contour (negligible risk) was at 600 meters. The implication was that the risk level at the nearest houses (one kilometer distance) was not above acceptable levels, however certain offices and companies were closer[99]. The discussion of alternatives for the intended activity, finally, was not very productive either. Many possible alternatives were 'written off' on the basis of qualitative arguments, specifically that they had 'disadvantages'. Arguments for not selecting alternatives included too high investment costs or lack of track record of alternative technologies in terms of operability. Such statements of course, reflect the expertise of Ecoservice and the consultancy firm that had prepared the assessment document. However, it is interesting to note that some of these statements were in contrast with the assessment report that was produced by North Refinery. Pyrolysis for example is written off in the Ecoservice report as a less operable technology because it was said to be susceptible to break downs. North Refinery's assessment report concluded differently. I found it remarkable that many of these aspects seemed to receive little attention and the assessment report was relatively quickly approved. The Commission for EIA did have some questions regarding odor and external risks from the facility, questioning some of the assumptions made in the assessment report (however not on operating capacity)[100], but the Commission judged the report sufficiently informative for further decision making.

Licence procedures follow

Licence applications were declared valid by April 1993, but the energy somehow went out of the process. By 15 March 1994, only a so-called 'concept draft-licence' under the Environment management act had been discussed with the company, but no public draft, let alone definitive, licences had been issued and the company wrote a personal letter to the member of the provincial executive responsible for environmental affairs. As a decision on the licence applications should have already been taken by September 1993, the company notified the province of the fact that damages were being

[99] Although not exactly spelled out in the risk assessment report, it seemed that individual risk for employees of the nearby company and a nearby office was within acceptable standards too.

[100] 'Toetsingsadvies over de inhoud van het milieu-effectrapport opslag, be- en verwerking van chemisch afval te Dordrecht door Ecoservice B.V.', Utrecht, 22 July 1993, no 407-71.

incurred and that the company was considering legal action against the province[101]. In the end, the draft licences of the Ministry and province were issued in March 1994 and put up for public inspection from the 5th to the 19th April 1994. Before the inspection period was over however, the company appealed the deemed refusal of the application to the State Council. The company requested compensation for damages and provision in the matter by the court in the sense that the refusal be quashed and the province be ordered to decide the matter by a certain date[102]. It seems that a mistake by the province had triggered this action by the company; the province had forgotten to place the draft licence under the Chemical Waste Substances Act from the Ministry with the documentation for public inspection. Such an error would be fatal to any final decision, and therefore the public inspection period had to be redone, causing additional delay. As licences for the existing plant were about to expire and requests for extensions seemingly lost in the provincial bureaucracy, the existing facility almost came to operate without a licence under the Chemical Waste Substances Act. The situation was restored to 'normality' with a 'gedoog licence'.

Contentious issue

In terms of content, the most salient parts of the draft licences are perhaps to be found in the licence by the Ministry of the Environment. A basic assumption underlying this decision was that the facility was not new. The Ministry noted that Ecoservice had been treating liquid wastes with mobile equipment and that part of this activity would be replaced by treatment at the new facility. As environmental precautions at the new plant would surpass those present at the sites where the mobile equipment operated, allowing treatment at the new plant was in the interest of the Dutch waste treatment infrastructure. Ecoservice largely obtained what it requested for, and the objections of its competitors were not really awarded. However, the creation of secondary fuel for cement ovens was an 'inferior treatment' method in the eyes of the Ministry, as the cement ovens where Ecoservice wanted to transport the wastes did not meet Dutch waste incineration standards. Therefore, this part of the application was refused.

As in the case of North Refinery, Ecoservice decided to incorporate various changes in the design of the facility after it had filed applications. The company wanted to add a storage place for diesel fuel, use different fire fighting equipment, and install a different air treatment system[103]. The

[101] Letter Ecoservice to the province, 15 March 1994.

[102] Letter Ecoservice to the State Council, 12 April 1994.

[103] Instead of a bio-gas cleaning system, the company was now proposing a bio-filter installation, which would use natural organic material and micro-organisms in it to clean

province chose to regard the company's request as a 'report of change to a facility', and notified the company that such reports could only be filed after a licence had been issued[104]. The provincial approach, in my opinion, was better than what the province of Groningen had done with North Refinery, but is still legally incorrect. This is because it was obvious that Ecoservice would not construct a facility that was in accordance with the original application and licences. The facility was indeed never constructed as had been outlined in the licence application and publicly discussed during public hearings. Instead, the company reported several changes to the then as yet to be constructed facility, which were then incorporated into the licence without a public procedure. The Water Board, taking the 'Groningen route' of the North Refinery case, had by then long accepted the change and consequently changed the PSA licence. This is an action that would normally require restarting the procedure as we have seen with North Refinery. Although arguably the changes were not detrimental from an environmental perspective, they created legal vulnerability to the licence.

Appeal

The only actor that filed an appeal against the licences was Ecoservice itself. The company, planning to sign the lease on the land in question in January 1995, had already filed an appeal against non-determination of its application. The company had already requested award of costs and now appealed the expiration date (after five years) of the Chemical Waste Substances Act licence as well. An important part of the company's case for requesting a licence that would be valid for longer than five years was a pending change in the Multi-year Program on Hazardous Wastes. At the request of the provinces, the Program was about to allow issuance of licences for 10 years in exceptional cases. The State Council proved sensitive to the argument. In its verdict[105] the Council noted that Dfl. 35 million was a considerable sum. Since the Ministry and province had both indicated they had no objections to a ten-year licence in extreme cases in discussion on the Multi-year Program, the State Council decided to declare the appeal valid and quashed the licence as far as it indicated a validity of only five years. The issue of award of damages was ruled in a separate verdict. During the hearing before the State Council in February 1995, the province suggested there was no proof of damages and that seven months

---

various organic and inorganic wastes and a moderate amount of water mixed with silicon-oil to wash remaining organic contamination. The proposed change is remarkable as the bio-filter option had been discussed in the environmental assessment report and was rejected there.

[104] Letter province to Ecoservice, 10 May 1994.

[105] 'Raad van State Afdeling Bestuursrechtspraak. Uitspraak dd. 10 April 1995, no. E03.94.1328'.

was too short a period to decide such a complex application. In addition, the province suggested that delay of the procedure was the Ministry's fault, that the delay was an entrepreneurial risk, and that Ecoservice itself had waited several months with construction activities after it had received a licence. The State Council disagreed with just about every argument provided by the province[106]. The Council was convinced that damages had been incurred as the shift of treatment activities from the old to the new plant had been delayed. The Council therefore awarded costs against the province, but as the amount of damages had not been accurately assessed, the judge ruled that further study be made of these[107]. After the court ruling, the issue of damages was settled outside the court between the Ecoservice and the province's insurance company.

Actual operation and change of government policies on secondary fuel

Construction activities were started in March 1995 with festivities and in the presence of prominent local politicians, who suggested that Ecoservice was an important asset to the municipality as it fulfilled an important societal task, pollution clean up, and as the company offered employment to the community[108]. The actual facility was opened late May 1996. Commercially, the facility proved to be a limited success until Watco, an emerging power on the Dutch waste market, took it over in late 1997. Ecoservice indicated to me that the take-over was the reason why the company was doing better and could actually expand[109]. A change of policy, allowing the creation of secondary fuel and using it in cement ovens will also have been helpful. By February 1999 the company expanded operations by introducing double shift operations, a move which had been portrayed as impossible in the environmental assessment report. The implications of the expansion on the level of external risk from the facility was not a factor in this decision as far as I could detect. In fact, risk seemed to be of very limited concern to the local authorities as they approved the presence of a boat which would function as living quarters for tens of refugees waiting for completion of the procedure on their final admittance to the Netherlands. This is the more remarkable because there had actually been a fire in both the old (August 1995) and the new plant (October 1997). The latter fire was quickly extinguished by company personnel using the anti-fire equipment[110], but

[106] 'Afvalverwerker verhuist later en wil geld', Dordtenaar, 22 February 1995, p. 7.

[107] 'Raad van State, Afdeling bestuursrechtspraak, Uitspraak dd. 31 Maart 1995, no. E03.94.0579'.

[108] 'Ten Veen slaat eerste paal voor Ecoservice', Dordtenaar, 11 March 1995, p. 14.

[109] 'Ecoservice breidt activiteiten uit' Dordtenaar, 23 February 1999, p. 5.

[110] See 'Vinkenswaard uren afgesloten vanwege brand', Dordtenaar, 25 August 1995, p. 1 and 'Geen gevaar na korte brand Ecoservice' Dordtenaar, 8 October 1997, p. 3.

still.... The province was clearly more concerned as it started enforcing the licences very strictly after the fire in the new facility.

5.3.3 The role of the various institutions and discourses

Institutional pattern

Again the law structured the debate quite well in this case. The decision-making process was largely a technical affair between company and authority experts, with some but not a whole lot interference from the courts. There was an attempt on part of the authorities to make Ecoservice justify their site selection, but the company largely resisted this. My impression is that the company actually produced quite a good assessment report otherwise and that the authorities overseeing the report were therefore somewhat mild in their review of it. Certainly there was no great resolve to interfere in the site selection process of the company. The municipality of Dordrecht had (and has) a very old zoning ordinance in place (adopted in 1932) and this ordinance basically allows any activity in the area, which of course does not enhance a siting debate either. One reason why the Commission for EIA especially asked about site selection seems to be related to questions asked by a group named 'Association Environment Group'. This group participated in the procedures on the terms of reference for the environmental assessment report. The group was not at all from Dordrecht but from Moerdijk, which is about 15 kilometers to the South, and interestingly the group apparently wanted to lobby for locating the facility in Moerdijk (where there is indeed quite a concentration of waste industry and other heavy industry already)[111]. The group, perhaps indicating the involvement of the waste industry in Moerdijk[112], suggested that the safety and environmental measures proposed at the plant should be compared to what had been required of similar facilities so that no unfair competition would result. The group only responded once, as did another group, the Zuid-Holland Environment Federation. This group did not really engage in the technical debate about facility design, but actively opposed the production of secondary fuel for export because it considered such export not to be in the interest of the environment. The Ministry of the Environment subscribed to this view and therefore, the group could safely become on-lookers after its contribution.

Generally speaking 'the community', that is, the population of Dordrecht, was not at all concerned about the development. Public hearings on the

[111] Letter 'Vereniging Milieugroep Moerdijk' to the province of Zuid-Holland, 14 March 1992.

[112] I have attempted to locate this group but to no avail. I therefore do not know about its composition.

environmental assessment report or draft licences were not attended, and only written responses were received. The tradition of heavy industry at the site, the proximity of the already very intrusive motorway and the relatively large distance to housing will all have played a role here. As far as there were reactions from ordinary citizens, they came from the other side of the river, where an odor problem in relation to Dordrecht industry existed[113]. A licence for Ecoservice plans would only be granted if it were absolutely certain that the facility would not cause additional odor problems. Later this person added that the facility should be operated by qualified personnel only and that the exclusion list (wastes that could not be brought to the facility) was too short. These remarks had limited impact. The same goes for a group of seven competing waste treatment firms that objected to Ecoservice's plans as they feared it would create over-capacity. They also suggested that Ecoservice's facility was not a comprehensive facility, which was apparently favored under ministerial policies at the time - to reduce transportation of wastes. The Ministry rejected these comments, largely because Ecoservice was already treating a certain amount of wastes, whereas treatment would improve because of the new facility. Finally, there was response to Ecoservice's plans throughout the process from their prospective neighbors. An industrial company, also located on the peninsula, had a range of concerns, amongst others about traffic to the site, but more importantly about the risk posed by Ecoservice's activities to its personnel. The company argued that the assessment report had underestimated risks to its employees, as it had not taken into account the risks related to human error in operating the plant. Its personnel were sometimes working outside, and odor emissions could potentially affect them. The company requested that measures be taken which demonstrably made the risk to its personnel negligible. It also pleaded for further risk assessment to take place[114]. These comments had no effect on the outcomes of the process whatsoever. In the case of all the participants, they had no desire to go to court on the case and therefore the only appeal was Ecoservice's own appeal. Neither of them used communitarian discourse, but expressed themselves largely in technical terms.

5.4 Conclusions

5.4.1 Introduction

The approach of this concluding section will be similar to the one adopted in the UK chapter. Some conclusions stand out. First, in all three cases there does not seem to be great concern with hazardous waste

[113] Letter, Mr. W. Bol to the province of Zuid-Holland, 5 March 1992.
[114] Letter Grootint B.V. to the province, 19 April 1994.

proposals among the general public. In the first two cases there was a lot of resistance but this really came from either environmental groups or from a limited set of people. Second, the amount of litigation involved tends to be quite large, with even the relatively smooth Ecoservice case ending up before the State Council. Because the courts do tend to request expert advice on the issues before them, the discussion in court also becomes quite technical. The courts themselves tend to take a hard look at procedural errors. Third, the legislation covering the decision process structures the debate quite well, at least in the sense that language spoken by all participants corresponds to the model of relevant language in the law, which is a mixture of managerialism and conservative pluralism. There are various exceptions of course, e.g. the managerial approach to siting in the case of the Kaliwaal, but the courts have forced authorities to adhere to the formal model. No reference was made to the communitarian language in either case. Fourth, the record in terms of 'technical competence' of environmental groups and private citizens is somewhat checkered. Both environmental groups and private citizens have participated in all three cases, be it that environmental groups are present somewhat more consistently. Whereas one would expect a certain greater degree of technical competence with environmental groups, it does seem that this is actually not that great. In fact, the participants who seem to have strengthened the decision process in technical terms are actually private citizens (in the Kaliwaal-case). The expertise available among 'ordinary citizens' was quite a force there and was a factor in discovering deficiencies in the proposals.

5.4.2 Cross case comparison

There are some remarkable similarities between the three Dutch cases studied here, but also some differences. I will discuss both while discussing the role that the various institutions played and the language used.

If we look at the importance of the various institutions in the cases studied here, I think that private parties, experts, elected representatives and the courts play the greatest roles, with the community relatively absent from the decision process. When looking at the role of private parties, one can see that site selection is really left to these parties, as is the definition of what type of facility is proposed. Government officials essentially check these choices, and in concert with elected representatives, they approve the choices or not. Participatory elements in the decision process are really aimed at these public authorities, not at the market parties, and are only one form of input in the decisions of such authorities, without much influence in reality. The courts provide recourse for a range of actors, most notably for the proponents themselves, but also for concerned citizens and environmental groups. It is my impression that various characteristics stand out in the way this constellation operates.

The first is that political will seems quite important in determining the outcomes of procedures, despite all the reference to managerial language and use of technical terms. Take the North Refinery case. There, the political will of the municipality and the province involved, effectively decided that the facility should be allowed and approved. This support was so great that the formalities of licensing were essentially forgotten and the facility operated for years without a proper licence. In a similar vein, take the Kaliwaal case. The political will to approve this facility was completely lacking until 1994 or 1995. Then the province changed its mind and progress started being made in the sense that the project will probably be realized. Various qualifications and a remark must be made here. The first qualification is that the 'political will' I spoke of does not seem to stem from representative organs, but rather from the politicians who head the bureaucracy in their name. The actual involvement of elected bodies in the decision process is rather limited. The second qualification is that there is somewhat of a difference perhaps between the provincial and the municipal level, at least in the case of the Kaliwaal. There one local authority (West Maas en Waal) explicitly framed its opposition to the proposal in terms of the will of the people. The mayor of that town preached trust in experts and higher authorities, but the local executive in the end, with the support of the council, did appeal the licences because it said the population wanted it. The assumption of a difference between the municipal and the provincial level is somewhat difficult to gauge in this case, in part because there was so little opposition to the proposals locally, but it is perhaps telling that in the one case where there was a bit more pressure, local government opposed the developments. One remark, finally, is that the political aspect in the decision process is hardly ever emphasized. The fact that the political will may actually determine outcomes is continuously de-emphasized and decisions are couched in managerial terms. One working rule is that politicians do not attend public hearings, a rule that is only broken in a limited number of cases. In the cases I have studied, the provincial executive only attended public hearings twice, once in the case of North Refinery and once in the case of the Kaliwaal. On both occasions, the purpose of the presence of an elected politician was to stress the responsiveness of the administration to the public, and to communicate trust in the administrators. Politicians seem to take great care not to get mixed up in the consequent debate at the public hearing. In the case of the Kaliwaal, the member of the provincial executive delivered a speech and then left.

The second remarkable aspect is the role of the courts. As we have seen in chapter 3, the Dutch courts have a tradition of thoroughly testing government decisions, which is supported by their ability to involve technical advisors. Such technical advice was called for on a number of occasions in the cases I have studied, and when this happened, this advice effectively determined the opinion of the courts in the issue at stake, except for the procedural bit. The reason why the courts would request experts advice seems obvious, as they do not have the technical expertise to understand certain issues. Yet it does give a somewhat odd impression that experts prepare an environmental assessment report, that is checked by government experts, then by experts hired by the Commission for EIA, and then once again by experts hired by the courts. Several factors seem to drive this 'expertization' of the decision-making process. One is the fact that proponents, and the consultants they hired for performing environmental assessment, are often not considered unbiased and so the idea is that their assumptions and calculations need overview. Second, the comments of actors such as the Commission for EIA are public and tend to provide opponents to the plans with ammunition. The cases show that it is not beyond the reach of most environmental groups and private citizens to read environmental assessment reports, comments from the Commission for EIA, and find weak spots in the assessment documents. The real issue then becomes what should be the consequences of such weak spots, and this question is put before the courts and their advisors. One consequence is that it is hard to get closure of the issue, which is also partly related to the fact that the courts do not really foster consensus. The courts are both a strong and relatively weak institution in the decision process. They are strong because they have quite far-reaching authority in quashing decisions, but they are weak in terms of the speed of decision-making and in determining the course of action once a decision was quashed. This can be a problem for both proponents (Delgromij) and for opponents (Werkgroep Eemsmond). In either case, the courts have tended to function as upholder of the law, making deviations from the law impossible.

Regarding the language used in the decision process, I have already noted the fact that the importance of the political will in these processes is conspicuously de-emphasized. The impression one gets from all three cases is that practically every party sees the decision process as something rather technical or is 'forced' to do so (structuration). Only in the case of the Kaliwaal was a certain level of progressive pluralist language used (references to distributive equity by the surrounding community). Such references are allowed but not rewarded in any sense with having any real impact upon the decision. Even though Delgromij's plans in this case could be defended in progressive pluralist terms (as there was a certain level of compensation through nature development), this line of reasoning was avoided during hearings, which were focused solely on sludge disposal. The comments on distributive equity were listened to and then the debate moved on to the 'real issues'. In this specific case, it would have perhaps been easier to present the nature development plans as compensation if they had actually been developed in consultation with opponents. Certainly, Delgromij approached its opponents in informal sessions (just like North Refinery did) but such sessions were hardly effective from the perspective of the company. Note that conservative pluralist language was absent from the debates too. Choices and decisions were rarely defended by referring to the fact that they were essentially a private market choice. There are a few exceptions, especially in the cases of the Kaliwaal and Ecoservice. In these cases, there was a certain pressure to motivate the choice of location from a managerial perspective. The belief that this is possible seems limited however, as the very actors that require such a motivation (the Commission for EIA, licensing authorities) do not stringently enforce the requirements. In the case of Ecoservice, the company essentially refused to assess alternative locations, which led to few problems with the authorities. In the case of the Kaliwaal, the province could have pointed at its own failed attempts to scientifically identify the most appropriate sites when asked why the Kaliwaal was the only site under consideration. Instead, the province said that the proponent had only one location to offer and could not be required to study alternative locations.

The implication of both the lack of pluralist and communitarian languages in the debates is the increased importance of managerial language in defending outcomes and choices. This is somewhat odd, because it is clear that the final outcomes are really an amalgam of market, political and expert considerations, not just expert considerations. However, it is apparently difficult to accept other lines of reasoning, whereas the 'expert considerations' themselves are certainly not neutral and subject to change. It does seem odd for instance that an environmental assessment report like the one prepared by Delgromij did not really indicate a problem with the soil of the Kaliwaal. It is remarkable that a certain air treatment system is completely written off in the assessment report of Ecoservice and then becomes the best environmental option months later. And what makes underwater landfill of sludges more and more acceptable whereas landfilling of other wastes is being banned in the Netherlands?

5.4.3 Theoretical feedback

Introduction

The set-up of this final section of chapter 5 is similar to the set-up of the theoretical paragraph of chapter 4. I will first discuss the implications that the case studies have for the discussion on the relation between discourse and interests and their respective roles in explaining the actions of participants. Secondly, I will examine some of the empirically verifiable assumptions embedded in the various discourses.

Interests versus language

One point of interest, also in the study of the situation at the country level has been the question about the relationship between language and interests. I have observed that in the three Dutch case studies, the language most used is managerial, whereas in reality there is clearly a political and a market element in the decision process. The non-technical elements in the decision process are thus not really publicly addressed but they remain under the surface. If interpreted from the distinction interests-language, this can imply various things. For example, the actors involved in the decision process may all think that their interests are best served by maintaining a managerial line of reasoning, or they may have difficulty thinking outside the boundaries of this discourse, or there is an active maintenance system that enhances the dominance of this discourse. I think there are indications for each of these possibilities, be it interrelatedly. To start with some observations about a maintenance system: it is clear that the Dutch courts actively uphold the approach embedded in the legislation and make deviations difficult. The courts for instance will not entertain arguments about the unfairness of certain proposals to the community. The courts adhere strictly to the law, and the law is there to protect the environment. Therefore, challenges to decisions must be framed in the words required by the law, because they do not stand a chance to win the case otherwise. It seems that the stricture of the managerial discourse is somewhat less tight during public hearings, and indeed it is on such occasions that progressive pluralist story lines are sometimes used and not before the courts. However, there too, arguments about distributive justice tend to fall flat and are mostly avoided by those present at the hearings.

But this still does not automatically mean that the managerial discourse is used purely because people think it serves their interest best. That is because communitarian and progressive pluralist language can supposedly also be used to mobilize the population and therefore influence the political will that is of so much importance. So despite being of little use within the legal decision-making system, the other languages could still help in achieving goals, especially for opponents. Indeed, from the perspective of an opponent, it is relatively unattractive to enter the technical debate, mainly because they have much less resources and information that the proponent. There appear to be two reasons why this is not tried. First, mobilizing the community if that community is not at all concerned (Dordrecht, Delfzijl and to some extent West Maas en Waal) is an unlikely strategy. Secondly, opponents themselves have almost always received scientific training and think themselves that the debate must be done on managerial terms. In this sense, the existence of competing discourse coalitions has not revealed itself in the three cases studied.

Feedback into some assumptions

As I have done in chapter 4, I will discuss the implications of the case studies for some of the assumptions embedded in the three discourses. I will do so by discussing the role of the various actors in the decision process and drawing a comparison with what was expected from these actors.

(1) The role of experts

The three expectations about experts were, on the positive side, that they are objective and neutral and will separate values from facts. On the negative side, the expectations were that experts would not be able to separate values from facts, and secondly that they will deliver the results that the persons paying for them want. It is my impression that the three Dutch cases do contain some evidence about the role of experts. The use of expertise in Dutch practice is widespread, but so is the use of counter expertise. I think that the degree of suspicion about the independence and neutrality of experts that is embedded in the use of counter-expertise is warranted. Especially the practice of environmental assessment, which is a process operated by experts, commissioned by either a proponent (North Refinery, Delgromij, or Ecoservice) or by an authority (Province of Gelderland, Rijkswaterstaat) gives indications to that effect. The experts performing environmental assessment seem susceptible to the influence of the actor commissioning the research. Take the different reports on the value of the Kaliwaal for birdlife. The municipality's advisor suggests this value is great, whereas Delgromij's advisor say it is very limited or would not be harmed. Take pyrolysis as a technique for waste treatment. Experts hired by North Refinery, intending to use the technique, describe it as very practicable and dependable. At roughly the same time, the experts employed by Ecoservice, intending to use a different technique, describe pyrolysis as undependable.

I have encountered several formulations in all environmental assessment reports that I have read for my case study work that testify to an approach of doing biased research and then writing the report slanted towards a certain outcome. For instance: why were practically all the test drillings that Delgromij made of the Kaliwaal in the side of the pond where the sludge layer was thickest. Why did the environmental assessment report consequently arrive at calculations of very favorable numbers on seepage (the assumption was that the sludge layer was not present on 5% of the pond)? Why did the environmental assessment document on Ecoservice's facility assume only a one-shift operation, whereas in reality it quickly became a two-shift operation? In part, the answer to such questions must be that the experts performing the assessment work were biased towards the proponent. The supervision by other experts is thus warranted, but these 'counter experts' are very much bound by what a proponent presents them with. Proponents can be pushed for additional work, but at some point this process is at a halt as not everything can be studied. This is true in practice and is probably the reason why the Commission for EIA is critical up till a certain point only.

Thus, I find there are indications that experts are not as objective and neutral as the managerial discourse expects. Another tenet of that discourse, which is the separation between facts and values is, ironically enough, corroborated to some extent by this finding. This is because many experts lend themselves apparently to substantiating the opinions that are provided to them externally. In this sense, Frank Fischer's (Fischer, 2000) arguments about technocratic decision-making being full of values are very true. Especially assumptions about how a plant will be managed (e.g. how long will the working hours be) are essentially requests for trust in the fact that the assumptions will come out. The idea that experts will deliver the results that the people who pay want is thus confirmed by experts who are paid by the private and the public sector. It is somewhat less true, at least I have no indications to that effect, about experts in the service of the Commission for EIA or the courts. However, the importance of assumptions applies to these experts as well. Take the advice of the Advisor on Environmental Appeals on the appeal against the first licence under the Nuisance Act for Holding Bos. This advisor essentially assumed that Delfzijl would make the licence more stringent, reducing external risk for instance. In reality however, the licence was made less stringent. And an issue not identified in my chapter two is the fact that the common opinion among experts (e.g. towards underwater disposal of sludges) can actually change quite rapidly. The process in which this occurred largely involves the hiving off of sludges from the 'ordinary' waste regime, the blocking of alternative options by budget restraints, and providing reassuring news about the safety of such disposal. I would strongly hesitate to call this development value neutral.

(2) The role of market parties

The assumptions formulated on market parties were on the one hand that they would know their own interest best and will defend it if need be in court or in politics (pluralism). On the other hand, it was assumed that they define their own interest so that it includes the interests of others (pluralism - the harm principle). More negative assumptions were that market parties will have difficulty in defining their interest in complex issues and will not think of anybody but themselves.

In regard to the cases of interest here, I will distinguish between proponents and others (e.g. landowners). For certain, the proponents of the facilities that I have studied did know their own interest quite well and most certainly defended this interest in court. I think that to some extent, they were open to the interests of others, take e.g. the inclusion of nature development in the case of the Kaliwaal and the reference to the creation of employment in the case of North Refinery. However, such references are very much ad hoc and instrumental. Noor (1999) for instance, on the basis of interviews and document analysis, comes to that conclusion about the plans for nature development in the Kaliwaal. Proposals that take the interests of others into account are thus best qualified as enlightened self-interest. Other parties than the proponents fall in various subgroups. Speaking of landowners, it is certain that the acquisition of land was not a major problem to the proponents. I have not studied the motives of these owners, but it is clear that in Dordrecht the municipality itself had 'derelict land' that it wanted to put to productive use. The concern for others in the area was not so great that they were consulted on the sale. The Kaliwaal was largely owned by Delgromij already, but the landowner of adjacent lands (in agricultural use) was also happy to sell. I do not know whether this involved consultations with the actual users of the land, but is certain that the tenants objected to the waste disposal plans and the nature development plans. They objected to the licences and indeed appealed them to the State Council, so they did defend their interest.

(3) The role of elected politicians

Elected politicians played only a moderate role in the decision processes just described. That is, elected politicians in general. In neither case was there much discussion in councils or legislatures on the various proposals, although an incidental question was asked here and there. Only in the case of the Kaliwaal did the proposal become somewhat of an issue during local elections, with all local parties speaking out against the proposals. More important than elected members, it seems, were the senior politicians, the ones that head the local or provincial bureaucracy. The Kaliwaal and the North Refinery cases are good examples of this. The provincial executive long made realization of sludge disposal impossible, whereas it became possible after this body its changed opinion. In the case of North Refinery, the proponent had so much political support from the local and provincial executives and even the Minister of the Environment at some point, that it could overcome many legal problems. The assumptions about politicians that were formulated in chapter 2 included one that predicted politicians would be accountable to the people because they need their support. The second however said that politicians will be accountable only to 'special interests'. The third was that politicians will 'misuse' hazardous waste decisions for political gain, even if it is strictly a technical issue.

It is hard to say whether politicians were indeed accountable to voters because of various reasons. The first is that politicians, as I said, were not very active in the decision processes. The second is that 'the voters' do not seem overly concerned with the proposals, certainly in Delfzijl or Dordrecht. Only in Druten (the Kaliwaal) was there a certain concern, but this occurred extremely late in the decision process. Finally, the connection between voters and elected members is not very direct in the Netherlands (politicians do not represent a certain district but the entire municipality). There is thus on the one hand no sharp divide between parts of the locality that stand to loose and parts that are indifferent, but on the other there is no active representation of the particular area in question either. That being said, the proposals were probably among the most contentious political issues available and thus offered an opportunity for establishing a political profile. One party invited a member of the opposition group 'No Poison in the Kaliwaal' to be a candidate for council. In this sense, they were responsive to a perceived local opinion. Were elected politicians accountable to special interests, as managerialism suggests? There is only limited evidence on this matter. Note that on the one hand two municipalities (Ecoservice in Dordrecht, North Refinery in Delfzijl) were more than happy to welcome waste facilities and the associated employment, whereas a similar sentiment was present in the Provincial Legislature in Gelderland (Kaliwaal case). The desire to have the facility sited was clearly greatest in Delfzijl, where it resulted, de facto, in an illegal situation. I think it is fair to say that the executives here were sensitive to special interests. In the case of the Kaliwaal however, the provincial executive clearly tried to stop the proposals on the basis of arguments that were somewhat suspect in light of the applicable legislation (hence the court verdicts against the province). But to say that this was done to serve a special interest would probably be incorrect. Rather, the uncertainty about underwater sludge disposal that had existed in the 1980s was still felt in the provincial offices. The idea that the technology should only be tried on an experimental scale seems to have long permeated the acts of the province. The decisions have not been used, it seems, for political gain (critique of managerialism), except perhaps in the case of the Kaliwaal. In fact, a more remarkable element of the decision process was the attempt of politicians to remove themselves somewhat from the decision process, stressing the fact that technical considerations were paramount in their decisions. This was even the case with North Refinery, where the provincial executive publicly stressed that the countless legal problems were not related to the technical merits of the facility, but to futile legal issues.

(4) The role of voters

The more positive assumptions about voters derived in chapter 2 indicate that voters will hold politicians accountable if they do not deliver the right decisions and that they will form interest groups to influence decisions in the right direction. The three negative assumptions were logically opposite. The third assumption was that voters would not have an interest in decision-making and would not be sufficiently educated to form an opinion. The fourth assumption was that voters would behave selfishly, favor economic development above the environment and will express 'absolute values'. The last assumption was that voters would form interest groups to further their own particular interest and prevent achievement of the public interest.

The position of voters in the Netherlands has been discussed to some extent already, and the impression is they were not very active in the three cases described although there are ample indications of interest groups being formed. There is not much evidence in the three case studies of voters holding politicians accountable through punishing election results. The dynamic behind this has already been explained. There is however a rich pallet of action groups available (mainly environmental groups) that do take an active interest in this type of decision process. Many of these groups are permanent (Werkgroep Eemsmond, provincial environment federations, WWF) and are engaged both in broader policy debates (e.g. about water management, see WWF) and local discussions (e.g. the Kaliwaal). Apart from these permanent groups, that tend to receive government subsidies to maintain their organization, only in the case of the Kaliwaal was their group formation specifically for the proposal in question ('No Sludge in the River Foreland' and 'No Poison in the Kaliwaal'). All these groups do invest energy in the decision process by reading the available documentation (including the environmental assessment report), by attending meetings, and by going to court if need be. Individual citizens are present in the debates as well, but seem to play a minor role. They tend to use hearings more for the purpose of information-gathering rather than influencing decisions.

Managerial and pluralist lines of thought consider citizens incapable of engaging in technical debates and expect them to behave selfishly, thereby preventing the public interest to prevail. There is mixed evidence on these assumptions. It does seem that certain citizens are quite well capable of engaging in technical debate. Take the group 'No Poison in the Kaliwaal' for instance. Some of its members are trained scientists, with expertise in the field of sludge. It does seem that some of the issues they brought up, especially the condition of the soil in the Kaliwaal, were valuable to the decision process. The Werkgroep Eemsmond was also seen as quite a competent group by most others engaged in the decision process, be it that the group itself admitted that it had little expertise in hazardous waste treatment and felt somewhat short-handed. On the other hand, the few ordinary citizens who participated do seem to be overwhelmed with the technicality of the debate and tend to quibble about issues from the environmental assessment report and request clarifications (e.g. what does it mean to have a certain quantity of odor particles in the air at some point?). Also, it does seem that certain issues are somewhat difficult to interpret for the general public. Take the issue of 'piping' in the case of the Kaliwaal. Piping, simply put, implies that rising water levels translate into higher water levels at certain points behind the dike. It looks as if water from the river surfaces behind the dike. Base on that assumption, people tend to think that pollution from the river system could be 'piped' to the land behind the dike. However, most experts in the field contested this. Their argument is that there is a column of water in 'the pipe' that moves up if the water level in the river rises and moves down if it lowers. There is not much 'migration' of particles from the one side of the column to the other. Hence there is little transport of pollution either. This notion is however extremely hard to believe and 'piping' continued to be an issue throughout the decision process. There is a certain degree of selfishness in the process in the sense that the people living closest to the proposed sites tend to be more inclined to participate than others, with the exception of the environmental groups that work mostly on a regional basis. Geerts (1999) studied the risk perception and willingness to act against Delgromij's proposals for the Kaliwaal of inhabitants nearby on the basis of a questionnaire. He found that the people North of the river (the opposite side of the Kaliwaal) had greater feelings of safety and were less willing to act against proposals than those South of the river. Also, those who thought they lived 1 kilometer or less from the site felt less safe and were more willing to act than those who thought they were 2 kilometers or more from the site. Less well studied, but seemingly the same is the pattern for the Ecoservice and North Refinery cases. The only ordinary people participating in the decision process there were people who live nearby. But are these people selfish then? I think that such an impression of selfishness could be countered if those people who oppose plans consequently also enter the debate about alternative solutions for the plan they are fighting. This could be by pointing at better policy alternatives (e.g. incineration not disposal or vice verse), or by pointing out better locations. Such arguments are present indeed. The group 'No Poison in the Kaliwaal' for instance, has established links with those that present alternative sludge disposal techniques and support these in their attempts to convince public authorities to treat sludges instead of landfilling them.

However, such more generic arguments are rare and interest group activism seems almost always directed towards opposition and much less towards very constructive efforts. Opponents tend to mostly resist a certain site without thinking too much about alternative locations or changes in policy. A case in point is the Werkgroep Eemsmond. Still I would hesitate to call this group selfish because they do not have a personal interest in the decision process.

(5) Parties in a court case

Regarding the parties in court cases, it must be observed that the single most important factor in the courts' decisions is the advice of the expert advisors they hire. In this sense, the outcomes of the court case do not so much hinge on the activity of the parties to a hearing. Even the terms of reference for such an expert are set by the court, and they may be wider or narrower than the parties would like to see. Especially under the pluralist discourse, parties in a court case are expected to improve decisions by making the strongest possible case for themselves and thereby support truth finding. A criticism of pluralism is however that parties in a court case make the strongest possible case for themselves and thereby *hinder* truth finding. A third assumption derived in chapter 2 was that parties in a court case would not rest until the last legal resort has been tried, prolonging the decision process.

Do the parties make the strongest possible case for themselves and thereby hinder truth finding? I have found hardly any indications to that effect, although there is a certain degree of radicalization of opinions in court cases. As the province of Zuid-Holland, while their officials were convinced they would have to pay damages for slow procedures, tried to argue before the court that such damages would not be warranted. In the case of the Kaliwaal, the authorities that wanted to issue licences for the facility, now started cooperating with Delgromij to make their case as strong as possible. In that particular case especially, the opponents started to radicalize as they came with an ever-growing amount of documentation, leading to more and more radical conclusions, amounting even to a certain degree of conspiracy theories. In all three Dutch cases, there is an enthusiastic use of the courts to have one's way. Especially the case of the Kaliwaal is extreme in that respect. It does look like the opponents there won't rest until they have tried every option they have.

(6) The role of the community

As stated, 'the community' plays hardly any role in these decision processes, or it must be by silently agreeing to the plans (Dordrecht and Delfzijl, and the majority of the population around the Kaliwaal). The first assumption about the community was that community members are engaged in coercion free relationships and are in dialogue with each other. Another, also positive, assumption predicted that members of the community have local knowledge and that such knowledge is crucial for the decision process. Far less positive assumptions include the one that community members will be parochial in their outlook and not look beyond their own community. In addition, two more assumptions are that community members do not have scientific knowledge, which is crucial for the decision process and that they do not want to participate in the decision process because of a lack of motivation.

In the cases discussed in this chapter, I found few indications of 'coercion free' relations within the community or the presence of dialogue. The few indications I have gathered are, however, not very positive about the prospects of communal decision-making. Take the case of North refinery. This plant was sited in a climate where the majority of the population was probably in favor of the plans. When some, relatively new, members of the community were somewhat concerned about the expansion plans, they received negative feedback (coercion?) from other members who depended on the plant for their employment. A similar approval of the plans existed in Dordrecht. It thus seems that in these two municipalities, the plans were perceived to be in the interest of the community by the community itself. And in the case of the Kaliwaal, there were opponents but these largely bypassed the local community and fought the proposals in managerial terms. In all three cases it seems fair to say that 'the community' was not so much against the proposals. On the issues of local and scientific information: I have noticed only a limited role for local knowledge. In the case of North Refinery the environmental assessment report had overlooked the presence of a certain population concentration near the site (this concentration was not on the maps) and this was obviously brought forward by these people themselves. In the Kaliwaal case local knowledge about 'piping' was brought forward but seems to have been incorrect (see above), although in a sense only experience will tell, of course. In neither case was local knowledge crucial for the decision process. More remarkable was the high level of technical skill that some members of the general public or environmental groups had. The idea that citizens do not have scientific knowledge per se must be rejected, although it seems fair to say that most of them don't have the skills for very technical debates. My impression is that this need not be a problem to participate as common sense and a critical view can greatly help too.

Chapter 6

Canada: Community-based siting

The three Canadian cases I have selected are special in the sense that an attempt was made to site waste facilities outside the ordinary decision procedures. The first departure from the normal approach to decision-making occurred in the early 1980s when attempts to find a facility in Alberta became extremely controversial. A careful study of siting failure on the North American continent followed, and its result demonstrated the need for a greater emphasis on the community. This conclusion was not easily accepted everywhere, certainly not among the actors involved in normal decision-making processes, but did prevail because of support in the bureaucracy. The process failed initially, but after making it somewhat more manipulative, it worked in delivering a site for an incinerator. The Alberta example proved inspiring as another province, Manitoba, and the federal government also wanted to apply it, with modifications. In both cases the process did deliver a volunteer community, but in the federal process, the demands for compensation of this community were too high.

It is important to realize that community-based forms of siting in Canada were pragmatically inspired, not by certain discourses. The mix of factors responsible for putting more emphasis on the community included the presence of actors who believed that it would 'work', a desire by politicians not to burn their hands on the issue of hazardous waste siting, and (after Alberta) the example of success. It is also important to realize that the intention of the process, to some extent, was to have a rather technical debate at the local level. The discussions were not to be about the interest of the community, but about safety and the identification of technically suitable sites. Although Canadian citizens appeared capable of such a debate, it proved difficult for those setting up the community-based process not to influence or even manipulate the debates.

Figure 1. Cases in Canada

6.1 Alberta: development of an innovative approach to siting

6.1.1 Introduction

The first case I describe[1] in this chapter involves the attempts of the province of Alberta to locate a hazardous waste incinerator in the province. Initially the provincial ideas about siting had been along conservative pluralist lines as it waited for the private sector to develop and defend proposals. This worked to some extent as indeed one private company developed a proposal for a hazardous waste facility. The problem however was that the proposal resulted in a large degree of opposition, which made the private company retreat and the province reconsider its approach.

The new approach that consequently emerged gave a more central position to the province, which took up the role of proponent even though it made clear that it still did not desire to actually operate a facility but just wanted to find a location for one. Various committees and expert bodies consequently gave their opinion about how the province was to go about

[1] Helpful comments on an earlier version of this case report were received from Mr. T. Fernandes (Alberta Environment), Mr. G. Latonas (BOVAR), and Mr. R. Sherbaniuk.

siting, with a process called 'invitational siting' as a result. Invitational siting implies that the community could play a greater role than normal, indicated by the fact that only 'volunteering communities' were contacted and by the fact that the decision process included a plebiscite about the proposal connected to a veto right for the local community. The decision process was not only centered on the local community however. 'Normal' licensing procedures had to be applied after local approval, essentially implying that Alberta's regulators and politicians also had a veto right.

Invitational siting has been enthusiastically embraced as a way out of siting gridlock, showing the way towards socially responsive and technically competent siting. I will discuss whether such enthusiasm is in place. Issues of attention are for instance whether the aura of 'voluntary siting' that is often associated with the Swan Hills case is warranted, i.e. whether the siting process was really community driven or not. I will also discuss whether the greater role of the community has had an impact on the quality of the decision, and if so, in which direction.

6.1.2 The genesis of invitational siting

Introduction

In the early 1970s, the government of the province of Alberta became aware of a potential hazardous waste problem as part of a more general developing awareness that waste disposal practices in the province were not environmentally responsible. In response, the province commissioned various studies into the issue of hazardous waste disposal, which indicated that large quantities of these wastes generated in the province were not treated responsibly. Lots of wastes simply disappeared, whereas others went to relatively unreliable landfill sites. The studies, although somewhat alarmist, did not attract much public attention and did not lead to much response from the Alberta government.

Market initiative fails

Whereas the provincial cabinet agreed to the recommendation that the province needed a disposal facility specifically for hazardous wastes, it saw responsible hazardous waste management as a private sector responsibility and thus waited for proposals from private companies. The first company to actually pick up the challenge of providing the needed treatment facilities, Kinetic Contaminants of Canada Ltd., came forward near the end of the 1970s. The company saw a market for hazardous waste treatment in Alberta and, with the help of certain local councilors who considered the potential employment a great benefit, had selected a site at Fort Saskatchewan. When a meeting was held to inform the local public of the proposals, about 300 people attended, most of whom came to express their opposition. The gathering was a public relations nightmare. Furious citizens quickly created an intensely hostile environment for the speakers and failing slide projectors hampered presentations. In response to questions why Kinetic Contaminants had selected the site and did not go elsewhere, the company promised it would discontinue its plans if the local public opposed them. This promise was not lost on some of the attendants. An opposition group started an extremely well-organized and effective campaign against the plans, which for instance implied that one opponent would phone the Prime Minister's office every ten minutes over an extended period. Kinetic Contaminants made good on its promise and moved elsewhere in the province with its proposal, only to experience a similar reaction. Members of the provincial legislature from the communities where the debate took place saw the proposals as a threat to their career, and pressured the government to take over the decision process. The government was publicly committed to responsible waste treatment as part of its strategy for industrial development and could therefore not easily refuse.

Cool down period

The provincial government decided for a cool down period during which the extent of the provincial hazardous waste would be studied and insights gained into the dynamics of the siting process. One advantage was that the issue would be out of the hands of politicians who were all afraid of what could happen to them in the next elections.

Hazardous Waste Management Team: social concerns first

Thus the province established a Hazardous Waste Management Team that partly consisted of high-ranking bureaucrats, partly of people from outside government, to carry out the desired study. One of the activities that the Team undertook was the commissioning of a report on hazardous waste siting experiences elsewhere. This report (Krawetz, 1979) argued that in order to achieve siting success, social concerns and acceptability of proposals to the public should be placed at the forefront. This would for instance imply that there should be a clear local benefit associated with the facility for the community. In addition, the report advised that success required that the proponent of the facility was a trusted organization and that the facility in question should preferably only be intended for treatment of Albertan wastes. In terms of concrete interaction processes with the public, the report advised not to use public hearings as opponents often succeeded in overtaking them and turning them into opposition rallies. Although some of the members of the Hazardous Waste Management Team initially had great reservations about the advice, one member (who had actually commissioned the report on siting) was adamant that a more community-based approach towards siting was necessary, and slowly the other members started to agree. The negative experience with Kinetic Contaminants proposal gave credence to the report.

The Management did not just focus on siting, but studied many more issues. Besides a siting process, the Team also calculated Albertan waste arisings in the future and discussed waste treatment technologies. The Team had visited Germany and was greatly impressed by one of the integrated waste disposal facilities (incineration, recycling, treatment and landfill combined) it had seen there. The team proposed a similar facility, with a certain capacity based on projected waste arisings, for Alberta.

Hearings and further study

Indicative of the sensitive nature of the topic, there was no explicit political response to the recommendations. Instead, the Minister of the Environment decided that, in order to gain public understanding of the issues facing the province and to start building a certain level of social acceptance for future proposals, the report should be discussed during a series of public hearings. The responsibility of organizing these hearings was put in the hands of the Environment Council of Alberta (ECA), a government funded expert body. Based on the hearings, the ECA would advise the provincial cabinet on hazardous waste policies, including an approach towards siting. By 1981, after holding an extensive number of public hearings, the ECA came forward with the conclusion that the province needed not just one but two facilities for hazardous waste treatment. These facilities were to be located within a 100-kilometer radius from the major cities, Edmonton and Calgary, where most of the waste was being generated. The exact spot for these two facilities was to be found by a Site Selection Committee, which would identify a site and then engage in public discussions. The ECA report once again did not result in much public endorsement by elected politicians nor by the provincial Cabinet, although Cabinet declared its continued preference - counter to the advice of the Hazardous Waste Management Team and the ECA - for limiting its role to site selection. The Cabinet indicated it would not want to operate the future facility.

A first attempt at implementation

Despite this lack of endorsement, a Hazardous Waste Implementation Team was formed in January 1981. The Implementation Team was to select two sites, as envisioned by the ECA, but was in addition expected to devote some time to the development of more concrete site selection criteria and the pre-selection of four municipalities as suitable locations. The Implementation Team set out to further develop criteria for site selection and then apply these to identify four municipalities. However, by the time four municipalities had indeed been identified, immediately causing a huge outcry among the respective populations, the Implementation Team was no longer the forum where the real discussions were taking place. These discussions had moved towards the Siting Task Force, a group of officials appointed to assist the Implementation Team. Within the Task Force, the pre-selection of four particular communities was rejected. The search for a location should use the entire province as a search area, with certain a priori determined exceptions and with the purpose of siting just one facility. The facility was not to be constructed close to the waste generators per se, but in communities that would accept the facility. Only after a certain level of community interest had been expressed, would detailed technical issues be studied. The term used for this approach was 'invitational siting'.

A bit of a struggle erupted as the Implementation Team resisted the Task Force's ideas. The Task Force won this struggle as some of the higher-ranking public officials in the Task Force had the ear of the Minister of the Environment. In fact, the Minister even disbanded the Implementation Team after hearing the different opinions on the further siting process. Many observers looked at the choice of the Minister of the Environment in disbelief. According to one member of the Task Force, once permission for invitational siting was given, '*The attitude among the powers that be was, give them the money, let them try and they will fail*' (Sherbaniuk, 1992: 118). It appears that the price that had to be paid for allowing an experiment with the new approach was an assurance to the minister that the approach would be successful: it was made clear that a site had to be found by March 1982. Many reputations were thus at stake in what followed.

6.1.3 Invitational siting is redesigned and applied, twice

New design

After wresting away control from the Implementation Team, the Siting Task Force devised a ten-step procedure for identifying volunteer communities. The unique feature of this process is that the authority rules were expanded from what the law of Alberta prescribed at that time. Instead of having the Minister of the Environment and responsible planning authorities decide practically exclusively on the acceptability of a site, a local veto was added. How and by whom such a local veto should be exercised was not perfectly clear. However, the intention of the Task Force was that in cases of local opposition, they would retreat. In fact, they would not start discussing a possible incinerator if not invited by the local council. Hence the term 'invitational siting'.

Because the government of Alberta would retain the final say on the facility, invitational siting was not to be seen as something that overruled the ordinary methods of deciding. Instead, it was an added feature to the standard decision process, increasing local control. The process was designed to be divided in several phases that progressed from a very abstract level (was the local council willing to have its area studied to identify potentially suitable sites?) to an ever more concrete level (the final phase would focus on the acceptability of a specific site). Every phase of the process would have to end with an affirmative response from the local community that it wanted to continue; otherwise, the process was to be terminated[2].

The Siting Task Force moves ahead: process nearly collapses

The Task Force started implementing the process by organizing a large number (67) of informative community workshops. These were held with the assistance of the Rural Education and Development Association (REDA) a well-known organization in the Albertan countryside. The result was that 46 requests for free 'regional assessments', studies of the suitability of certain regions, were received and numerous local workshops to discuss the outcomes of the assessments. The local 'assessment workshops' subsequently resulted in an invitation to the local communities to send representatives to a provincial workshop by October 1981. A respectable number of 55 delegates attended the workshop, although some of these were from communities that had already decided to cease further participation in the process. REDA asked the delegates to start local discussion groups, and imagined that a dialogue in and between the various communities would lead to consensus on either acceptance of a waste facility or some other approach to hazardous waste issues. REDA saw its own role as a facilitator of community dialogue and had no specific interest in a particular outcome of the local debates. It appears that REDA's approach led to serious disagreement between REDA and the Task Force, because the Task Force wanted to keep the local debates focused on site selection. The two organizations parted ways, and REDA then publicly suggested that the Task Force was not serious in its commitment to a voluntary process and was focusing on siting instead (Krawetz et al., 1983: 62). In its turn, the Task Force indicated that REDA had not properly organized the local debates.

It does seem that REDA had a point with its comments. Even though many communities were still contemplating their continued participation in the process, the Siting Task Force had moved into those municipalities where the council wanted to proceed and which looked most promising from

[2] The prototype process consisted of ten steps, which I will not describe here (but see Krawetz et al., 1983: 52-53).

the perspective of the ECA's criteria. Although the Task Force had been invited to perform test drillings of the soil by the council in these municipalities, opposition arose when the local population noticed the work of the test drilling teams. The first addressees for the critique of opponents were the local councils. Statements by one public official associated with the Task Force that local communities should perhaps be compensated for acceptance of a facility had not gone unnoticed and opponents started accusing their councils of trading their citizens' health for provincial dollars. The Task Force had avoided public hearings on their activities and had set up various informal meetings with limited numbers of community members to present its proposals. Although these meetings were apparently going well, opponents used the fact that they had not been public as an argument against the Task Force. The Task Force was portrayed as an unreliable organization which used '*nothing but sneaky tactics*' (Krawetz et al., 1983: 67).

Particularly in one county, the situation escalated. A group of between 200 and 300 people invaded the county office during a meeting of the council. During that meeting, councilors were discussing a land use bylaw that had been prepared to authorize a hazardous waste facility at a specific location in the county. The group of citizens, headed by a local doctor, demanded that a referendum be held on a land use by-law that had been prepared to authorize the facility in the county[3]. They did not leave the county offices before a motion to that effect would be approved, which the council did. The subsequent referendum resulted in an overwhelming majority against the plant and the proposed change of bylaw was effectively killed. The fact that the provincial Attorney General started an action to have the council's motion repealed did not matter much. As a consequence, the invitational siting program was in shambles. The larger participative effort by REDA had stopped and in the communities that had looked most promising, opposition had effectively disrupted the Task Force's approach.

Revamped siting process

The Task Force came under tremendous stress but continued to receive support from the Environment Minister. The mandate of the Task Force was extended for one year and Burke Nagle, a nearly retired agricultural sociologist, was hired to study and revamp the process. His analysis of what had gone wrong focused specifically on the tactics that opponents had used and how these could be effectively responded to. One 'story line' that opponents had tended to use was the disappointing record of waste

3 In Alberta, a hazardous waste facility is compatible with heavy industrial uses. It seems likely that the by-law did not explicitly specify a hazardous waste treatment plan as permitted use, but just heavy industrial use.

management up until then, which did not command a lot of faith indeed. Nagle noticed that the Task Force had often replied by indicating that the new facility would be different and better and by downplaying the past record of the waste industry. He suggested it would be better to openly admit the mistakes of the past and then present the facility as a method to mend the situation. He suggested that the Task Force show the movies about mismanagement of wastes that the opposition had previously used in order to convince the public that a better facility was needed. Also, opposition groups had demanded referenda, especially in situations where local councils had been cooperating with the STF, without the knowledge of the majority of the population. Despite the fact that these did not fit in Alberta's legal system, Nagle suggested that the Task Force itself should offer to hold a referendum in order to preclude any questions about the motives of local councils. The response, despite the rather drastic nature of recommendations from an institutional perspective, was positive. The Minister had by then '*become desperate enough*' to accept the idea of a binding referendum in the communities[4]. Therefore the Minister accepted the proposals for change.

Core of the revised siting process was a new information program of three seminars. During the seminars, films would be shown that presented a balanced view of hazardous waste management, that is, did not ignore the bad practices of the past. After the films, an introduction and presentations about various technical topics would follow by communicatively able experts. Attendance of the seminars was to be controlled and limited to 50 people at most. During the seminars 'impression management' would take place, amongst others by approaching the most vocal opponents during a meeting and proposing to them to continue the conversation bilaterally. Also, there was increased attention to establishing trust and rapport. This was especially achieved by engaging the energies of local leaders. To find out who these opinion leaders were, people on the street were asked who they would turn to if they had questions about hazardous waste ('power structure analysis'). This led to a list of about ten people. It attempted to convince them of the need for the facility and the safety of its operations, and then ask them to sit on a local citizens committee (Community Liaison Group) that was to discuss the proposals. The idea was that these people would become active proponents of the facility and give a local voice to those in favor of it. To avoid the impression of bias, no councilors were approached.

4 Interview with N. Krawetz, formerly with Alberta Environment and for CEO of the Environment Council of Alberta, Edmonton, 9 July 1999.

Three new potential volunteer communities

This approach was indeed tried in various communities, which had expressed interest but had previously been ignored because of their distance to Edmonton and Calgary. These communities were the Special Areas, Ryley and Swan Hills. One might wonder why these communities had been interested. It seems fair to say that in each of these communities, the desire to create new employment, to stabilize the community, and to diversify economy, played important roles in their local council's willingness to participate. Swan Hills for instance was largely dependent on the oil industry and was an outgrowth of a trailer park of people working in that industry. The council desired more facilities, more permanent houses, a less transient population, and a more diverse economy. Besides its involvement in the siting process for a hazardous waste incinerator, the town had also issued a bid for a provincial prison and had lobbied the province for medical facilities.

In two out of the three communities, Ryley and Swan Hills, the new siting process was a success in the sense that the population widely attended the seminars, voted in the plebiscite on the topic, and voted in favor of accepting the facility. The communities - especially Swan Hills - even felt they were competing with each other for the facility, which had thus become something to vie for, a positive asset. Especially once Ryley had accepted, other communities started looking at the facility in wonder: had they been missing something that Ryley had seen? Some decided they had indeed missed out on an opportunity and decided to jump aboard. Such attempts were resisted however as the Siting Task Force presented the Provincial Cabinet with two options after Swan Hills had also agreed. Members of the Task Force had not indicated a preference to the Cabinet, but did hand a motivation for either choice so that the Cabinet could proudly announce and defend its choice, whatever it was. Despite not indicating a certain preference, it was clear that Task Force members considered Ryley the better site, mainly because it was closer to the waste generators than Swan Hills (a 'technical' reason). However, Cabinet decided otherwise and announced that Swan Hills would be the location of the new facility. The announcement was greeted with a party in that town, and with fury in Ryley. The losing town protested the decision and accused the Cabinet of making the wrong decision. Especially the Member of the Provincial Legislature from the Swan Hills area, and later Environment Minister, was seen by some as the bad genius behind this decision. The subsequent licensing procedures went through without much trouble and licenses were issued without any objections. No environmental assessment was required in this case because of the extensive public involvement programs that had already been undertaken[5]. Figure 2 depicts the site that was eventually selected.

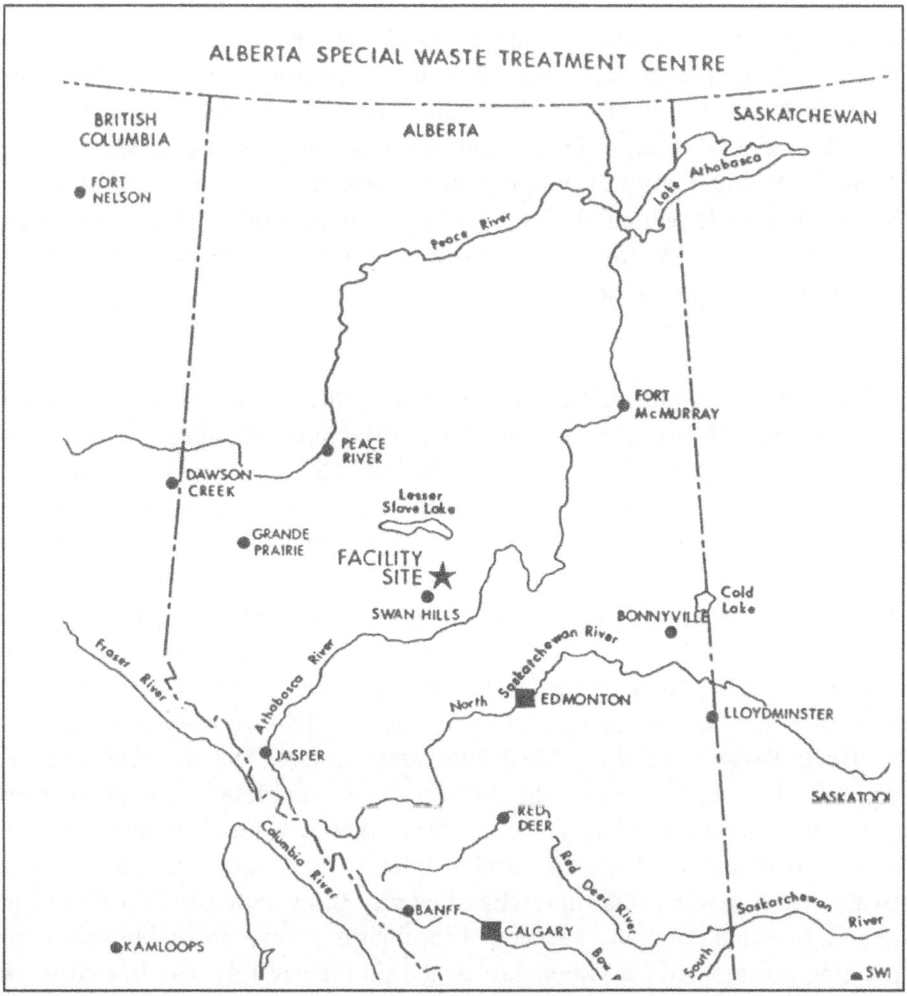

Figure 2. The Swan Hills site. Adapted from the license from Alberta environment.

6.1.4 Developments after facility construction

Unpredictable market

After the site selection process was completed, hazardous waste regulations were being phased in and wastes started being stored at the facility that was being constructed. It was only then that a first real

[5] Personal correspondence with Mr. G. Latonas (BOVAR).

indication was received on the need for hazardous waste treatment in the province. The required capacity turned out to be less than had been predicted by the Management Team years ago. The supply of wastes was so low in fact that a decision was made to operate with new and untried incineration technology that could destroy many types of hazardous wastes but would not require a large feedstock ('rocking kiln incinerators') and was therefore supposed to provide flexibility. Even after the plant opened however, waste streams continued to be different from what had been predicted and the two installed rocking kilns never operated to satisfaction. A small rotary kiln incinerator supplemented the rocking kiln incinerators and started doing most of the work[6]. A transformer furnace for the purpose of treating PCB-transformers was also added.

Controversy

The facility is controversial in Alberta in various respects. One reason for that was the fact that the facility cost considerable amounts of money to operate (estimates are that about 400 million C$ of provincial money has been spent over the years). Part of the reason why such an impressive amount of money needed to be pumped into the facility is related to the contract with the operator of the plant. After the site had been found, a Crown Corporation (akin to an arm's length public firm) by the name of the Alberta Special Waste Management Corporation was founded for the purpose of dealing with Alberta's hazardous waste. People associated with the Task Force considered an extensive role for this Corporation crucial for the effective operation of the hazardous waste system. The idea was that this firm would actually operate the facility. The provincial Cabinet decided against this however, mainly on the basis that it wanted to terminate its dealings with the facility and hand it over to a private operator. Such a private operator was found in CHEM-Security (now called BOVAR Inc.) in the end, and this company succeeded in obtaining very profitable terms for operating the facility ('a sweetheart deal'). Some say this was in return for donations to the election campaign of Alberta's ruling party (see Sherbaniuk, 1998 and Nikiforuk, 1996). Various people I spoke to have explained to me that the contract guaranteed Chem-Security a minimum rate of return on its investments. Thereby the company had an incentive to invest in the facility (the plant was really a Cadillac facility) and this forced the province to subsidize the facility to an increasing extent despite the fact that the facility was a limited commercial success (ibid.). The facility was hampered by its odd location (a three to four hours drive from the main waste generators in the province at least) which increased operating costs. Also, the capital costs for the facility were relatively high, certainly compared with landfill

6 This was a CE Raymond kiln facility with 8,000 tons per annum capacity.

facilities. Despite that being the case, Chem-Security succeeded in presenting a case for enlargement of the facility in the early 1990s with a large rotary kiln incinerator[7]. The company's parent, BOVAR Inc. was to pay for 60% of the costs, whereas the Crown Corporation would pay 40%.

The issue was controversial and came before the Natural Resources Conservation Board (NRCB). After hearings, this quasi-judicial body decided that the extension should be allowed. The waste arisings that were predicted by Chem-Security at that time of the hearing have never materialized however. This was mainly because the predictions were premised on the tightening of waste regulations that was in the pipeline at the time. The province had such regulations prepared, but backed away from them after pressure from the oil industry. The stringent waste regulations would have forced this industry to use the Swan Hills facility and the facility's future was somewhat uncertain. Instead they could continue landfarming and injecting wastes in deep soil. The dynamic of the situation then changed to Chem-Security proposing to start accepting wastes from elsewhere in Canada and somewhat later also from abroad, despite agreements with Swan Hills that forbade this. The local community did not resist waste importation by then anymore, but environmental groups succeeded in forcing the provincial Cabinet to yet again request a NRC hearing into the issue. Waste importation from other Canadian provinces was formally authorized in the end, but resulted in complicated issues related to the subsidy that Alberta was giving the plant, whereas other provinces were not contributing and refused to start doing so. By 1999, Chem-Security threatened to not renew its 10-year contract with the province if subsidies would not continue and waste importation would not be allowed. After complicated negotiations the company indeed stopped operating the facility at the end of the year 2000 and returned it to the province. The province had no longer wanted to subsidize the plant, because opposition parties were effectively using the facility as a way of demonstrating the failings of the government. Also, newer members of Cabinet had little sympathy for public subsidization of the private sector. These developments could be seen coming for a long time and Chem-Security had tried to make the facility more profitable by reducing the staff. Already shortly after I visited the plant in 1999, about 70 of the 170 people working at the plant were fired. Currently, the province has contracted out the operations of the facility to Sensor Environmental Services Ltd., while waiting for private bids to operate the facility.

7 This was a FBD incinerator.

Quasi-judicial hearings before the NRCB

The NRCB hearings, especially the ones on the expansion of the plant in the early 1990s, demonstrated that opposition had started to build up against the plant. This happened not so much within Swan Hills, but within communities nearby and with native Canadians (Indian First Nations) at about 20 kilometers from the plant. Nearby communities were not benefiting from the plant like Swan Hills, but were dealing with a certain extent of waste transport. Although the facility had its own trucks that were much safer than the ordinary trucks used for these transports (mainly because they had a double tank surface) and no accidents had happened with these trucks, this was different for trucks from other provinces. One truck carrying PCB-wastes from another province had had an accident relatively soon after the plant opened, causing leakage of wastes. Because of this, communities on the road to Swan Hills had become sensitive due to accidents with waste transports. The ban on the import of wastes that was initiated in response did ease this only to some extent. In addition, the First Nation people up North thought their hunting grounds were being polluted by the facility. There was indeed evidence to this effect. The transformer furnace was found to be leaking PCBs in 1996 and that part of the facility was consequently shut down and has not operated since. Subsequent sampling for PCB-levels in the immediate area of the facility consequently found elevated levels in plant and wildlife, resulting in a health warning against hunting and fishing in a 30 kilometer radius around the facility[8]. Together with environmental groups, the First Nation people acted as an opponent to the expansion of the facility and the import of wastes from elsewhere.

6.1.5 The role of the various institutions and discourses

The role of the market, elected representatives and experts

The role of the market is relatively limited in this case, although at the periphery some remarks can be made. First, it was a private sector initiative that got things started. Alberta's politicians had laid the groundwork to some extent by analyzing the waste market and several of the larger waste disposal firms had checked whether a facility in Alberta could be profitable. Most firms however did not think the size of the market was sufficient to sustain operations. Kinetic Contaminants, a relatively small company, had a different opinion however and tried. Without success and with the emergence of more provincial studies and invitational siting as a consequence. Only when this succeeded, private market parties came back into the picture again to do actual facility design and operation of the

[8] Mr. G. Latonas (BOVAR) reported to me that samples in the years since have demonstrated PCB levels near background levels.

facility. The way the facility was subsequently operated demonstrates various issues. The first is that the market for hazardous waste incineration is an 'artificial' market in that government regulations determine what is defined as hazardous wastes and what must be done with it. Clearly the province itself had embarked on site selection and a subsidy scheme for the facility, thinking that wastes from the oil industry would go to the facility. However, the province gave in to pressure from waste generating industries and these could continue some of the practices (e.g. deep well injection) which should actually be terminated. The second is that private market parties should perhaps not be given a guaranteed rate of return, because it places the risks associated with investments in the lap of the public sector, while keeping potential profits in the hands of the private sector.

The role of elected representatives in the case study was quite large. It must be noted that the normally major role for private sector initiative in Alberta was largely terminated because members of the provincial legislature feared that their political future was at stake if the decision was to go ahead. Coupled with a private developer (Kinetic) that did not want to go against the popular opinion in the towns where it was trying to locate a facility, and with a very effective opposition campaign, this laid the groundwork for invitational siting. One may wonder why the provincial Cabinet was not minded to leave matters as they were, and the answer seems to be a real concern with responsible waste management, public pressure generated by Kinetic's proposals, and the perception that such waste management was a key part of Alberta's economic development strategy. The study of the hazardous waste problematique and hazardous waste siting strategies was, from the perspective of the provincial cabinet, a good way of getting a contentious issue to cool down a bit. The members of Cabinet and the Provincial Legislature could thereby avoid burning their hands on the issue. Only the Minister of the Environment was taking a certain (limited) risk in siding with the proponents of invitational siting as most onlookers considered the chance of succeeding quite low. The initial attempts to start invitational siting indicated however that local councils were risking all the more burning their hands, mainly because some councils saw the facility as a good economic development opportunity and rapidly moved forward in their dealings with the Task Force. The perception of the facility being an economic development opportunity was clearly not the right perception in the eyes of the local population. Especially the rather intimidating act of occupying the council offices with hundreds of people seems to indicate a certain readiness of participation outside the legal system that I have not observed in any of the other case studies. It does seem that this act later helped in largely de-emphasizing the role of local councils. Although they were the ones inviting the Task Force to town, council members would no longer participate in the so-called Community Liaison Groups that were founded as a forum for deliberation in the various communities. Also, their decision-making powers were suspended until a local referendum showed the local population supported the facility. The councils of the jurisdictions from where such approval came all agreed to the outcomes. The decision between the two communities that volunteered, Swan Hills and Ryley, was a political decision once again, at least there was a certain sense that the lobbying of members of the provincial legislature played a certain role.

The role of experts is relatively modest in this case, and the type of expertise used differs somewhat from what we have seen so far. Although technical and experts were responsible for various steps in the decision process, such as market assessment, facility design and licensing, the technical suitability of the proposal was not concern number one in this case. Several people have told me that social acceptance was the dominant criterion after Kinetic had failed with its proposal. The technical difficulties related to facility design were considered to be solvable. The idea was thus that the facility would be designed around social acceptance, within certain broad limits. The broad limits related to a minimum set of criteria that had to be met, e.g. concerning the soil condition and the absence of sensitive nature reserves. The 'constraint map' that resulted from the application criteria indicated the moving space of the various communities that wanted to cooperate with the Siting Task Force. As the community progressed in the decision process, the data on suitability would be tested at some point, for instance by drillings, and the results were discussed at public meetings. The Task Force largely determined the exact sites of such drillings, but citizens were asked to speak out as well.

The type of expertise employed by the Siting Task Force for constraint mapping and some other tasks was technical. 'The' expert of hazardous waste incineration that the Task Force used during hearings was a chemistry professor from the University of Alberta, who inspired a certain level of confidence in audiences across the province. This professor, Walter Harris, had been a member of the Hazardous Waste Management Team and had been greatly inspired by the idea of invitational siting and the German system of integrated waste facilities. He had continued to be present in the decision process, for instance by becoming a member of the Implementation Team, and continued to press forward the ideas of the Hazardous Waste Management Team, i.e. invitational siting and an integrated treatment facility. His opinion on the risks associated with hazardous waste incineration was (crudely summarized) that '*adverse effects on health have been highly overstated*' (Collections Editor, 1987: 3). Although there are very different opinions about this issue, such alternative sounds were not presented to audiences throughout the province. What's more, people closely associated with the Task Force noticed that they sometimes lacked the required expertise on hazardous waste incineration, which might have led to misinformation. '*We had people in the community talking about processes and technologies who were not qualified to talk about them. I was there when the plant was being built and promises were made about no runoff, zero effluent and such - it's not possible to have zero effluent. I remember one promise was that it would be a "fully enclosed pristine plant". That's ridiculous - it's a very lovely looking plant, but I wouldn't use a word like pristine*' (Sherbaniuk, 1992: 225). Such a statement reflects, above all, the emphasis that was placed on social acceptance. An important criterion in the selection process of Task Force members had been their communicative abilities, more so than their technical abilities. In addition, social scientists ran the Team. To be fair, the Siting Task Force based itself largely on the reports written by the Hazardous Waste Management Team and ECA, which broadly indicated suitable technologies for Alberta. However, the waste market and waste technologies changed rapidly at the time, for instance waste minimization became more popular. Officials at Alberta Environment responsible for hazardous waste could have given warning about this development and they attempted to do so. However, their warnings were not heard by the Task Force, which had developed into a very strong subgroup of the Department, operating under close political leadership of the Minister. Somewhat bitterly, these officials told me that '*Technical input was neglected at any phase of the project*' and that the Swan Hills facility '*is not a system that reflects the needs*'[9]. It should be noted that in the case of Swan Hills, another attempt was made to provide 'counter expertise', this time by Greenpeace. However, by the time this organization came to town, most citizens had been informed already and a community opinion in favor of the facility had been formed. Greenpeace was allowed to hold a meeting by the chair of the Citizen Liaison Group, but only on the condition that proponents of the facility would be given an opportunity to respond to its arguments. The Swan Hills did not provide a very hospitable environment for the environmental group, which consequently left the decision process. Some of its members had spread the message that the incinerator would result in a massive increase in illness among the population, but when people from Swan Hills tried to approach Greenpeace to back this statement, Greenpeace did not follow up. The consequence was that Greenpeace came to be seen as a 'scaremonger' and an unreliable organization.

[9] Interview with A. Fernandes, Alberta Environment, Edmonton, 17 June 1999.

The role of the courts

The courts were not involved in this case. In fact, the licensing procedures that followed after invitational siting had become a success, were largely pro forma events. The critique of the plans within the bureaucracy did not lead to a rejection of license applications and participation in the procedures was minimal and did not result in any objections, let alone appeals.

Later on, especially when the facility was to be expanded and when waste imports were to be discussed, the issue ended up with the quasi-judicial Natural Resources Conservation Board. During the hearings, the Edmonton Friends of the North Environmental Society, the Alberta Trappers Association, and Indian Nations from the Lesser Slave Lake region and the Indian Association of Alberta all wanted to participate. Along with them came the Swan Hills Expansion Review Coalition (SHERC). Uniquely and in line with the earlier decision process, Chem-Security, the operator of the facility and proponent of its expansion, gave SHERC and the Indian First Nation group intervenor funding in the order of C$ 60,000. Also, First Nation groups were given additional time to prepare themselves for the hearing at the request of Chem-Security, after they indicated they had not been informed in time (the NRCB had not wanted to grant extra time). During the hearings, there was much discussion about the need for the expansion. Opposition groups used some of the intervenor funding to have legal counsel and experts represent them. SHERC applied a certain pressure to broaden the agenda of the NRCB hearing from a discussion of the environmental, social and economic implications of the expansion (the NRCB focus is not just on environment) to a discussion of hazardous waste policy. Such pressure was resisted by the NRCB however. Although Chem-Security was in many ways a model participant, the company refused admittance to data on incoming waste streams at the facility (the basis for its waste projections). This seriously hampered the productivity of the debate because the hearing now became dependent on Chem-Security's figures. The odd thing is that the NRCB had refused the First Nation group intervenor funding that the group had wanted to use for assessing the market.

The NRCB hearings into waste importation show a somewhat more litigious atmosphere than the first set of hearings. Although essentially a policy issue, pressure from environmental groups that referred to agreements made during the invitational process that precluded waste importation had resulted in bringing the issue for the NRCB. The NRCB was to determine whether the incineration of wastes from other provinces would be environmentally, socially and economically acceptable compared with continued incineration of Alberta waste only. The difference with the first set of hearings is that opponents were not offered intervenor funding by

Chem-Security nor by the NRCB (the groups had requested C$ 400,000). The groups asked for a leave of appeal by the Alberta Court of Appeal to be able to contest this decision, but the Court refused. Consequently, most groups withdrew, except the First Nation group, which sunk C$ 300,000 in the hearings (Kruhlak, 1995: 2). The group argued that the risks associated with waste transportation would be unacceptably high and that there would be no demand for the facility's services because it was so expensive. The NRCB approved the importation of wastes in the end, but not after the effective interventions of the Indian First Nations, and various reports about government interference with the hearing, had attracted considerable media attention. To start with the latter, there were two incidents during the hearing that may have given the impression that the government sought to influence the Board. In particular, the Premier of Alberta indicated that the import of hazardous wastes would imply '*tremendous opportunity*'[10] and that it would '*be a problem - a real problem - if we don't get approval*'[11]. These statements are clearly not in line with the practice that silence must be maintained while a case is under consideration by a (quasi-) judicial body and may be taken to be a message to the members of the Board, all government appointees. Furthermore, there were allegations that the president of the Alberta Special Waste Management Corporation (the Crown Corporation) had contacted a consultant employed by the First Nations for the NRCB hearing, and had threatened this consultant with the loss of government business should his report be too critical of the expansion. The First Nations chief indicated that the report had indeed been stripped of some critical passages after the director's intervention[12]. The ethics commissioner of Alberta was asked to investigate these reports[13] but this was inconsequential to the decision process. In the end, Chem-Security started collaborating with the First Nations in setting up a monitoring system for measuring the concentration of pollutants around the facility.

The role of the community

Invitational siting stresses community acceptance to quite a degree. This could potentially imply a large role for the community in the decision process. I will make two remarks here about the community. The first relates to the drawing of boundaries between the relevant community and other members of society. The second relates to the question of approval by the

[10] 'Shipping toxic waste to Alberta 'makes sense' premiers say', Calgary Herald, 21 May 1994, page unknown.

[11] 'Province can dump waste plant', Calgary Herald, 31 August 1994.

[12] 'Liberals call for judicial inquiry', Calgary Herald, 20 October 1994, page unknown.

[13] See for this story 'Waste plan meddling to be probed', Calgary Herald, 19 October 1994, p. A-3.

community and its meaning. Starting with the definition of the relevant community - mainly an issue of defining the relevant boundary rules - one can observe that this definition does sometimes result in certain difficulties. Effectively, the Siting Task Force departed from the assumption that the relevant community existed of inhabitants of the towns nearby potential sites.

In the case of Swan Hills, this implied that inhabitants of the town could determine whether they wanted to accept the facility at eight or nine kilometers from their own village, a distance that led to a certain passivity among the population in the UK and Dutch cases. Not so in Swan Hills however, where the initial reaction towards the council decision to embark on invitational siting was wholly negative, mainly because citizens considered their health under threat despite the relatively large distance to their houses. Perhaps the perception of distance is very different in Alberta after all, because the First Nations that participated later on were in fact located at a distance of 80 kilometers to the North (be it that their hunting grounds were more to the South). Despite their activity however, much of the risks associated with the facility are borne by other communities that are on the road to Swan Hills. This is one reason why these communities are not overly happy with the choice of Swan Hills and don't look very favorably upon the town. More generally, Swan Hills has suffered somewhat in reputation because of its choice, since many Canadians associate the town with the treatment facility, which has experienced operating difficulties (especially the escape of PCBs). In the case of Ryley, the second town that responded positively to the request to site a facility, the tension with the surrounding communities was greater from the start. That is mainly because Ryley is part of the county where the county offices had been stormed by 200-300 citizens and where the task force had already been defeated by referendum. Perhaps indicating a certain attempt to influence the outcomes of the county referendum, Ryley had been excluded from voting in that plebiscite, and now acted on its own. The groups that had organized the opposition at the county level tried to target Ryley now, but they failed, mainly because most inhabitants of the town had by then reached consensus. Ryley's mayor in response to calls of the opposition leader said '*As far as I'm concerned, it's none of his damn business getting into our affairs*' (Krawetz et al., 1983: 73). This despite the fact that Ryley was completely surrounded by county land and communities that were hostile to the facility and despite the fact that the site required for the facility was twice the site of the village. So, if the proposal had gone ahead, the village would have had to annex land from the county. It does seem that restricting participation in the invitational siting process to citizens from Ryley points to a certain mismatch between the affected and the deciding community.

More generally it does seem that the subtle difference between how the Siting Task Force named its approach, 'invitational siting' and how others have later called the approach, 'voluntary siting', is important. Whereas voluntary siting has an aura of local democratic decision-making, invitational siting suggests something else, particularly that there is a certain proponent (the one who is invited) who has an agenda and does not just leave it to the community, so to speak. This difference demonstrated itself already in the conflict between Task Force and REDA, but also surfaces if one looks at the way the decision process proceeded, for instance in Swan Hills. Under 'voluntary siting' one would expect complete openness of information and free opinion forming. Under invitational siting one can expect less openness and perhaps an attempt to influence the public opinion. This is indeed what happened in Swan Hills. The description of the 'revamped' siting process already indicates a certain manipulative element, or as people associated with the Task Force called it, 'manipulation with integrity'[14]. Especially the power structure analysis can be seen as a covert attempt to manipulate local opinion. Indeed, the Siting Task Force did not tell ordinary members of the public that their opinion leaders had been 'turned around' by the Siting Task Force and subsequently started influencing the opinion of others (Sherbaniuk, 1992: 97-98). One of the community leaders later recalled: '*The way this process works is manipulative as hell. It was a real lesson in blatant manipulation, but it was honestly stage-managed. There was no effort at deceit, but we did everything we could to convince people and bring them onside*' (Sherbaniuk, 1992: 92-93). Besides not being open about the power structure analysis, the Siting Task Force was not very open or even honest about another issue. It appears that the Task Force created a situation where the public assumed a high level of compensation from the province. One citizen told me: '*There was a statement that we would never be short of money. Council expected that the streets would be paved with gold*'[15]. Members of the Task Force have said they did not make any promises and that the local council was responsible for any such promises. However it is certain that the Task Force was aware that there was an impression that economic benefits would accrue, and failed to counteract (Armour, 1990: 194). In addition, experiences of one community were not communicated to another one. Thus, whereas Ryley had already signed a contract for receiving a levy for every ton of waste taken at the facility, no one told the town of Swan Hills of this option or assisted them in drawing up such an agreement. The explanation for the diverging treatment of the two towns could well be that Ryley was first to

[14] Interview with W. Harris, Edmonton, 15 June 1999.
[15] Interview with J. Butler, Swan Hills, 21 June 1999.

say yes, whereas Swan Hills was second. Indeed, among the people I have interviewed in Swan Hills, there was a sense that they had been competing with Ryley at the time and should not ask for too much[16]. I am sure that such an impression would be actively counteracted in a community-based decision process, but not in invitational siting.

Some actions of the Task Force are thus best understood from the perspective of invitational siting and *not* voluntary siting. But even then one can legitimately ask questions relating to the deliberations and motivations of the people of Swan Hills. For instance was the consent of the people of Swan Hills an informed consent? Certainly they knew that the facility they were consenting to would involve hazardous wastes and incineration. In that sense there was little problem, although one may wonder why different visions about the health implications of hazardous waste incineration were not communicated. It seems that in terms of health risks, the relatively large distance of the incinerator site to the community and a great degree of trust in the Siting Task Force helped overcome uneasiness that was present at the outset. The fact that trust was more important than actually knowing what would be built and what would be the consequences is also corroborated by the fact that the citizens of Ryley and Swan Hills had little information on a range of issues. What the facility would look like, who would operate it, and what would be the benefits for the community was really uncertain. This is partly due to the fact that the provincial cabinet had not moved forward with its hazardous waste policy and was yet to condone any of the outcomes of the decision process. The consequent pressure upon the Task Force was great and actually forced them to anticipate some of these decisions. '*In the face of public opposition to the sites, the absence of government decisions and the perceived need for a facility, they advocated what was the only type of facility acceptable to the public, i.e. world class. In other words, they became staunch advocates consciously selling not only the site but also the plant*' (Krawetz et. al, 1983). The Task Force had learned that the public wanted a 'Cadillac facility', a first class facility, and that is what they promised the public. The 'Cadillac facility' was a comprehensive system for collecting and treating wastes according to the newest technologies, including first-class, high temperature incineration. Whether or not the government of Alberta would actually build such a Cadillac facility was however yet unknown.

Whatever the critique one might have, it is a fact that the population of Swan Hills was overwhelmingly in favor of the facility after the decision process. The Swan Hills referendum showed that an amazing 79% of the voters (69% turnout) was in favor of the plant. The commitment to the plant

16 E.g. interviews with Mrs. J. Butler, Swan Hills, 21 June 1999 and Mr. H. Junck, Swan Hills, 22 June 1999.

was still great when I visited Swan Hills in 1999. One expression is the continued presence of a Community Liaison Group that meets with the management of the facility to discuss its operations and which provides the company with feedback from the community. The Liaison Group, I have been told, was especially effective in influencing the company when provincial subsidy was available. The extra-safe transportation trucks for instance are an idea of the Group. The Group does not seem to function as an active means of enforcement, although occasionally news from employees on the functioning of the plant reaches the Group.

The role of discourse

The choice to use invitational siting came about in a unique situation, involving few politicians daring to go against the public, a private developer refusing to press on in the face of opposition, a government committed to the development of waste disposal capacity, and a very strongly organized opposition that takes very bold actions such as the occupation of council offices. The decision to turn to invitational was never a principled one as pragmatic arguments dominated. Especially politicians did not want to burn their hands again on the issue: '*By our process they could avoid taking the blame*'[17]. The basis of the innovation lies in a thoughtful report on the factors driving siting processes and advocating a greater role for citizens' concerns. The thrust of that report was hard for many to understand however, especially those with a technical background that advocated a more or less managerial approach. Professor Walter Harris, specialized in chemistry, told me that his first reaction towards more socially responsible siting was wholly negative and that the Hazardous Waste Management Committee had to be convinced of its soundness mainly by the one social scientist on board, who had commissioned the study of hazardous waste siting elsewhere[18]. The Management Committee in the end embraced the idea because most of its members had seen that even the public in the relatively scarcely populated province of Alberta could get very upset if siting was done in the traditional way.

In the eyes of Professor Harris, the step forward that had been made in advising the government to adopt an invitational approach was largely lost when the ECA, a body consisting mainly of scientists, advised the government on the Management Team's report. Indeed, if one looks at the ECA report, one can see that it endorses community-based siting, but then only to a very small degree. Instead the ECA proposed a more or less managerial siting approach, with a progressive pluralist hint as some

[17] Interview with N. Krawetz, formerly with Alberta Environment and former CEO of the Environment Council of Alberta, Edmonton, 9 July 1999.

[18] Interview with W. Harris, Edmonton, 15 June 1999.

compensation is recommended. So, whereas the ECA wrote that public acceptance should be 'the number one criterion' in site selection, the council suggested at the same time that a Site Selection Committee should be established, which '*should consist mainly of technical experts, since a sound technical decision will facilitate public acceptability*' (ECA, 1980: 152). The Site Selection Committee was to first select a site, and then engage the public in discussions. Through discussions, the public would come to understand the nature and need of the operations. In combination with an 'open' process, which could include negotiations about compensation, it was expected that acceptance would result. However, if this were not to be the case, then '*it should be possible for the province to establish a site for facilities even if a local municipal authority objects*' (ibid.: 152-153).

The Implementation Team, an amalgam of Management Team and people associated with the ECA, set forth the strategy outlined by the ECA. The Committee was developing very refined siting criteria and started applying them. They had already selected three municipalities as potentially suitable, when the decision process was taken from their hands. This decision was not so much taken because of principled reasons, but because of pragmatic considerations, that is, a fear of the opposition that was building up again. In the end, those who favored invitational siting 'won' the debate, in part because they believed so strongly in the approach that they promised to deliver results relatively quickly. But even then, the Task Force met with strong criticism. Take for instance the decision process that took place in the Special Areas, one of the last three communities to participate. Opponents there occupied the discursive space that had arisen because the province had left it. Their opposition came in the form of a managerial attack on the underpinnings of the Task Force's approach. Opponents there effectively used the criterion of social acceptability against the Siting Task Force. The Task Force, said one opponent, was '*trying to pick and choose what fits into their political structure or programs in Alberta. (. . .) We suggest that they shouldn't do that - that this plant has to be placed regardless of the political pressures involved and must be placed using "logistics"*' (Krawetz et al., 1983: 80). The report of the ECA and its emphasis on proximity to the waste generators was an inspiring resource for opponents. The Special Areas were outside the search area as defined there, just like Swan Hills actually. The evaluation by the Task Force was that the information seminars had missed a large part of the population as these had taken place during harvest season. Particularly when the town council requested drilling at certain locations there was an angry response from those who had not been aware of what was going on. In response, the series of information seminars was repeated but the by now angry citizens refused to play by the Task Force rules. Led by two 'opinion leaders' who had been overlooked in the power structure analysis, more than 100 people showed up at one session, most of them not pre-registered. Although the Task Force wanted attendance limited to 50 people, all 100 were let in and negative emotions flew high. The effect of the opposition here was that the Task Force was discredited. The referendum that was held showed only 21% of the electorate to be in favor of the facility.

6.1.6. Review

This case study has resulted in various interesting results. The first is that an apparently counter-intuitive idea of more community control in siting may in fact work to help site facilities. This in itself explains the great amount of attention that the case study has received and the hopeful message for hazardous waste siting on the North American continent that some derive from this experience. The second is that the description of this case as an example of local democratic decision-making along communitarian lines (emphasis on local knowledge, on deliberation, etc.) is not completely correct. The case is better described as an example of invitational siting than of voluntary siting because there are clear attempts to influence local decision processes from outside. The third is that invitational siting, while solving the problem of siting gridlock may create new problems. It does seem that the fact that social acceptability was placed at the forefront has had certain impacts on the economic viability of the incinerator, mainly because the community willing to accept it was quite far away from the source of the wastes. Another 'new' problem is related to 'border drawing', that is the identification of the relevant community. The impacts of a facility this size do affect a large area. One can think of the emission of pollutants and the risks associated with transportation of wastes. The communities affected by these impacts have not had a say in the decision process and do not benefit from the facility. Finally, the combined effect of competition between two or more communities and the manipulative character of invitational siting make the process vulnerable for a race to the bottom, and creates ethical questions. If Swan Hills were already hesitant in requesting certain favors in return for acceptance of the facility because there was one competitor, what would have happened if there had been even more competition? Sure, there were minimum technical requirements to the sites that could be proposed and therefore there was a certain level of protection built into the decision process, but still the idea that the most economically needy communities should compete for a facility like this does raise some ethical questions. The manipulative part of the siting process and the apparent favoring of one community over another do not help to alleviate these concerns. On the other hand, the facility has provided much employment to Swan Hills since it opened and did help stabilize the community.

6.2 Manitoba: competition between rural and urban communities

6.2.1 Introduction

The second case study I describe[19] takes place in Manitoba, the province that borders Alberta to the East. Just like Alberta, Manitoba used a community-based approach towards hazardous waste siting. The influence of the Albertan example was great in the sense that Manitoba essentially followed and built upon Alberta's approach. The decision process involved ten communities at some point, but this number dropped relatively quickly to two: Montcalm and Winnipeg. Montcalm, being the smaller community by far, was much quicker in concluding its deliberations and said yes in a referendum at a time when Winnipeg still had a long way to go. The provincial Cabinet in the end decided to go with Montcalm. The deliberations in that municipality demonstrate various interesting dynamics, including fierce opposition from one specific village and a clear contribution from the ordinary public in the quality of the decision.

6.2.2 The development of a siting approach

Introduction

Manitoba encountered hazardous waste problems at about the same time as Alberta, although an important difference between the two provinces is that Manitoba has little petrochemical industry and therefore generated (and generates) much less hazardous waste than Alberta. Perhaps because the waste quantities were relatively small and the province fortunately controlled storage capability for those wastes that could not be processed by the limited existing facilities or exported, there was less pressure on hazardous waste policies and decisions on treatment facilities than had existed in Alberta. Because the demand for waste treatment was so small, the private sector had shown no interest in the development of proposals for treatment facility, and the province itself was mainly looking at possibilities for cooperation with

[19] Helpful comments to an earlier version of this part of the text were received from Mrs. B. Connell (formerly with the Manitoba Hazardous Waste Management Corporation, MHWMC), Mr. D. Foidart (former Chair of the Montcalm Community Advisory Committee), Mr. E. Kucera (formerly with Manitoba Environment), Mr. A. Richards (formerly with the MHWMC), and Mr. L. Strachan (Director of Environmental Approvals at Manitoba Environment).

other provinces, notably Alberta which was a bit ahead with its proposals. The hope of cooperation was dashed when the Alberta government took a position against cooperation and after construction of the Swan Hills facility formally committed itself to a policy of non-importation. Manitoba thus had to develop its own treatment capability.

The Manitoba Hazardous and Special Wastes Management Program

In comparison with Alberta, Manitoba was in an advantageous position because it had storage capacity, but also because there was no siting failure in the history books and hence no sensitized communities (Cooke, 1988). Manitoba learned much from the way Alberta had sited its facility, for instance through a symposium on hazardous waste management with national and international experts in March 1983. The meeting marked the official adoption of the Manitoba Hazardous and Special Wastes Management Program. This Program was to result in an integrated provincial hazardous waste treatment system in three phases. First an information-gathering phase would take place, then a design phase for a hazardous waste management system and the third phase would imply the actual implementation of such a system. The first phase involved an attempt to assess the hazardous waste market in the province by sending out confidential questionnaires to the relevant industries and by interviewing responsible officers in these industries. The fact that Manitoba choose to adopt hazardous waste legislation in 1984 (the Dangerous Goods Handling and Transportation Act) assisted the province greatly in getting a reliable image of waste arisings. The act started regulating both waste generation and waste transportation and required notification to the province in many instances. The first phase resulted in a report describing the extent of the problem, which was presented to the public through a set of hearings by the Manitoba Clean Environment Commission. During these hearings various issues were discussed, including the need for hazardous waste management and the role of the private sector in a hazardous waste management program. The hearings were not well attended by ordinary citizens, but most organized interests had been present. On the basis of the hearings, the Clean Environment Commission (CEC) concluded there was support for the institution of a hazardous waste management program and construction of a hazardous waste treatment facility (the shape of which was to be determined later). Industrial representatives and environmental groups had both warned about being very careful on the issue of waste projections as many developments were taking place in the waste market. Manitoba Environment heeded these warnings and decided that the data on waste arisings stemming from the first phase of the program would be updated annually. More generally, the decision was made to progress 'incrementally' from then on.

Design of a provincial system

After the hearings of the CEC, the Manitoba Cabinet took the decision-in-principle to construct a waste management facility. In addition, the Cabinet decided that a Crown Corporation, the Manitoba Hazardous Waste Management Corporation, should be responsible for the design of the system and would also act as the proponent for the new facility. The Corporation would thus play a large role in the second and third phases of the Manitoba Hazardous and Special Wastes Management Program (design and implementation). The reason why the Cabinet opted for a Crown Corporation was that such a corporation would be seen as an institution that worked for a public purpose and did not seek profits from hazardous wastes. At the same time, the Corporation would be seen as being independent from the government, specifically Manitoba Environment that would have to license the new facility.

The groundwork for the provincial hazardous waste program fell in the hands of the CEC again. Ordered to do so by the provincial Cabinet, the Commission organized a new set of hearings, which were to function as an input into the design process of the new system. On the basis of the hearings, that were once again not very well attended, the Commission suggested that there should be a central treatment facility for hazardous wastes in Manitoba. Given the small market for treatment, such a facility would not encompass incineration capability. Wastes requiring such treatment were expected to go out of the province. The Commission also recommended the set-up of a siting process that was largely based on the process that had been applied in Alberta. The Commission advised that the siting approach should be highly participatory and communities should be given a veto-right against proposals for the central facility in their area. In the eyes of the Commission, such participation served a dual purpose. Not only would resistance against the facility be avoided, but in addition, citizens would come to understand the waste problematique. The Commission added that economic and other incentives should be given to communities willing to host the facility, but that such incentives should not be allowed to corrupt the process by resulting in technically unsuitable sites. In order to guarantee that selected sites would be technically acceptable, the Commission suggested that a list of siting criteria should be developed. All sites receiving real consideration would have to meet the criteria. The Commission did not apparently have a rock solid faith in the participatory approach. It suggested that in case of failure of a community-based approach, the government should chose a site and override potential local resistance.

On top of the CEC-advice came a long-term strategic plan of the newly-founded Crown Corporation that described its goals and approach. The crucial goal of the Corporation was described as the provision of a universally accessible waste management program, with all treatment options present, a large emphasis on waste prevention, and low costs for the waste generator. The Corporation explicitly rejected an approach to the issue of waste management that would be rigid. Instead, the explicit purpose was to proceed incrementally, not only because waste arisings were rapidly changing, but also because technologies were changing from large scale fixed processes (such as incineration) towards source based, smaller and transportable technologies. Note that Alberta invested in expansion of its incineration capacity at the same moment.

Site selection process

The MHWMC learned a great deal from Alberta in terms of siting. Various people with knowledge of the decision process there came over to Manitoba and some of the lessons learned in Alberta would also be applied in Manitoba. Various documents (amongst others Cooke, 1988; Latonas, 1988) outlined the MHWMC approach. Like in Alberta, the authority rules guiding the process would revolve around a 'double veto': both a local community and the province could decide that a certain site was not acceptable. Like in Alberta, 'constraint maps' of the province were made, which indicated what parts of the province would certainly not be acceptable from a technical point of view, and which ones were potentially interesting. In addition to these maps at the provincial scale, regional maps were to be made that would reflect local needs and wishes. Partly based on the experiences in Alberta, the MHWMC was sensitive to possible disagreement between neighboring communities that might result from a community-based decision process. The choice was made to approach the decision process at the municipal level, but to leave open the option that several municipalities would collaborate. In essence, the MHWMC was hoping that the relevant community would define itself (see Richards, 1996: 3). Like in Alberta, a referendum did not really fit the political system of Manitoba. But on the basis of legal advice, the MHWMC suggested that the municipal council could in fact organize a legally valid referendum in the relevant communities.

The key terms used in the MHWMC documentation to describe the intended decision process were 'voluntary siting' and 'co-management'. Voluntary siting differs from 'invitational siting' (Alberta's term) in that it suggests a higher intended level of voluntariness. My interviewees, previously connected with the MHWMC, indicated that a power structure analysis as had been performed in Alberta was out of the question under voluntary siting. Co-management implies that the community's involvement with the facility would not stop after it had been sited. Like in Alberta, community liaison committees would be established and these would play an important role in the management of the facility.

6.2.3 Actual siting discussions

Open houses

The MHWMC went public with its new approach and organized a set of 13 'open houses' (exhibitions) in the month of November 1988 to present the siting process and the site selection criteria. The open houses were public events, but were specifically intended to raise interest among municipal councils. At the conclusion of the open houses, all municipal councils in Manitoba were asked to express their interest in the siting process. About 40 municipal councils did so. As a consequence an information campaign started in these municipalities.

The information campaign revolved around more open houses and some public meetings during which the local public could educate itself on the proposals. During this period the MHWMC learned some tough lessons and developed some rules of thumb that would be helpful later. One lesson learned was that public meetings were not a suitable forum for presenting the information and discussing the siting process with the public. Open houses were much more suitable as they allowed for opportunities for the public to interact with the MHWMC staff and discuss the issues on a one-to-one basis. Visitors were free to come and go during an afternoon or evening and view the exhibits, discuss their concerns or questions, and register their views about assessing the possibility of siting the facility in their communities. Whereas the few public meetings that the MHWMC held resulted in a very hostile atmosphere, the exhibitions tended to result in friendly interactions with an interested public. Because of this, exhibitions became the only way of interacting with the public.

Indicators of acceptance

But even open houses were sometimes not very successful and the MHWMC staff learned to recognize the indications for this in the questionnaires that all visitors of the exhibitions were asked to fill out. Initially, the analysis of these questionnaires was seen as a way to learn about the public reaction, but it soon transpired that the *number* of questionnaires was much more telling than the content of the responses. In some communities the results of the questionnaire were rather positive, yet the general public reacted in a negative way when the MHWMC consequently progressed with the decision process. After some experiences, the MHWMC staff concluded that unawareness of the MHWMC's activities in the community could quickly lead to a negative reaction. Furthermore, it was crucial that those who heard about the MHWMC heard about it from the MHWMC directly. As the exhibitions were the main way of communicating with the public, the MHWMC started using turnout as an indicator of the community opinion. The MHWMC adopted the policy of stopping its activities in a certain community when turnout at the exhibition had been low. As a rule of thumb, the MHWMC also decided to stop the process in the case of a negative response from those filling out the questionnaire and a quick and hostile reaction from the public, for instance in the form of a petition signed by a large part of the community within two months.

The elimination game

Based on the just discussed indicators, the MHWMC left 30 of the 40 communities alone after the first open houses. Ten communities remained in the process however, and serious attempts were made there to gain acceptance by establishing Community Advisory Committees (CACs). At this point a drawback of the incremental approach surfaced. In order to maintain flexibility, the province had retained an open mind about the way the central facility would look like, apart from the fact that incineration was excluded. As had been the case in Alberta, MHWMC representatives in the field were often pressed to tell which types of treatment would take place, what the facility would look like, and what would the impacts on the surroundings be. Such questions could only be answered in rather vague terms as the province continued updating its waste data until relatively late, March 1990. Based on the information that was available at that point a facility design was made, but this was too late as five further communities had withdrawn from the program. Three more followed somewhat later. Despite all the care and attention to public acceptance, opposition emerged in all three municipalities.

Two communities remain

In the end only two communities remained, which were rather different from each other. On the one hand there was Montcalm, a rural municipality, consisting of three small agricultural villages. On the other hand was Winnipeg, the capital and largest city of Manitoba. Alun Richards, one of the members of the siting team that later analyzed the differences between the two communities (1992), concluded that the siting process encountered greater difficulties in Winnipeg than in Montcalm. This was especially so because in Winnipeg the benefits of approval of a facility would be dispersed, but costs would be concentrated. In Montcalm on the other hand, costs and benefits would both be concentrated. In addition, Richards argues that in Winnipeg there were far less possibilities for continuous interaction between citizens, who had 'delegated' their involvement in local politics to public interest groups anyway. In Montcalm on the other hand, many members of the community were personally familiar with each other and could often interact. These factors indeed resulted in a different response, initially only in the sense that the decision process in Winnipeg progressed much more slowly than Montcalm's. Despite that being the case, the MHWMC continued to be very interested in a location in Winnipeg. This was mainly because the city generates most of the wastes from the province and less transportation of wastes would thus be necessary if the facility were to be located there in comparison with a location in Montcalm, which is some 80 kilometers South of Winnipeg. In addition, if Winnipeg dropped out of the process, there was a chance that other communities would develop a sense that Winnipeg was abusing them. Therefore, Winnipeg's role was of great symbolic value.

Winnipeg's city council had been very aware of its responsibilities, and took notice of the fact that many other communities had described the hazardous waste problem as a Winnipeg problem. The council continued to support the siting process even after most other municipalities had withdrawn. In order to speed up the process somewhat by avoiding problems with landowners, the council looked at various pieces it owned itself. This exercise resulted in 11 potential locations, all on municipal land. The MHWMC consequently studied these locations and determined that four were promising. However, public opposition mounted very quickly in three of those sites, and in response the MHWMC came to focus on just one site. Detailed assessment of this site took place by the MHWMC with seemingly promising results. Winnipeg council then established a CAC with a very limited mandate: it was to check whether the MHWMC had not missed anything in its analyses. Furthermore, the CAC was to explore public opinion on the issue and report its findings back to the council. Because Montcalm was far ahead of Winnipeg, the CAC came under enormous time pressure, ran into public opposition, and its report was published after the actual decisions on the facility had been made, in favor of Montcalm.

Montcalm: the desire to diversify

The population of the Rural Municipality of Montcalm, located some distance away from Winnipeg to the South, had already decided they were willing to host the facility through a referendum one month after the Winnipeg CAC had been established. In the municipality, a CAC had been established in May 1990, after open houses had indicated public support. The interest of Montcalm in the facility was somewhat surprising in the sense that the municipality had little experience with chemical industry: it was and is dominated by agriculture. However, the council of the municipality had decided that its economy was not diverse enough and that the economic base should be broadened. When the council decided to found a CAC they had had the wisdom not to appoint any councilors, since these were widely regarded as proponents of the facility. The council appointed Dennis Foidart, a science teacher at the local high school, as the chair of the CAC. Like him, most of the CAC were not outspoken opponents or proponents of the facility and most of them had no specific knowledge of chemicals, although the farmers had some experience with pesticides, of course. Most members of the CAC were interested in the protection of agricultural land and products, safety, and possibilities of establishing a certain level of control over the facility. Because there were relatively many issues to be studied, the members split up in subgroups that set out to study certain specific themes.

Added value of the CAC's work

The CAC's work on collecting information about the acceptability of the facility for the area largely focused on a study of the relevant siting criteria, the collection and checking of data for the regional constraint map, and the identification of specific sites that were particularly promising. Some of the information collected by the CAC had a great impact on the outcome of the decision process. This is especially true in the sphere of determining where the flood prone areas were located in the municipality. Because the village is near the Red River, it should be no surprise that large chunks of land had been considered unusable by the MHWMC initially on the basis of provincial geographical data on flooding. The CAC however interviewed the farmers in the area to see whether certain spots had really been flooded during the past decades. In doing so, the CAC found a spot that was flood prone according to provincial maps, but had not been flooded according to local farmers. The local knowledge was in the end accepted and the provincial maps were adjusted. The specific site was later selected as the best spot for the facility, also because the soil was practically impermeable and the owner was willing to sell[20].

Conflict

Like in some other communities, the test drillings that started taking place to corroborate the findings of the CAC and MHWMC had quite an impact on the local population. The CAC had identified various spots that seemed suitable. However, when some drilling started taking place, opposition suddenly reared its head. A very active action group formed, and it appeared that frictions within the municipality arose, specifically because most of the resistance was concentrated in one of the villages in the municipality. This prompted a discussion in the relevant constituency for the

[20] The selling of land was voluntary; owners had to offer their land.

decision process, but it was decided that the decision would be taken on the municipal level, not the level of specific towns in the municipality. The action group had written a petition against the facility. They criticized the CAC for tinkering with the criteria regarding flooding and also questioned the exclusion of nature reserves as the area for the facility and suggested that economic considerations were paramount in the CAC's work. '*Are humans, animals, meat and dairy products, cereal grains, lentils, corn, peas and sugarbeets not worth of protection against contamination???*'[21].

Although the opposition group had collected quite a number of signatures, the MHWMC did not immediately withdraw because there was a suggestion of manipulation with the action group's petition. In response, an opinion poll was held among the population, which indicated a high level of support. The process, nearly terminated by the MHWMC, survived in the end. After the CAC concluded its work with a positive recommendation on the proposals, a referendum was held in September 1991. The outcome was that about 67% of the voters in the municipality were in favor of the proposals. The opposition group admitted its defeat and accepted a place on the CAC that would now start negotiating the terms and conditions for the municipality's acceptance of the facility. As various communities in the decision process, including Montcalm, had expressed a preference for a written contract, the negotiations between MHWMC and the municipality resulted in a binding agreement. This agreement guaranteed the municipality, represented by a small group of citizens (the co-management committee), that they could veto fundamental changes to the facility and ownership changes. Apart from that, Montcalm had to wait for a while on the provincial decision on the municipality that would receive the facility in the end.

The province decides

Although the MHMWC tried to buy time so that Winnipeg could stay in contention, the Manitoba Cabinet announced that Montcalm was its preferred site in February 1995. Official licensing procedures were started on the basis of draft environmental assessment guidelines that were discussed with the CAC of Montcalm. Because much of the required information had already been collected by the MHWMC and the CAC, the assessment report could be largely based on already existing material and was quickly finished. Review by Manitoba Environment and public hearings by the Clean Environment Commission revealed few problems and a license was issued without any trouble. Because the municipality of Montcalm had practically

[21] Leaflet Citizens Concerned about Ecology and Environment, '*How concerned are you about the agricultural industry in the rural municipality of Montcalm?*', 1990. Underlining in the original.

no zoning bylaws at the time, there was no requirement to go through planning procedures. However, some of the employees of the MHWMC considered the lack of planning policies a bit of a liability in the sense that future development could not be steered away from the facility. With the financial assistance of the MHWMC, a zoning bylaw was drafted and accepted that would prevent such encroachment. Figure 3 indicates the location of the site.

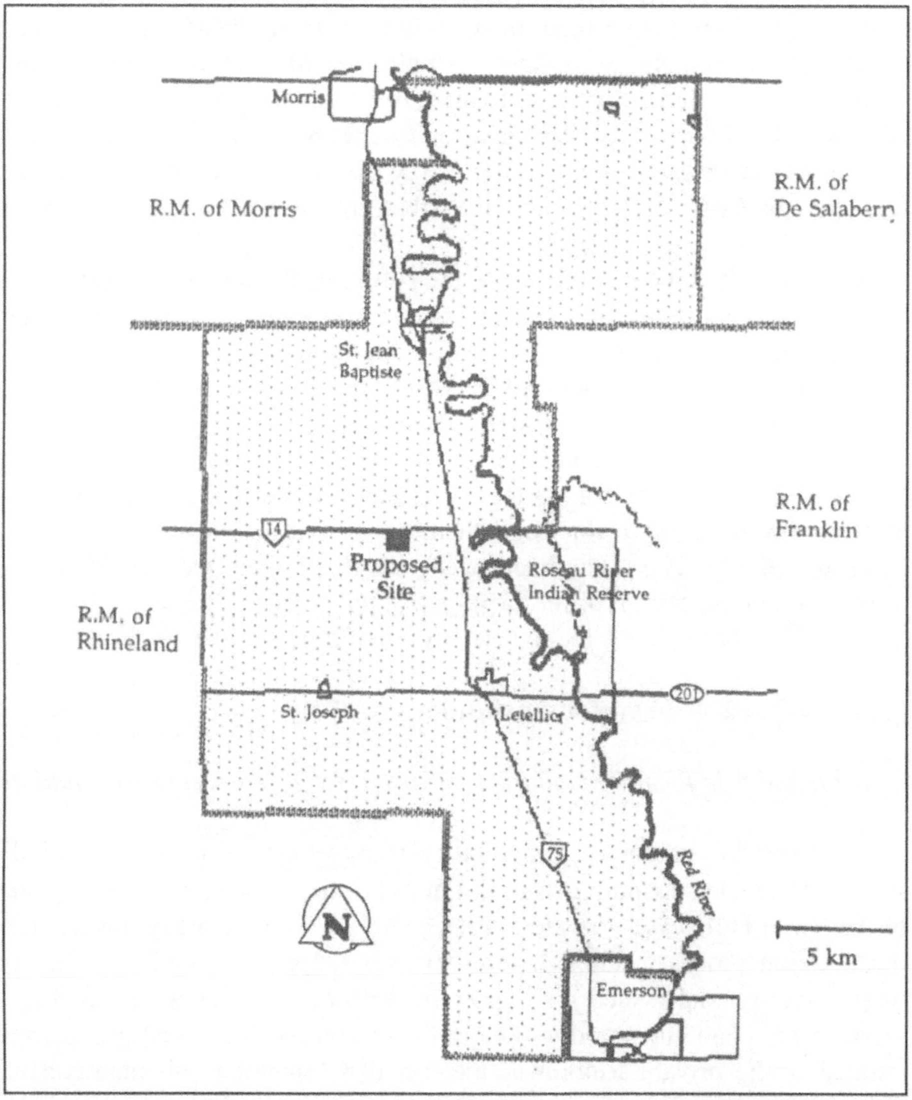

Figure 3. Location of Manitoba's special waste treatment station. Constraint map used by the MHWMC. The 'proposed site' later became definitive.

Interestingly, as happened in Swan Hills, the speed of the decision process was reduced considerably after the co-management agreement was signed and regulatory approvals had been issued. Instead of starting construction immediately, the province went looking for a private investor that could start operating the plant. The province had difficulty in finding private investors and several potential investors did not sign an agreement after some negotiations. This situation lasted for quite some time and by August 1994 still no private investor had been found. Instead, the MHWMC itself had started to operate a soil remediation facility and a small transfer station with an extensive monitoring system, actually designed for a much higher level of activity. Many people in Montcalm were actually disappointed by the level of employment that was being generated. It was to take until 1996 when a private investor was found, which has continued to operate the facility in a relatively low-key fashion. By most standards, the facility remains small and is best described as a transfer and pre-treatment station.

When I visited the plant in 1999, the soil remediation facility had ceased operating. Some of my interviewees who were involved in the plant's co-management committee were critical, but generally happy about the operations at the plant. They indicated that most people in the community were not aware of the plant anymore, as activity was slow. The plant was not a great concern, and as a consequence, it was difficult to find new members for the co-management committees. The new members they did find were also less aware of the history and therefore perhaps somewhat less critical about the plant[22]. There has been one great flood since the facility started operating, but the terrain of the facility was not flooded. Instead, it has been used as a safe haven for some materials of the farmers in the neighborhood.

6.2.4 The role of various institutions

A limited role for private market parties, elected representatives, and the courts

As described, voluntary siting and co-management in this case implied a somewhat greater deviation from the normal rules of decision-making than in the Swan Hills case. Private market parties were practically absent from the decision process, although industry was helpful in providing data on waste arisings and issued an important warning that the waste treatment market was changing rapidly. The main reason for the otherwise passive attitude of the private sector was the fact that Manitoba was uninteresting from a commercial perspective, given the fact that little waste was being

[22] Interview with G. Fontaine and L. Leclair, members of the Community Liaison Group, Letellier, 5 July 1999.

generated there. It was only after parts of the projected facility had been constructed that private market parties became rather interested. At that point, they ran into the co-management agreement that the province had made with the municipality and this curtailed their plans with the facility. The current operator had a significantly lower bid on the facility than some of its competitors, but the co-management committee had been tipped off about one other company that had been interested in the facility and that apparently only liked the landfilling capabilities of the site. The co-management committee opposed such plans and prevented this party from taking over the facility.

The fact that the siting process was set up the way it was, is of course in the end a political decision. Even though Manitoba had not experienced any siting failure, its politicians, especially the Cabinet, was convinced that siting failure would be a result if the decision process were to be structured 'normally'. The Alberta experience provided an illuminating example and it seems that the Manitoba Cabinet never considered an alternative approach. What may have helped is that there was not a single private initiative to develop treatment capability that would have required another approach and secondly that the province had a certain storage capacity for the wastes so that they had some time to waste so to speak. Throughout the entire decision process, the political climate remained supportive of the decision process, even though a different party came to rule Manitoba.

The involvement of local politicians was somewhat greater of course as they had to decide whether they wanted to extend an invitation to the MHWMC to enter their territory or not. In many cases, this was a relatively haphazard decision, even if the open house questionnaires had demonstrated great support for the siting process. The fact that 38 of the 40 municipalities that were initially interested had retreated in the end attests to that. A recurring theme, as it was in Alberta, is that many citizens equate the council with an economic development perspective and that they consequently do not trust the council with this decision. The presence of councilors in the debate at any point before the referendum had convincingly demonstrated a majority in favor of the proposal, was detrimental to the process.

In one of the last five communities to be involved in the process, Rossburn, the local council, which was on record as a proponent of the facility because it provided such a great economic opportunity, chose to have two of its members sit on the CAC. As a consequence, the CAC was not perceived as being neutral anymore by the public. In fact neutrality per se became an important issue of contention in that municipality. An opposition group formed which explicitly demanded that the CAC hire independent consultants to assist. The CAC members, one councilor included, refused to do so and indicated that its deliberations would be based on data and information provided by the MHWMC. This refusal only increased speculation on the independence and neutrality of the CAC. The issue then became very divisive and started influencing the next elections for the local council. The opposition group had targeted various councilors in favor of the plant, and some of them lost indeed their seat. When the new council then pushed for a referendum, even before the CAC deliberations were completed, the consequence was a head-to head collision between council and CAC. Despite protest from the CAC, a referendum was in fact held. The council adjusted the boundary rules and excluded the parts of the municipality in favor of the facility from voting. As a consequence the proposal was defeated.

Even in Montcalm, there was a serious challenge to the siting process at some point, even though local politicians remained far from the CAC. When the work of the CAC came under fire the reeve (mayor) of the municipality stepped in and prevented the process from collapsing. His exceptionally good standing in the municipality helped him achieve this result, as did his intimate knowledge of community interaction processes. He was effectively able to undermine the case of the protest by pointing out they had intimidated citizens. He claimed that '*at least 300 or more*' of the over 400 signatures were from people who '*didn't know what they were signing. They actually thought they were signing in favour. (. . .) They didn't know what they were signing or were misled. (. . .) We've had people tell us they actually signed just to get rid of the people. (. . .) There is a very general consensus among Councillors and the Committee people, that the petition doesn't mean a thing*'[23]. This intervention was effective and saved the decision process.

In this case, the courts played no role, and neither did any of the quasi-judicial bodies of the province, except the Clean Environment Commission, which has a largely advisory role, but does perform some quasi-judicial duties. The Commission played an important role through its hearings. Environment groups especially came onside after hearing and debating the extent of the hazardous waste problem in Manitoba. These groups have played a limited role since then.

The role of experts and the role of the community

Like in Alberta, the Manitoba decision process was driven in part by experts and in part by the community. Experts played a role that was somewhat more limited than they would normally do. It is interesting that, like in Alberta, there was no pretense that experts could select the most suitable site. Instead, they did provide the provincial constraint maps that

[23] 'Fears for agriculture arise over waste plant', in: The Scratching River Post, 21 January 1991, page 1 and 7.

demonstrated the maneuvering space for the communities. Experts also assisted CACs in assembling regional versions of the provincial constraint maps and experts also designed the projected facility and would license it. Like in Alberta, the MHWMC elected communicatively able experts to represent its case. Especially in Montcalm, where the CAC asked for already existing examples of such contracts in the USA, they were diligently assisted by the MHWMC. In Montcalm, the MHWMC acted as an ally of the CAC, but also appears to have corrected the CAC to some degree in its local approach to the siting issue. Specifically, the MHWMC representative urged the CAC to invite representatives of nearby municipalities. Although the CAC was somewhat hesitant towards the idea, it was implemented and remarkably, the support of all surrounding municipalities was received. This later paid off in terms of the standing of Montcalm with its neighbors[24].

It would be wrong to say that 'the' community played a great role in the decision process in the municipalities that the MHWMC visited in the course of a few years. The greatest role was of course for those who became members of the CAC, and this membership was limited to 10-15 members of the ordinary public. More passive forms of participation, such as the visiting of open houses, the signing of a petition, and in Montcalm the voting in the referendum was however very common.

There is another reason why it is difficult to speak of participation by 'the' community. This is because the decision process tended to create new or intensify already existing tensions within the communities. In Rossburn, the municipality where the CAC and council went head-to-head, feelings of inferiority of certain segments of the population infiltrated the decision process. The proposal was seen as being directed against the Ukrainian farmers in the area, which was unfair, certainly given the fact that the facility would not be sited near a nature reserve[25]. The issue became very divisive. A local journalist said: '*It's group against group. (. . .) They are both fighting so hard, it gets personal. It's hard to believe things like that would happen in a small town*'[26]. When the decision was made that a referendum would go ahead despite the objections from the CAC, the reeve said '*There's so much animosity and dissension over this. The rural council is caught between the devil and the deep blue sea*'[27]. Rifts and disagreements in communities were certainly not the exception. In another community, where the siting process

[24] In correspondence with me, Mr. Richards observed that the site that was finally chosen is in the geographical center of Montcalm, 'taking away the arguments from the nearby communities that it is "next" to them'.

[25] 'Bunch of Ukrainians say "no" to hazardous waste site', Winnipeg Free Press, 18 January 1990, page unknown.

[26] 'Waste site plan causes town feuding', Winnipeg Free Press, 23 October 1989, page unknown.

[27] 'Waste plant goes to vote', Winnipeg Free Press, 14 December 1989, page unknown.

was terminated before the CAC issued its advice, one opponent said '*I don't think the community could have lasted the summer not knowing whether it's on or off. The rift in the community would have been too great*'[28].

Restricting myself to Montcalm, it is clear that there was a certain potential to divide the community there was well. Its population is mainly French speaking, Roman Catholic, and is divided over three towns: St. Jean-Baptiste, St. Joseph and Letellier. Although Alun Richards considered the community tightly knit, there are also some indications of (small) divisions within the community (Castle, 1992: 21-22). In particular, there appears to be some competition between St. Joseph and St. Jean-Baptiste. Each town holds a small vegetable processing plant and these compete with each other. Also, there was some bitterness in St. Joseph over the closing of its school, and it appeared that the Roman Catholic religion and French culture was taken more seriously in St. Joseph (ibid.). It does seem that the opposition group that arose during the CAC's work was concentrated in one of these villages and thus had the potential of creating a rift. The intervention of the reeve and subsequent result of the referendum ended this potential however. I have asked one of the staunchest opponents of the facility, who later became a member of the co-management committee, why he had ceased opposition after the referendum. In reply he explicitly referred to the interest of the community that would not be served by contensions[29]. It does seem that bringing the staunchest opponents on board has strengthened the outcome of the decision process in the end. The community liaison committee, fueled by a certain level of suspicion by their new opponent members, negotiated very hard on the co-management agreement. The group was allowed to visit Swan Hills at some point, and especially the former opponents were able to stir up some critical reflection from some citizens there. The entire Montcalm delegation came back with the impression that Montcalm should get a far greater level of control over the facility than the citizens of Swan Hills[30], and this was indeed realized.

6.2.5 The role of discourse

Enlightened managerialism

Despite the fact that the decision process was largely community-based, one should not conclude that communitarian discourse had captured those designing the decision process. Instead, the people involved are perhaps best seen as subscribing to pragmatism or an enlightened version of

[28] 'Hamiota sets June for vote on waste centre', Winnipeg Free Press, 30 March 1990, p. 2.

[29] Interview with G. Fontaine and L. Leclair, members of the Community Liaison Group, Letellier, 5 July 1999.

[30] Interview with D. Foidart, Letellier 5 July 1999.

managerialism. The choice for the community-based approach to siting was purely pragmatic in the sense that Alberta has demonstrated that it worked. The Alberta example was closely analyzed and improved upon to some extent, although some lessons learned in Alberta (e.g. no large public hearings) were apparently not communicated to Manitobans. It is clear that there was a certain level of concern, especially with the Clean Environment Commission, about the chance of success and this body essentially advised to override community resistance in the case of failure of the siting process. In this sense, the communities stepping forward and having themselves assessed took a certain risk, of which they were well aware. This risk revolved around the possibility of identifying a technically suitable site in the process that would not be approved by the local population. The province would, in case of complete failure of voluntary siting, be aware of the site and might try to focus on that particular one. However, even the CEC, a body largely consisting of experts, mostly accepted the voluntary siting approach.

When I say that the discourse that captured the MHWMC is one of enlightened managerialism, I mean that corporation personnel saw the importance of delivering technically suitable sites, yet recognized the limits of what a scientific site selection process could deliver. The MHWMC for instance stated that: '*No simple or right definition of "the best site" exists, because the judgement of "best" will vary from individual to individual*' (MHWMC, 1992: 158). Also part of this 'enlightened managerialism' was a certain level of help for the communities in the race. Certainly in Montcalm, where Alun Richards functioned as the MHWMC liaison, the CAC was assisted quite diligently in its tasks and was in fact reminded of issues (such as the need for support from the surrounding municipalities) that would later be valuable to the community. Managerial in the sense that the MHWMC largely avoided the topic of compensation during the decision process. Although it was a fact that the facility would bring a certain level of employment, and many local councils (including Montcalm's) were driven to some extent by this 'reward', there was no reference of financial compensation during the CAC's activities. I have been told that this was at the request of the MWMHC, which anticipated strong negative local reaction if there was talk about compensation. Indeed, in some communities where the MHWMC failed, opponents had successfully alleged that the MHWMC was bribing the community. This occurred especially in communities where local councilors sat on the CAC or played an otherwise important role in the process.

In Montcalm, the CAC approached the issue mainly from a technical angle, meaning that unsuitable sites should be eliminated from consideration and fully suitable sites identified. To that end, the CAC could use a so-called 'drilling budget' for testing certain promising sites. In the process, there already was some attention paid to the will of local landowners. Potentially suitable sites, owned by people who did not want to sell their land, fell from consideration. The strong emphasis on technical suitability of sites dropped off even more when the siting process reached the level of negotiations on the co-management contract. During these negotiations, Montcalm secured a certain level of tax income from the facility and inserted a condition in the contract that obligated the future operator to hire local personnel. Especially the latter condition was a serious concern for the municipality, and was often followed up upon after the facility had been constructed and too little local personnel was hired in the eyes of the council.

Vulnerability

Enlightened managerialism implied a certain vulnerability from the communitarian perspective and also put a strain on the MHWMC in cases where the local population applauded the coming of the facility, but where it was hard to find a technically suitable site. To start with the communitarian critique, the MHWMC was sometimes attacked for influencing a free local dialogue. The Manitoba Eco-Network (1989: 5), a network of environmental groups, described the MHWMC's approach as follows: '*A philosophical or perhaps uninformed "no" isn't seen as valid, by the Corporation, or by the CAC's. Opponents are urged to wait and make up their mind after they consider all of the facts*'. The MHWMC should stay out of the local dialogue between community members: '*The Corporation should take care not to insert itself in the dialogue, however misinformed, between the community members, aware that it is a guest, not a resident. It should also recognize that with its superior resources and technical background responding to criticisms might appear to be clobbering the honest reaction of citizens*' (ibid.: 8). Such statements were made after the MHMWC had issued 'counter information' against some of the arguments made by opponents. The Eco-Network lamented this approach and stated that - in the interest of the environment - decisions should be made solely by the local public and environmental groups (ibid.: 1). This communitarian language remained without effect however. Throughout the decision process, the MHWMC continued to apply a policy of correcting 'false' information by either party in the debate. In this sense, local knowledge and dialogue were not allowed to flow freely.

Somewhat similar was the case of Pinawa. This municipality was a prime candidate in the eyes of the MHWMC for quite some time and the community was very willing to host a facility. However, it seemed that there was no location in the municipality that could meet the siting criteria, especially the ones related to geological conditions. The reason why the community was considered a prime candidate is that it is home to a large research facility of Atomic Energy Canada Limited (AECL). In fact, AECL had constructed the town to house its employees, and it was expected that their familiarity with chemicals would provide a receptive social environment for the facility. Alun Richards: '*Pinawa probably has the highest proportion of Phd's in the country. We felt that, as the local citizenry was familiar with the issues involved with hazardous materials, it would be more receptive to the plant*' (in: Castle, 1992: 6-17). This expectation appeared to be correct. Rick Cooke, the director of the MHWMV, said the municipality's reaction was 'fascinating' as the attitude of the town was exactly opposite to what could be seen elsewhere[31]. Certainly the mayor of the municipality was very receptive to the idea. He stressed that environmental studies would be required first, but was very enthusiastic about the economic consequences of the facility, which made 'good sense' and would improve the tax base of the municipality[32]. He indicated that most of the inhabitants of the town understood that treatment and disposal of hazardous waste did not have to be dangerous[33]. The surrounding communities did not voice any objections either and thus social acceptability appeared to be guaranteed. The problem however was that only one potentially suitable site existed in the community and the site would likely meet the provincial criteria. The locally available expertise was used to test this site. Drilling was undertaken jointly by AECL, the MHWMC and the municipality, with AECL personnel doing the actual drilling work. The tests revealed that the potential site was located in a very flat and poorly drained area. The clay and till layer at the site was fractured and site hydrogeology was very complex and difficult to characterize. The conclusion was that the site would fail to meet the requirement of restricting potential contaminant movement in perpetuity (MHWMC, 1992: 178). The mayor of Pinawa then indicated that he was still optimistic about Pinawa's chances to obtain the facility. He expressed hope that the Corporation would consider further drilling in the municipality because some of the C$ 10,000 drilling budget allotted by the MHWMC would still be in place after the initial drilling tests[34]. The MHWMC, however, grudgingly ruled Pinawa out of further consideration, a decision necessitated by the will to adhere to the siting criteria, and indirectly the will to remain credible in other communities where siting took place. The decision caused considerable anger on the part of the municipality. Again, a feeling of unfairness transpired, but this time

[31] 'Pinawa fearless about disposal site', Winnipeg Free Press, 2 November 1989, p. 2.

[32] 'Man. town hoping it gets waste site', Western Producer, January 1987, page unknown.

[33] 'Pinawa fearless about disposal site', Winnipeg Free Press, 2 November 1989, p. 2.

[34] 'LGD of Pinawa still in the running for waste site', The Leader, 19 September 1989, p. 1.

because the community felt that other communities were unrightfully favored over Pinawa.

6.2.6 Review of the case

In terms of institutions and discourse this case is largely similar to the Alberta case. The decision process can be viewed as a double veto, with the local community having a place side-by-side with the institutions that are normally involved in the decision process. This is, again, reflected in the outcome of the decision process, as the final location of the facility is located relatively far from the waste generators. In comparison with the Swan Hills case, the role of private parties in the decision process was even more limited in Manitoba because of the arrangements for co-management. Especially because the CAC in Montcalm found that Swan Hills had only limited control over the facility once it had been sited, they desired a continuing role after completion of the siting process. They obtained such a role and exercised it, for instance in determining who would become the private operator of the facility, once it got to the point where the province relinquished its role in operating the facility. The motivation for this rather remarkable approach to siting seems largely pragmatic. There was no private initiative towards development of the facility whatsoever, which implied that the normally reverent attitude toward the private sector could be abandoned. In addition, politicians feared the local reaction to siting proposals (this fear later proved to be correct) and were more than happy to be inspired by the Alberta approach, which had been successfully completed by then.

6.3 Deep River: low-level radioactive wastes

6.3.1 Introduction

This section describes[35] the third case I have selected for study in Canada. Again inspired by the Alberta siting success but also by the continued failure to find a solution for large quantities of low-level radioactive waste, the federal government of Canada decided to set up a voluntary siting process for these wastes. The approach was to be even more participatory in nature than Alberta's and Manitoba's processes, as local communities not only just determine the site, but also decide which technologies they deem appropriate for the wastes. In addition, there was very explicit attention to compensation for the volunteer communities. Although the process worked in the sense

[35] Helpful comments to an earlier version of this text were received from Prof. Mrs. A. Armour (formerly with the Siting Process Task Force and the Siting Task Force) and from Mr. D. McCauley (Canadian Department of Natural Resources).

that one community volunteered to accept the wastes in the end, the federal government decided not to accept this offer, mainly because the demands of the community were considered too expensive. Other comments that can be made concerning the outcome of the voluntary process relate to the technical suitability of some of the proposals that were made and to the role of the bureaucracy supporting the decision process.

6.3.2 The solution to the low level radioactive waste problem: a contentious issue

Introduction

Not only the province of Manitoba was inspired by Alberta's successful siting attempt: so was the Canadian federal government, and specifically the Department of Energy, Mines and Resources. The Department (now called the Department of Natural Resources) had and has the responsibility to deal with the legacy of the early nuclear industry of the 1930s, 1940s and 1950s. In the 1930s, pitchblend ore from Canada's North was brought to Port Hope in Southern Ontario for refining. Initially the ore was refined for radium that was used in medical treatment. In time, the ore was refined for its uranium content, which was used for military purposes first and then later for nuclear purposes. As there had been little awareness of the dangers associated with the wastes (including tailpipes and polluted soils from the refining process), such wastes had basically been dumped. Or worse: in some cases they had been used for construction and landfilling purposes and had thereby been dispersed over a wider area around Port Hope.

Attempts at better waste management

Although the uranium industry, specifically the refinery owned by Eldorado Nuclear Limited (later renamed to CAMECO), had began more careful management of its wastes since the Second World War, there were still problems. Eldorado had been landfilling wastes on its Welcome site, just outside of Port Hope, and on a separate location near Lake Ontario (Port Granby). It turned out that the last location was not stable and that there was a possibility that certain amounts of wastes would slide into the lake. Therefore, the Atomic Energy Control Board ordered Eldorado to find a new location for its wastes (500,000 cubic meters) in 1980. The intention was that the new waste facility would also accommodate wastes from the Welcome sites (650,000 cubic meters) and the 'historic wastes' that are present in Port Hope (350,000 cubic meters).

Eldorado started to develop a proposal and seemed to opt for a location near Lake Ontario again, which was strongly opposed by the local public. The site of the nuclear waste disposal facility became an item during the federal elections in 1984, when the opposition party expressed its disagreement with Eldorado's proposals and in fact won the elections. The sites that Eldorado had been favoring were withdrawn after intervention from the new government. The incoming government suggested Eldorado to find locations that were farther removed from Lake Ontario and also decided that environmental assessment would be required. Eldorado followed the orders of the cabinet at a relatively quick pace. Four months later, the company had already taken options on two properties in the municipality of Newcastle, both more than ten kilometers from the lake. The two sites in Newcastle had been selected because of their supposed technical suitability, i.e. mainly the presence of a thick clay layer and glacial till. The options had been purchased through a local real estate agent, without revealing the purpose of the purchase to the selling landowners. Neither the landowners that had sold the land nor the community had been informed about Eldorado's plans with the sites. When the news did become public however, there was a furious reaction from the community. The local newspaper described Elodorado's acts as '*surreptitious*'[36]. A spokesman for the corporation was critically asked about it and indicated it was standard practice in the industry not to tell landowners about the purposes of the purchaser, because land prices would be driven upwards otherwise. The issue of whether Eldorado had a moral obligation to tell the landowners of its plans was answered with: '*Questions of morality should not be directed at Eldorado*'[37]. Besides the two sites in Newcastle, Eldorado also purchased land for waste disposal in Hope township.

Environmental assessment

Eldorado's intended approach to site selection revolved around various elements. The first was a comparative environmental assessment of the three sites by an independent panel, which would result in the identification of the best site. Secondly, Eldorado tried to influence the opinion of the responsible local authorities in a positive direction. The company offered the two involved municipalities a certain degree of compensation for the fact that they would have to make costs in association with the proposal, e.g. by the hiring of expertise to study the plans. Both municipalities refused this money however and actually interpreted the offer as an attempt to bribe them into cooperation. One of the local authorities involved consequently adopted a zoning ordinance that prohibited the disposal site. In both communities, local opposition groups quickly formed and a strong lobby of parliament undertaken. The purpose of protest groups had been to obtain representation

[36] 'No national dump Eldorado rep says', Port Hope Evening Guide, 22 September 1986, page unknown.

[37] 'Eldorado eyes Tyrone as possible nuke dump', Oshawa Times, date and page unknown.

on the assessment panel, but this desire was not met and that decision threw more oil onto the fire. Members of the assessment panel received death threats and the work of the panel was made physically impossible. The mounting and intimidating resistance was a reason for the federal government to postpone the assessment work until it could focus on one site in specific, which would calm the resistance in the two other locations. When one site was then selected the opposition there became even fiercer if possible. In response, the federal government decided to take away the decision process from Eldorado and stop the process.

Voluntary siting: cooperation as the solution

The government was trapped between its election pledge to solve the problems in the so-called source communities and the impossibility to site a facility. In order to calm down emotions somewhat and to see whether it was possible to escape from siting gridlock, the cabinet appointed a Siting Process Task Force. This Task Force, as its name suggests, was not set up for actually siting a facility, but for devising a decision process that would work. The responsible minister was partly influenced by some of the members of the assessment panel, who had heard about the Alberta and Manitoba siting approach and believed that a community-based approach could possibly work for radioactive waste as well. The minister now started working under the assumption that previous attempts to site the facility had failed because the social concerns of the involved communities had been ignored. The Siting Process Task Force was to devise a process that would be different. The change in perception of the whole problematique was already apparent from the composition of the Task Force, which did have local representation this time.

The Siting Process Task Force started its work in January 1987 and published its report one year later. This report was a straight condemnation of the approach to decision-making that had been followed until then, describing it as a technocratic effort to override local concerns. Instead, their idea was that siting should become a matter of cooperation (hence the report's title 'Opting for Co-operation'). The report offered a very well-elaborated structure for an alternative approach, almost to the level of a blueprint. The approach was largely similar to Manitoba's, except that the idea of compensation was rather more embedded in it and that the local communities were given various differing waste disposal technologies to choose from. On the issue of compensation, the Task Force said: '*It is both reasonable and fair to require that any hardships or difficulties that a community might experience as a result of siting should be compensated, either directly or indirectly, using both financial and non-financial means*' (Siting Process Task Force, 1987: 58). The purpose of compensation was not to be limited to offsetting the negatives, but also to ensure that the community that accepted the facility would be better off with the facility than without it. Such compensation was not to be regarded as a bribe according to the Task Force, since the reward would be well deserved and not based on betrayal of trust. At the end of the decision process, there would be negotiations on compensation between the Task Force and the host community.

The receipt of this report in the communities that had been confronted with Eldorado's proposals - and the source communities - was positive. Both the fact that an independent organization would lead the site selection process, and the large role envisioned for ordinary citizens were positives from the perspective of these communities. A spokesman of the so-called Granby Monitoring Committee indicated that an all-expert task force would not be warranted. '*If people can be tried in court by a jury of their own peers than there is no reason why laymen shouldn't be able to decide the best means for locating a waste site*'[38].

6.3.3 Implementation

Province-wide start: information sessions; numbers drop quickly

The minister accepted the recommendations of the Siting *Process* Task Force saying that there was a need for '*a much more socially responsible way of finding waste disposal facilities*'[39]. He established a Siting Task Force, which started its work in November 1988. The Siting Task Force's first concern was to create better awareness of the size of the problem and potential solutions within the province of Ontario. The Task Force organized various information sessions, and invited a representative from each of the 850 municipalities in Ontario to attend them. About 400 of such municipal representatives indeed came to the information workshops, but this resulted in relatively little interest in the rest of the process: only 26 municipalities requested further information. In these municipalities, the Task Force approached the local council and gave them a long presentation on its work. After these presentations, 15 more municipalities dropped out and only 11 communities remained, although the three 'source' communities later also decided that they would become potential hosts, thereby expanding the number somewhat. Just like Winnipeg in the Manitoba case, the source communities felt compelled to enter the process largely to create the impression of burden sharing. It appeared that this was an important point for many of the potential host communities.

38 'New Approach!', Port Hope Evening Guide, 3 November 1986, page unknown.

39 'Communities should have veto over nuclear dumps, study says', Toronto Star, December 1987, exact date and page unknown.

Motivation for participation

For most of the potential host communities, their participation in the process should be seen in light of their council's desire for economic development. The two communities with the deepest interest in the facility were Deep River (collaborating with surrounding communities[40]) and Elliot Lake. Both have an interest in the nuclear industry, upon which their economic livelihood largely depends. Deep River has a large nuclear research facility and largely depends on the nuclear industry. Elliot Lake is home to uranium mining and, at the time, had a large quantity of improperly stored radioactively contaminated mine tailings stored in the lake, which occasionally resulted in airborne emissions of radionuclides. Elliot Lake's idea of a solution was simple. The mayor called for an end to 'the dog and pony show' and suggested that all wastes should be added to the already existing wastes in the lake, thereby creating a massive repository for radioactive wastes. The Task Force refused to entertain this idea because disposal in open water was not among the nine possible disposal techniques and thus an inappropriate technique in their eyes. The proposal therefore did not get very far.

Feeble support

In all of the 14 potential host communities, Community Liaison Groups (CLGs) were established with the purpose of studying the waste problem, potential waste disposal techniques, the siting process, and performing a broad analysis of their area to identify possible suitable locations. Of the 14 communities, 11 withdrew after their Community Liaison Groups had reported to the council, leaving only three communities in the process: Deep River, Geraldton, and Hornepayne. Both the support from the local council and from the federal government for the siting process proved to be somewhat feeble at the time. This was partly related to the fact that local elections were being held in the middle of the process. In Hornepayne, opposition groups had formed that specifically targeted councilors who had favored a continued participation in the process. They succeeded in unseating some of them and gaining a majority against continuation of the decision process. Something similar had happened in Geraldton, but there the smallest possible majority of the council remained in favor of the siting process. What did not help the siting process either was the fact that the federal government now took quite a while to ponder upon a further continuation of the siting process. The decision to actually start the next phase of the process, the analysis of specific sites, was only taken after a full year of contemplation. Somewhat to the surprise of the potential host

[40] Chalk River, the United Townships of Rolph, Buchanan, Wylie and McKay and the Townships of Head, Clara and Maria.

communities, which had developed a certain level of trust in the Task Force, the government appointed a new chair. This chair then decided to completely alter the membership of the Siting Task Force, which took another few months and led to sour comments from the CLGs in the only two remaining possible volunteers because the members of the STF whom they had come to trust were now removed. The CLGs pressured the chair and as a consequence, one of the members of the previous Task Force returned.

A certain change of philosophy

The mandate of the new Task Force reflected concerns that must have arisen in Ottawa over the direction the process was taking. The new Task Force was pressed to continue the siting process in the most cost-effective fashion and to select only the most cost-effective solutions for waste disposal. The Task Force was warned not to spend (waste is the verb used) millions of dollars on detailed technical analyses before it had assured community support for the facility. The technical work at prospective sites in the two host communities was to remain at a relatively general level. Extensive test drilling of the soil at specific locations was to be postponed until after the referenda in the communities. The implication was that the philosophy of the program, that 'informed consent' on the part of the local community, was slightly altered. A referendum would now be held in the possible host communities at a time when it was not 100 % certain that the potential sites were technically safe.

Problems and tensions

The new Task Force experienced great difficulty in its work because of the delay that had been caused by the federal government's slow decision to proceed. Much resistance had built up in the year that had gone by and the momentum, for instance in the form of community awareness of the process in the communities, had slowed (STF, 1993). Added to this were problems between the Task Force and the Siting Task Force Secretariat, a group of government officials that was to assist the Task Force. The chair of the Task Force quickly withdrew after the start of the new Task Force, as did two ordinary members later. These resignations were partly related to tensions between task force and secretariat, hinging on their different definitions of success. For the Secretariat, finding a volunteer that would accept the wastes in return for a relatively low sum of money (cost-effective) was crucial. This implied amongst others that Secretariat personnel stimulated a certain level of competition between the various communities, for instance by telling them that other communities were further along in the process. Instead, for some of the Task Force members, the interest of the communities had been the number one priority. Indeed, one of the ordinary members who withdrew told me that her resignation was related to such differences[41].

One candidate lost, one gained ... and lost again

Deep River became the only host candidate when Geraldton also withdrew from the process. The problems in Geraldton, apart from a certain level of opposition and a divided council, were legal in character. The site that the town was offering to the Task Force was not in their jurisdiction. The legal proceedings to incorporate the site into Geraldton territory would normally take quite a while, but even longer in this case as these proceedings would likely become a battlefield for opponents and proponents of the site. This would imply that the council or population of Geraldton could not determine that the site was to be used for low-level radioactive waste storage, even if it had wanted to do so. As a consequence, the town withdrew.

Deep River was the only volunteer community for some time, and this fact forced the source communities, which had also dropped from the process, to rethink their position. Especially Port Hope now started thinking about hosting a facility to store the wastes as well. The site of the uranium refinery became the center of attention. The council, in association with Eldorado, developed a plan that would solve the problem of the historic wastes and at the same time provide room for the wastes that would be created by the eventual closure of the refinery. The idea was to create caverns in the rock beneath the refinery that would then be filled and left there forever. This plan was at odds with the basic approach of the Task Force towards siting, because the proposed approach was not one of the nine possible waste disposal technologies that the Process Task Force had identified. The crucial difference between what was being proposed now and the nine technologies envisioned by the Process Task Force related to the retrievability of the wastes. The Process Task Force had concluded that public fears could be allayed - to some extent - by making the wastes retrievable so that there was a possibility of intervention in case of failure. Figure 4 depicts an accessible cavern, identified as a possible storage option by the Siting Process Task Force.

[41] Interview with A. Armour, Former member of Siting Task Force, Toronto, 11 June 1999 and personal correspondence since then.

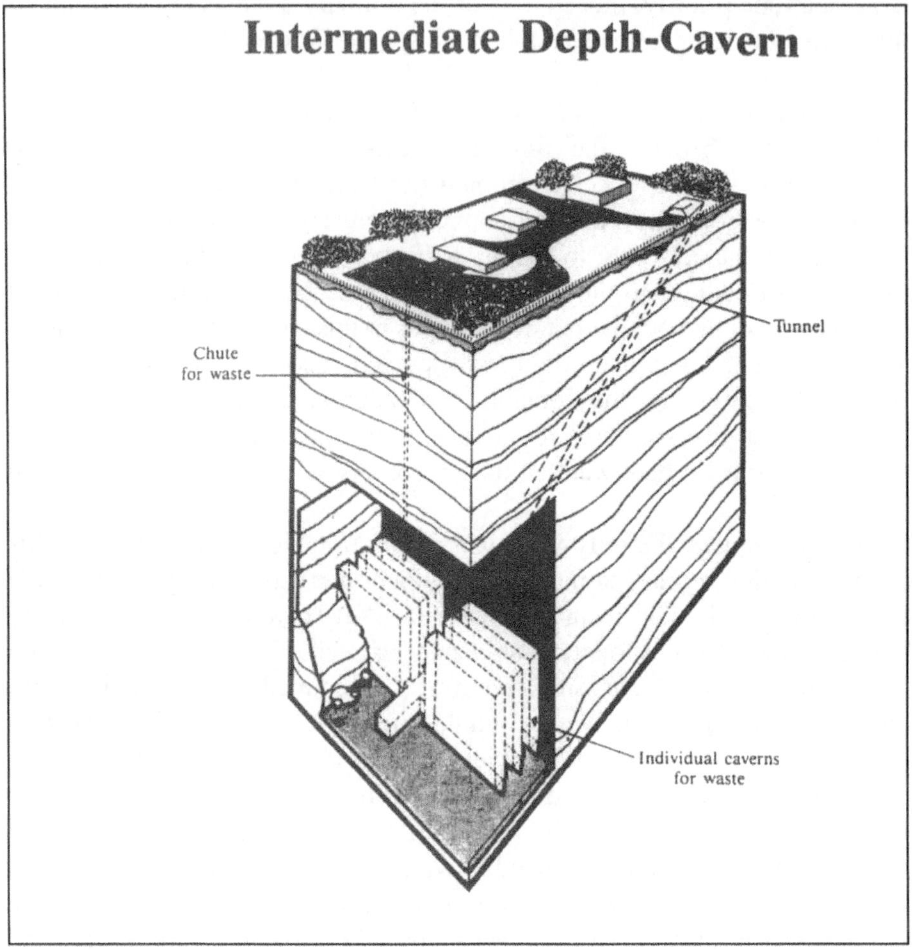

Figure 4. Cavern proposal as envisioned by the STPF. Acces is through a tunnel; depth is only a few hundreds of meters, rather than a few kilometers. Adapted from STPF-documentation.

All nine suitable technologies had the possibility of access to the wastes, but the 'cavern proposal' by the STF and Port Hope town council offered no such possibility. The public reaction was as the STPF had expected and the cavern proposal ('Crazy Caverns' said opponents) stirred up quite a controversy in Port Hope. Despite strong resistance, the council decided to press on, provided that certain conditions would be met. One of these conditions was a guarantee from the federal government that value of property surrounding the site would be guaranteed. The potential loss of property value was large however[42], and the federal government indicated it

[42] The potential costs were estimated at around C$500 million. The STF was willing to offer C$ 8.75 million.

would not be prepared to do so, effectively ending the negotiations with Port Hope.

The local reaction was one of relief and there were many negative opinions about the STF. One local paper wrote in its editorial: '*True or not, the perception was that this federally driven process was not all it was made out to be by the government. Meant to be a co-operative approach with much consultation from the local community, it looked like a well-funded arm of the government bureaucracy which did everything to promote its own agendas while frustrating the needs of the local community. (. . .) We applaud Port Hope council ignoring the threats nothing would ever be done about the problem if it opted out of this flawed process*'[43]. After completion of the decision process, very critical news on the technical quality of the cavern idea came forward. After about one year, an engineer from elsewhere in Ontario used data that had been released by the STF *after* Port Hope had taken its decision. He claimed that the data had been withheld from the CLG because they suggested that the cavern option would have not been safe[44]. The STF would not comment, leaving a somewhat suspect impression.

Progress in Deep River

After Port Hope's retreat, Deep River was once again the only town still participating in the program. But the process was not problem-free there either. Specifically, the Deep River council was having problems in maintaining its good relations with Atomic Energy Canada Limited (AECL). The company was the major employer in town and had been very supportive of the process initially. In fact, it was AECL which had suggested that the municipality should perhaps seek participation in the process. AECL however had an agenda of its own, and the company demanded that this agenda should be accommodated. Indicative of the skepticism that the Siting Process Task Force report had met in Canada's nuclear energy community, AECL demanded that if its land was to be used for waste storage, the company and not the community should determine the storage technique. AECL, in line with expert consensus, argued against retrievability. In addition to control over disposal technology, AECL also demanded that it should be allowed to co-dispose its own wastes with the historic wastes from Port Hope. The proposals were once again counter to some of the basic ideas underlying the siting process, but both the Task Force and CLG agreed to the AECL's dominant role. Despite the fact that AECL's demands were being met, AECL's commitment to the process waned considerably, however, after its bold start. The dealings with the CLG

43 'A sigh of relief', Northhumberland News, 1 August 1995, page unknown.

44 'Proposed dumpsite wasn't stable, engineer charges', Port Hope Evening Guide, 13 May 1996, pages 1 and 2.

were delegated to lower ranking officials and when the CLG started becoming very specific about its preferred locations on AECL land, the company reacted very negatively as if it had been unaware about everything that had been going on until then. This resulted in a certain level of friction between the town and AECL. Figure 5 demonstrates the area where the facility was broadly planned.

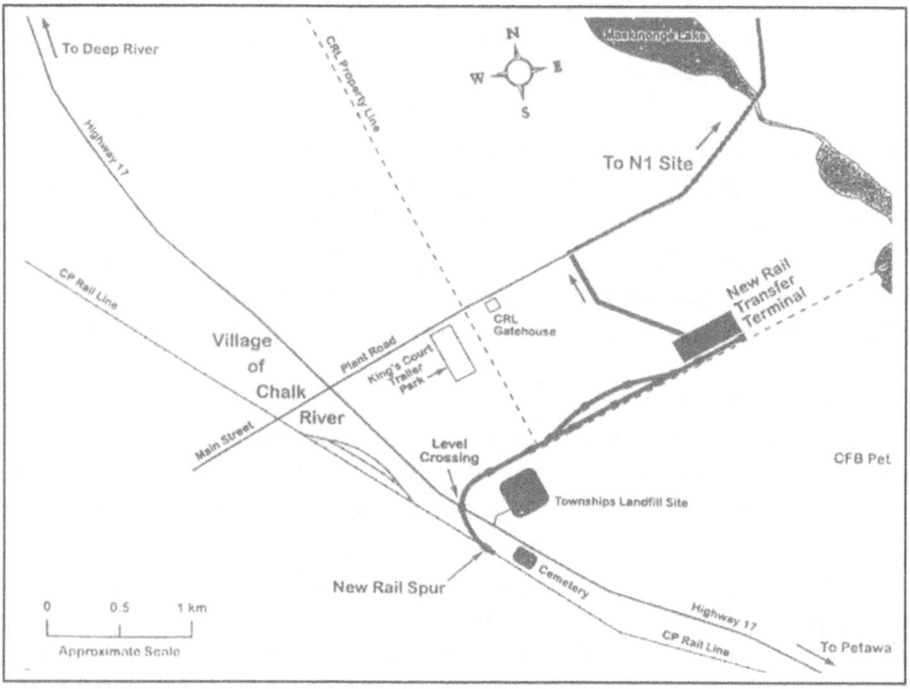

Figure 5. Search area for site on AECL property. AECL property indicated by the dotted lines and 'CRL property'. A few kilometers to the North-east is Deep River. Chalk River is to the South. Adapted from STF-documentation.

Motion to withdraw withdrawn

When the CLG determined its most preferred site after many meetings that were often attended by AECL personnel and on the basis of data that were partly provided by AECL itself, AECL objected to the CLG's choice. Their argument was that the selected site would not fit their business plans. When I interviewed the former chair of the CLG to ask for an explanation for this, he suggested that AECL's wavering may have been caused by a fear that detailed assessments of the sites would indicate unsuitability for waste disposal[45]. Although it would not be disastrous for AECL not to receive the historic low level wastes on their site, a finding of unsuitability would also

[45] Interview with former CLG-chair D. Thompson, Ottawa, 21 July 1999.

imply that the wastes already at the site were not safely stored and AECL's problems might be great[46]. This line of reasoning, although speculative, does not seem completely illogical, as the media started reporting AECL's own waste storage at the site had been leaking for years with emissions of radionuclides to surface water as a consequence[47]. AECL had chosen not to make these emissions public. For the mayor of the town, the waffling by AECL resulted in great feelings of frustration. He decided to draft a motion for the town council to withdraw from the process. Opinions change quickly however. One week later, the motion was later narrowly defeated in council, in part because the mayor had voted against it himself. The mayor had changed his mind on his own motion because AECL had changed its tone and now suggested it wanted to cooperate with the CLG.

CLG resigns

Because of the events just described and the news about the siting process in Port Hope, the CLG developed a certain level of distrust in AECL and the STF. Such distrust was fueled by questions from the public that attended the public meetings of the CLG. The amount of expertise on radioactive waste was impressive as many inhabitants worked at AECL and had obtained scientific training to a high level. Many critical questions were asked, for instance, regarding seismic activity in the area and the proximity of the proposed location to a river body. It appears that many of these concerns could not be addressed by the STF, even after long pressing by the CLG. Because of this, the CLG resigned by October 1994. The CLG felt it was irresponsible to hold a referendum in March 1995 (as was the planning) because so much information was lacking.

Council presses on

The CLG resignation did not imply the end of the process, because the authority rules described the CLG as just an advisory body to the council. Deep River council decided to ignore the opinion of the CLG and indicated that it would continue to work towards a referendum by March 1995. In the course of the preparations for this referendum, a very big public hearing was held, where the chair of the CLG played the role of opponent to the facility. He warned the citizens of Deep River that '*This thing is going to leak like a sieve. The AECB will never allow something like that.*'[48]. Despite the fact that STF reports did not report any problems with the soil of the selected site, the chair told the public that he had been present when the drilling work

[46] Interview with D. Thompson, Ottawa, 21 July 1999.

[47] 'Chalk River's dirty secret', Ottawa Citizen, 28 May 1997, p.1.

[48] 'Lack of technical soundness at Deep River site may nix plans to build radioactive waste dump', Port Hope Evening Guide, 22 February 1996, page unknown.

was done and had noticed how the rock on site was fractured and porous. In response, a spokesperson for the Task Force indicated that the problems with the selected sites could be engineered away, a statement that comes close to admitting problems that had actually been denied by the STF until then. The STF now played out an important discursive joker, their critique of managerialism. They indicated that there are inherent tensions between technical suitability and public acceptance. According to the STF '*getting the support of the public is more important than anything else when trying to find a place to put something like a radioactive-waste dump*'[49]. Therefore, in Deep River, '*If there are any problems with dumpsites they can often be resolved through engineering work*'[50]. In reply to questions about the Port Hope siting failure, the STF now said the site in Port Hope had been excellent from a geological perspective but no public acceptance was achieved.

Despite the opposition from the CLG, the town council decided to hold a referendum in the Fall of 1995 and there was a fair chance that the population would support a facility. An important consideration for many people in Deep River was the fact that the town council and STF had reached an agreement on a level of compensation for the municipality. The STF had agreed that the jobs in the nuclear industry in the town should be guaranteed in return for Deep River's cooperation. These jobs were very much under threat from the pending reorganizations at AECL. Therefore, such a promise was very welcome to most inhabitants of Deep River. On the other hand, such a guarantee would imply strong interference on the part of the federal government in AECL's operations and would potentially require quite a bit of money. It was thus highly uncertain that the federal government would actually agree to the negotiation result. The council sought to gain assurances that the federal government would back the deal but received no affirmative reply.

Referendum

Although the actual implications for the town of a decision on the facility were uncertain because of the federal silence on the issue of compensation, a majority of 72% of the voters agreed to the facility. Interestingly, there was little reply from the federal government except that a four-year study of the rock on AECL property would be necessary. Before those studies could be completed however, Deep River withdrew its willingness to be host for the facility. The main reason for that appears to be that the federal government sent a negotiation team to the town, trying to break open the agreement that

49 'Lack of technical soundness at Deep River site may nix plans to build radioactive waste dump', Port Hope Evening Guide, 22 February 1996, page unknown.

50 Ibid.

the town had reached with the STF. One of the things that the new federal negotiators were demanding was storage of extra wastes. The town responded negatively and negotiations collapsed, only to be briefly reanimated in 1997. In 1998, the federal government began new discussions with the source communities on local long-term management including in situ isolation. Those discussions have been successfully completed.

6.3.4 The role of various institutions and discourses

Introduction

This particular case resembles Alberta's experience in several ways, amongst others, in the sense that siting attempts in the ordinary fashion had failed and there was an impression that 'co-management' would be the only possible way to get a facility sited. In line with that, community acceptance, compensation and a certain relativation of the managerial language were necessary. There was a clear political necessity for solving the problems in the source communities as promises had been made during the elections. It does seem however, that the political will weakened considerably over time. Especially a concern about the costs of the decision process itself and a concern over the costs of the final solution seemed to trouble the relevant decision-makers, who consequently interfered in the decision process and started stressing the 'normal' conservative pluralist and managerial concerns with cost effectiveness. It does seem that the potential host communities indeed asked for a hefty price in return for their co-operation, which may in part be in response to the explicit discussion on compensation by the Task Force. In many of the communities where the decision process took place, local councils and Citizen Liaison Groups had a problematic relationship. In addition, the case of Deep River demonstrates a potential tension between various communities affected by the proposal. In this section, I will devote no attention to the courts and private market parties because they played no role in the decision process.

The role of elected representatives, experts, and the community

The role of elected representatives in the decision process was great. In their capacity as supervisor of Eldorado's activities (as a state company) they could stop the company from taking further steps on the path it was taking. The Minister in question, under pressure from his Prime Minister who had promised a solution to the source committees, started believing in the more community-based approach, mainly on the basis of the advice from some members of the assessment panel who could point to the positive experiences of Alberta and Manitoba. Over time however, more 'ordinary' concerns such as the cost-effectiveness of the process returned to the forefront. The changed political will was greatly influential on the process. The reconfirmation that voluntary siting was still the desired approach took quite long, so long in fact that the potential volunteers started questioning the seriousness of the process for the minister. The changes in membership of the Siting Task Force did not help either. The final decision to reject the Deep River proposal was also taken at the ministerial level, and this was clearly negative. The role of elected politicians at the local level is so interwoven with the role of the community, that I will discuss it there.

The role of experts in the decision process was large, even though there were attempts to lessen it somewhat. Like in Alberta and Manitoba, the siting process in this case was intended to introduce a double veto, with the final say on the proposals for the government and the Atomic Energy Control Board (AECB). Experts, especially the ones from the AECB, would play a role in the checking of sites and proposals that would be voluntarily generated. It is interesting however that the Siting Process Task Force deviated from the standing expert consensus on waste disposal in the nuclear industry. That standing consensus was that waste disposal should be permanent and not dependent on human institutions as such human institutions would not survive long enough to guard the wastes permanently[51]. Permanent and inaccessible disposal, out of sight out of mind, was the dominant idea in the industry. The Siting Process Task Force, far more inspired by the public concern and supported by various technical consultants from outside the nuclear industry (but with knowledge of disposal techniques elsewhere in the world), pleaded for retrievable disposal. These dramatically different visions would have likely collided if the AECB would have had to license a disposal facility along the design of the Task Force. Instead, this collision came much earlier, namely during the interactions between STF and Eldorado and between STF and AECL. In both cases, the members of the nuclear industrial complex (so to speak) demanded, in return for the use of their land for waste disposal purposes, that the proposals would meet their conditions, among which the fact that waste disposal should be permanent. The reason why the Siting Task Force gave in to such demands relatively quickly appears to be that the people who thought out the process had not retained control over the program and had been replaced with others who had less knowledge of the historic background.

The status and neutrality of the STF and their supporting consultants as 'experts' was criticized to the point of becoming suspect during the process. This was especially the case with the 'cavern proposal' in Port Hope and to some extent as well with the Deep River proposals. Some of the people I

[51] See e.g. AECB regulatory document R-104, 'Regulatory objectives, requirements and guidelines for the disposal of radioactive wastes - long term aspects'.

spoke to brought the apparent mistakes in risk calculations in connection with the definition of success that the STF was applying. I was told that for the STF, success equated with actually siting a facility and that this implied a certain tendency to try and sway communities into acceptance. Such a tendency would seem to be only human, but could potentially imply a threat to the impartiality of the STF's experts. Norm Rubin of the environmental group Energy Probe was sometimes asked by certain communities to come and provide counterexpertise. He described the work of the STF as follows: '*They behaved as guardians of the status quo, telling people not to worry about radioactive risks. That's why some of the communities said No - they knew they weren't hearing the whole story*'[52].

The role of the community was great in this sense that a referendum would guarantee the population a veto right in the decision process. Most ordinary citizens could remain relatively inactive throughout the process, although their coming to CLG meetings was stimulated. These CLGs functioned as community representatives, albeit only within the framework of a mandate given to them by the local councils, which also selected the members. This mandate tended to be technical and informative. The CLGs were to study the technical aspects of radioactive waste disposal and then inform other citizens and their municipal council, with the help of STF experts. It appears that the CLGs often collided with their local council. In the course of the decision process Community Liaison Groups were established in 14 municipalities. When the time came to decide on the continued participation in the decision process the CLG advice to the local council was counter to the preferences of the local council in the majority of the cases. In three of the municipalities that decided to continue after having studied the issues at a relatively abstract level (Hornepayne, Deep River, Geraldton), two did so against the recommendation of the CLG. The opposite situation also occurred; whereas three CLGs had recommended continuation, only in one of those cases had the council concurred. All in all, there was only one town, Geraldton, where both CLG and council were interested in continuation of the process. The Task Force, following its own process decided to go with the local council in each case. There was only a limited negative response to the councils' decisions, and if such a negative reaction was visible, it was largely constrained to former members of the CLGs in question. In one case, the town of Hornepayne, one former member of the CLG sought publicity with very negative messages expressed about the operations of the Siting Task Force (Currier, 1993). She alleged that the Task Force had sought to control the information supplied to the CLG by preventing them from hiring independent experts and had tried to pressure

52 'A messy legacy of radioactive waste', Maclean's, 23 September 1996, pp. 52-53.

them into a positive recommendation by asserting that many other communities had already decided to go ahead. She started effectively organizing the opposition against the council's decision to continue the process. The item became an issue at the next municipal elections, where several councilors lost their seat and were replaced by councilors who opposed the siting process. The majority of the new council consequently decided to opt out.

The case of Deep River is especially interesting, because the council there even neglected the preferences of a CLG twice. The first time came when the council decided to volunteer after the general study of the issue, the second after the debate on a specific location. In both cases the CLGs had advised against continuation. The decision to ignore the advice of the first CLG cost the municipality much goodwill in the region. This is because Deep River had entered the decision process in combination with the surrounding municipalities and communities and each of them had had representation on the CLG. When the CLG issued a negative advice on the proposals, all councils withdrew, except (to the amazement of the others) Deep River. Deep River motivated its decision by rejecting the CLG's argument for terminating the process. In line with the thoughts underlying the siting process, the CLG had polled the population of the various communities and had concluded that a large majority of the population saw nothing in a continued involvement in the process. The mayor of Deep River criticized the opinion poll and would not accept the (negative) outcomes because the CLG was not the dominant actor. He used managerial and progressive pluralistic language to defend his choice: '*We kept reminding them that they were not elected. They did something very unfortunate (. . .) sent out a questionnaire. Gave everybody a ballot that could be dropped at a box in the post office. It's hardly a scientific way of doing it. I took nothing from the exercise, thought it was badly done, and told them so. They told us it was a significant outcome, but I didn't accept it*'[53]. The council overrode the CLG and decided to continue its involvement on two conditions. The first condition was that the effort of the source communities (where the wastes were at that time) should be increased. The motivation for that was a certain level of skepticism in the population of Deep River towards the idea that low level radioactive wastes were really a problem, and secondly a degree of fiscal conservatism among the population. The second condition that Deep River wanted imposed was that the wastes should be transported by rail. The reaction towards the Deep River's council decision was mixed. The surrounding communities initially wrote a letter of support recognizing that Deep River had a lot of local expertise to deal with the problems and

[53] Interview with L. Smith, former mayor of Deep River and former employee at AECL, Deep River, 21 July 1999.

rightly wanted to continue. Representatives of a citizen group that had been active in the process were negative however. According to them '*The efforts of many residents of this community who involved themselves in Phase Three activities have been swept aside in favour of a closed process conducted under highly questionable circumstances*'[54]. They were especially suspicious of the fact that Atomic Energy Canada Limited (AECL) was the employer of many Deep River councilors. The continued role of Deep River, especially at the stage of specific site assessment, started to greatly annoy the nearby towns. The fact that trains with radioactive waste would cross their jurisdictions and thereby impose certain risks upon the local population caused unrest. The nearby towns sought representation on the second CLG that had been established to perform an assessment of the proposed site. Their pleas were successful to some extent, but Deep River managed to reduce their influence to a minimum by ensuring that CLG members from the surrounding communities would be outnumbered by Deep River members (10:4). Deep River also decided that a great majority of the population in the nearby communities could not cast a vote in the possible referendum. This decision was motivated by a somewhat odd reference to the sovereignty of the other municipalities. The mayor of Deep River argued that '*This council has no right to interfere*' with other councils and that '*It would be presumptuous to try and force them into a referendum at this stage*'[55]. A councilor said: '*They clearly expressed their unwillingness. I think you have to respect that*'[56]. The surrounding municipalities consequently held their 'own' referendum, which resulted in a 80% majority against Deep River's plans.

The role of discourse

Although the ministerial choice for a cooperative siting process was in the end much more pragmatic than inspired by a certain discourse per se, the Siting Process Task Force did motivate its proposals on the basis of a strong and eloquent rejection of the managerial approach and an embrace of a mixture of progressive pluralist and communitarian ideas. The Process Task Force's analysis of the Eldorado failure was that the approach had been too technocratic and had not sufficiently taken into account the interest of the involved communities. The report contains a rejection of the standard approach to siting. This approach, argues the report, is rather straightforward and involves narrowing the scope of study from a large area to a short list of

[54] Letter by D. Foster and I. Teilheimer to the Siting Task Force, 15 August 1990, found in the Deep River municipal archives.

[55] Final vote for town only, says council, North Renfrew Times, March 1995, no exact date, page unknown.

[56] 'Townships might ask for reversal', North Renfrew Times, 15 March 1995, p. 3.

sites on the basis of specific siting criteria. '*The most significant feature of this process is technical rationality. Its primary aim is to find the "optimum" solution, the one that represents the best match of ends and means*' (Siting Process Task Force, 1987: 46). Three 'implicit' premises underlie the approach, according to the Task Force. The first is that scientific and technical criteria should be the proper basis to assess siting options. Locations are therefore assessed on the basis of 'hard' data, rigorous technical analysis, quantitative assessment of costs and benefits and systematic application of the scientific method. The second premise is that socio-political factors should not be the key determinants of site selection, but environmental factors; the technology-environment relationship is the central concern in the siting process. The third premise is that the public's role in the process is secondary to that of the expert. Experts are to make judgements on technological options, evaluation criteria, analysis of consequences and the rankings of alternatives in isolation from affected parties. Public consultation is only intended to inform citizens of the issues being addressed and to obtain their reaction. The Siting Process Task Force argued that the standard approach to siting was appealing because '*According to conventional wisdom, such technical rationality is better than approaches to decision-making because it leads to recommendations that are objectively sound and ultimately in the interests of the "common good"*' (Siting Process Task Force, 1987: 47).

The description of 'traditional siting' closely resembles my earlier description of managerialism, and it is therefore perhaps no surprise that the critique formulated by the Siting Process Task Force against the 'standard' approach is similar to the critique of managerialism. The Task Force concluded that there are a number of problems that cast doubt on the efficacy of this approach. In particular, the presumed objectivity of the process is considered suspect by people who recognize that many of the technical decisions made in the decision process contain assumptions which are related to the subjective judgements of the experts involved. Secondly, the standard approach does not consider the concerns of the potentially affected communities until candidate areas are selected. The criteria that are considered by the experts and the kind of trade-offs they are willing to make are not in agreement with what the community would prefer according to the Task Force. A final point of critique on the standard approach to siting was that it inevitably leads to one outcome, which is the imposition of a decision. As citizens have not been involved in the early phases of the decision process such as establishment of the criteria for the decision, they feel the decision is imposed upon them and they will feel victim of technocratic decision-making (ibid.: 47). The Siting Process Task Force considered the standard approach unacceptable.

I have not been able to study the reaction to the new approach in all communities that the STF approached or even all the communities where a CLG had been set up. Only for Hornepayne, Deep River, and Port Hope have I collected some evidence on the way the proposals and process were perceived. I have already said something about the reaction in Hornepayne, which is perhaps best summarized as a plea for far more local democracy than the STF was offering. The CLG there felt that the STF did not leave the local decision process alone enough and started effectively challenging its presence. In addition and in a communitarian vein, the CLG there did not feel comfortable with the competition between the communities that was being stimulated by the STF as well. The former member of the CLG who started an opposition group after leaving the CLG said: '*Just before I left the CLG, our groups received a letter from the Siting Task Force stating that, of the ten communities left in the process, four were willing to take ongoing as well as these historic wastes. (. . .) I don't think it's too far off the mark to say that this letter was the task force's way of saying, "if you want to stay competitive in the process, you'd better be willing to take other wastes as well". Such magnanimous gestures on the part of these towns are not mentioned in the phase 4 report (. . .)*' (Currier, 1993: 40). She also contends that the STF told the citizens' group that as Hornepayne is located quite far to the North of Ontario and transportation costs would thus be high, there was little chance of obtaining a facility with 'all the bells and whistles'.

The reaction in Deep River was very different and seems to have been largely conservative pluralist and managerial, but there was some progressive pluralism present as well. In that community, beside a certain desire to keep the nuclear industry viable, there was doubt about the usefulness of the whole exercise and the cost-effectiveness of the operation. The idea of compensation for the local community initially met with hesitance. Deep River's local newspaper wrote that '*we are in the midst of an expensive and complicated process to deal with a non-problem. As local residents, we can, of course, take the attitude that if money is being spent, however foolishly, we might as well be on the receiving end. Surely, though, it is better to say that in this community we are comfortable dealing safely with radiation and are willing to consider a storage facility if necessary, but that, as Canadian taxpayers, we cannot endorse such a huge expenditure when it is so clearly unnecessary?*'[57]. Continuing its managerial comment, the newspaper compared the millions of the Siting Task Force with the budget cuts that had been made to the Science Council, which was '*the body*

[57] 'Bonanza II', North Renfrew Times, 22 April 1992, page unknown.

most equipped to advise (. . .) on such matters'[58]. The paper thus found that the real experts should decide.

Both in Deep River and Hornepayne, there was a fair bit of sourness over the fact that wastes would be transported from the industrialized and wealthy South to the North of Ontario. People in Deep River rejected '*to ship it up to the north and forget about it instead of looking after it where it is*'[59] and had demanded a bigger role for the source communities. These were not just the concerns of the local newspaper, but also the opinion of the local council. For instance, the council, already relatively early on in the decision process, pointed out to the STF that a certain degree of waste separation at source would be very profitable. Certain polluted soils were not very radioactive and could perhaps simply be capped or go to normal landfill. Some of these suggestions were actually followed up on, and the quantities of wastes to be brought to Deep River reduced. It does seem however that later on, the municipality started singing a different tune. Especially in the negotiations on the compensation package, the municipality representatives tried to get the most out of the decision process.

In Port Hope, where the proposal was to create permanent storage in caverns, a large part of the discussion centered on the ability of engineers to design a repository that would contain the wastes forever, with many arguing that this could not be done. The opposition group in Port Hope made a tactically very smart move of inviting Mr. David Suzuki, a well-known environmentalist on the North American continent, for a lecture. Although Mr. Suzuki did not know the specifics of the cavern proposals he pleaded against them on the basis of anti-managerial thinking. According to Mr. Suzuki, the proposals came from a technocratic mind, '*incapable of imagining itself mistaken, yet knowing that it cannot be held responsible if it does prove mistaken*'[60]. The lecture had a major impact on public opinion and the uncertainties associated with the proposal were a major point of contention. Opponents came to portray their fight as '*a struggle between ordinary people outraged at what is going on, and a task force of professional persuaders using millions of taxpayers' dollars*'[61]. The opponents argued that the STF did not care about '*the best interests of the people*' and called upon the residents of the town '*to say no to the technocrats*'[62]. Such statements had a mobilizing effect on the community, but did not in the end influence the decision process much. The single most

[58] Ibid.

[59] 'From Hornepayne', Timmins Daily Press, 24 August 1991, page unknown.

[60] This summary is from a resident of Port Hope, 'Safety not guaranteed for proposed cavern option', Port Hope Evening Guide, 20 July 1995, page unknown.

[61] 'A classic democratic grassroots struggle', letter to the editor, Northhumberland News, 2 July 1995, page unknown.

[62] Ibid.

important reason why the council and STF did not agree in the end was that they could not agree on a contract that would guarantee stable property values. When the news came out that the cavern proposal might have been a lot less safe than the STF had indicated, a spokesman of the former opposition group lashed out at experts. '*We learned that experts are people with impressive degrees who will say whatever they are paid to say. We learned that ordinary people who take the trouble to get informed really do know better than the experts what is best for them. We learned that grass-roots democracy can still work, even against the overwhelming odds*'[63].

6.3.5 Review of the case

The low-level radioactive waste siting process was perhaps the most ambitious of all three decision processes studied here as far as the level of local control was concerned. This is in part because the designers of the process could greatly benefit from Alberta's and Manitoba's experiences and wanted to avoid some of the pitfalls experienced there. Especially the blueprint of the process that was made beforehand, the fact that compensation was so clearly part of the approach, and the fact that ordinary citizens were expected to choose from a certain range of waste disposal techniques (instead of just one) were indicators for the high ambition.

It was apparently very difficult to actually attain that high ambition level. The reasons vary, but seem to be largely concentrated around the fact that political support for the co-management method was somewhat feeble. This was then again related to a declining degree of commitment to the process in Cabinet and increasing concern over the costs of it all[64]. Cost-effectiveness became a very important criterion in the decision process but also for the decision process itself. This implied for instance that the possibility of CLG supervised test drillings was largely cut, until the final stage. One other explanation for the difficulty in attaining the high ambition level is that it was apparently quite hard to maintain the role of neutral assistant to the local community. People working for the STF almost unavoidably developed a certain desire to actually locate the facility and this seems to have affected the way they presented things and perhaps even some of the outcomes of their studies.

The concern for cost-effectiveness does seem warranted to some degree as both Port Hope and Deep River had steep demands, which could result in huge financial obligations for the federal government. It does seem that such

[63] 'When experts are shown to be fallible human beings', Letter to the editor, Port Hope Evening Guide, 3 May 1996, page unknown.

[64] The siting process itself cost tens of millions of Canadian dollars. The actual disposal site would cost hundreds of millions.

demands were made in part because the process was so explicitly structured on the notion of compensation and the fact that 'negotiating a compensation package' was an explicit step in the process. The demands made by the two Ontario communities were clearly much greater than the ones made by Swan Hills or Montcalm (where compensation consisted basically of employment plus some tax income benefits in Montcalm). There are various possible interpretations for the failure of the negotiations between Deep River and the federal government in this respect. One could perhaps argue that the explicit mentioning of compensation created a certain level of greediness in the local communities. In that case, it was logical, if not required, that the federal government did not reward this behavior. More neutrally put, one could say that the federal government, from the outset, had the possibility of judging the outcomes of local decision-making in the comparative perspective of its entire budget. If this explanation is correct however, then the next 'co-management' process would greatly be served by a more explicit indication of what it all may cost. A certain link must then be made with the federal political process, not to constrain the local municipalities, but to be fair to them. A different vision of the failed negotiations would revolve around the fact that local costs associated with waste disposal facilities (i.c. deprivation of property values) were now - and rightly so - expressed in the process, whereas such costs normally remain hidden. If this is the case, then the decision of the federal government of Canada must be seen as a refusal to take its responsibility.

The Siting Process Task Force had undertaken a frontal attack on the idea that waste repositories for radioactive waste should contain the wastes eternally. Such repositories cannot be designed according to the Task Force, and therefore why not openly admit it and have the wastes stored in sites that are accessible if need be. This opinion was strongly against the common opinion in the nuclear waste sector and it does seem it never stood a chance, certainly not when the AECL got involved[65]. This 'automatically' implied that a certain degree of maneuvering space for the CLG was taken away, but this happened without much protest.

The relations between the various communities involved are interesting in this case. Deep River's motivations were in the end largely economic, directly (job guarantees) or indirectly (supporting the nuclear industry) and the municipality was not minded or stimulated to take the concerns of other

[65] It is interesting that there was at the same time doubt about the seriousness of the waste problem in the host communities (low-level radioactive wastes were considered almost innocent) and a requirement that the wastes should be disposed of in almost exactly the same way that high-level radioactive wastes are disposed of. Ironically, the solution chosen after voluntary siting revolves around covering the wastes ('capping'), which is nowhere near as advanced as the disposal methods proposed by the Task Force.

municipalities into account. Deep River unilaterally terminated the initial cooperation with other communities, initially without a very negative reaction from the others. When Deep River remained in the process even though many questions arose about the technical quality of the proposals, and when it became clear that transportation of the wastes would be through other communities, these communities became concerned and tried to influence the decision process. This was to limited avail and as a consequence, the standing of Deep River in the region decreased.

6.4 Conclusions

6.4.1 Introduction

There are various issues that stand out in the three Canadian cases. An important finding is that the legal framework only partially structured the decision process in all three cases. Whereas hazardous waste facilities are normally realized by private market parties (a state firm in the case of radioactive waste), their control over design and siting was lifted and replaced with control by a specially appointed Task Force. There are a number of reasons why this happened, and these differ somewhat among the three cases. An important factor was the opposition from local communities, which had either materialized already (Alberta, Ontario) or was expected to materialize (Manitoba). Under the Canadian regulatory system such resistance must simply be overridden, but it seems that especially in Alberta, the elected politicians did not dare to take such a decision. The fact that in that province, the local public was very well organized and at the same time motivated almost to the level of hysteria will have helped make a strong impression. In Alberta, the public was actually willing to start exerting physical violence (occupation of the county offices) and in one case simply forced the county council to write out a referendum. Such resistance made hazardous waste siting a political hot potato that politicians would rather not touch, even though they were honestly committed to more responsible waste management.

The start of an alternative approach lies in a thoughtful analysis of North American siting experiences performed at the request of the province of Alberta, which influenced high-ranking officials there and helped design a more community-based process. This approach produced results in the end. The Alberta example proved inspiring and the federal government and the province of Manitoba both made good use of these experiments, somewhat modifying the approach, which evolved from invitational siting to co-management and voluntary siting. These different terms already suggest a difference in approach. Whereas Alberta's process is best described as 'manipulation with integrity', Manitoba's approach is much less manipulative, and the federal process was explicitly intended to avoid manipulation (but failed in that respect to some extent).

In terms of decision quality, there are various issues that stand out. The community-based approach tends to place great emphasis on community acceptance, which implies a great amount of attention to the social quality of the proposals. At the same time, the maneuvering space of this process is set by certain pre-determined technical criteria, which are intended to protect the technical quality of the decision. In all three cases, the technical criteria were in fact very important in determining the location of the facility and largely structured the terms of the debate with discussions of compensation (if present at all) postponed until the final part of the process. It appears that citizens are quite capable of having such a technical debate, and can actually contribute new information (Manitoba: flood risks; federal process: radioactive substances and risk assessment). On the other hand, it is not the entire community that is engaged in such a debate, and some citizens were delegated (by the council) to invest more time in the process than others. In all three cases, there was a certain tension between the technical and social quality criterion, for instance, because some communities wanted to volunteer without having a suitable location. What's more, it seemed very difficult to pay attention to the 'economic quality' of the proposals in all three cases. Alberta sort of forgot to update its data on waste arisings and therefore projected a facility that was rather large for the amount of wastes actually offered at the facility. A private operator was hard to be found and the province needed to hand out subsidies to actually get such a operator. Manitoba learned from this experience and set up its own incremental approach. This proved difficult too, as the province was not able to tell local populations what kind of facility was to be expected. The province then designed a prototype of a facility, but this prototype was never realized in reality because of the waste market. In the case of low-level radioactive waste, the amount of compensation demanded by the community, and obtained from the Siting Task Force, proved too much to swallow for the federal government in the end because of cost considerations.

6.4.2 Cross case comparison

Institutions

The population of the three provinces involved consistently showed a large level of distrust in the role both private companies and their local own elected representatives played in hazardous waste decision-making. Citizen opposition, coupled with a desire to stay the siting of hazardous waste facilities, implied that these two institutions were largely de-emphasized during the decision-making process, as were the courts that did not play any role and that therefore need not be discussed here. It is interesting to see that in situations where councilors did try to influence the process, e.g. by sitting on the citizen liaison committees, the process always failed. In place of these institutions came a larger role for the local community, which obtained veto rights through a referendum (uncommon in Canada) and had people appointed to the CLGs[66]. The referendum was initially not part of the siting process, but became necessary after the first experiments with invitational siting in Alberta backfired. In many cases (especially in the federal process) there was a certain level of tension between the council and the CLG as they had different ideas about the acceptability of the proposals. In such cases, going against the CLG involved a certain risk for councilors because the CLG members could start driving the opposition. In various communities this happened and councilors in favor of the siting process were unseated. The fact that CLGs had a mandate that derived from their local councils had another drawback. In all cases, the implication was that the municipal (or county) borders became the formal delineation of the relevant community. Especially in Alberta and the federal siting process (note the tension between Ryley and the surrounding county, and the tension between Deep River and surrounding communities) this delineation was not especially fair to others who wanted to participate as well. To some extent this implied that the community benefiting from the facility (so to speak) partly 'exported' some risks (mainly related to transportation) to other jurisdictions that had not consented to the facility and were not benefiting in economic terms.

The role of experts was different from what it would have been normally. Surely, experts were the ones that would design and license the facilities in question in the end, but the circle of experts that was normally involved in licensing was largely left out. In their place came 'communicatively able' technical experts and social scientists, who supported the community (Manitoba, federal process) or sold the facility to the community by reassuring and somewhat manipulating it (Alberta, federal process at a later stage). Especially in the low-level radioactive waste case, and to some extent in the Alberta case, the Siting Task Forces distanced themselves from the ordinary regulatory circuit in order to be seen as independent, but also because of some different insights in appropriate waste disposal technologies. Especially in the case of the federal Siting Process Task Force, the attack on normal disposal practices was quite fundamental and an attempt was made to change the philosophy of waste disposal practices in general. In Alberta, it does seem 'ordinary' regulatory officials had hesitations about the projected waste facility because they did not see the need for the proposals developed by the Task Force. The Task Forces

66 I am using the term CLG here for style purposes. The actual term in Manitoba was CAC.

meanwhile, were under pressure from local communities, a pressure that expressed itself, for instance, in the need to 'fill in' gaps of what the future facility would look like, even though this was not certain yet. The reality check in Alberta came after the province introduced hazardous waste regulations and started developing an image of actual waste arisings. The projected facility was then adjusted to smaller waste streams. In the federal siting process the reality check did not come so much from the ordinary regulatory circuit, but from elected politicians. These ruled that the deal negotiated between the town and STF was unacceptable in financial terms. More generally speaking, the desire of provincial and federal politicians to stay out of the decision-making process had certain drawbacks. This is because it added to the uncertainty already embedded in the fact that the 'ordinary' regulatory circuit had to approve the proposals after agreement had been reached locally. It seems that politicians only have a certain level of patience with the process. The Manitoba process was terminated before Winnipeg completed its assessment, in Alberta siting success had to be guaranteed, and at the federal level there was great concern about the costs of the process.

There are indications that the phenomenon of communities making high if not impossible demands is not intrinsic to community-based siting, however. To realize this, one must look at one importance difference between the Alberta and Manitoba process on the one hand and the federal process on the other. In the federal process, only one potential volunteer remained in the end. This 'monopoly' implied that demands could be relatively high. In Alberta on the other hand, Swan Hills felt it could not demand too much because it saw itself as competing with Ryley. The comparison between Swan Hills and Ryley suggests to some extent that the winning over of the first community necessitates greater investment than the winning over of other communities (although the STF members in Alberta also seemed to have favored Ryley because of technical reasons). The 'game' implies that a first community volunteers, sets a certain price, and that the second community must set lower demands in order to win the competition. If this is maintained, without free exchange of information between the potential hosts (as was the case in Alberta where the STF did not help Swan Hills in guaranteeing a profitable siting result in some respects), the result is a driving down of community demands, and consequently the driving out of community needs as the most important determinant of the compensation package. There is however one other aspect to this dynamic, which is the degree to which compensation is actually mentioned as inherent to the decision process. It does seem to some extent that the relatively big demands of Deep River should also be interpreted in light of the explicit references to compensation in the federal process. In

Alberta and Manitoba such references were much less open. It does seem that the mentioning of compensation discredits the siting process somewhat everywhere. One needs to remember that although the three siting processes have often been set as an example for other countries, the process did not produce siting results in many communities. This is partly related to the fact that the Siting Task Forces were often seen as offering bribes.

Discourse

The choice for invitational siting and co-management seems to have been largely based on pragmatic considerations in all three cases. Especially a report on siting failure in the rest of the continent and its causes which was produced for the Alberta government (Krawetz, 1979) seems to have resulted in an experiment on a different basis. This experiment was based on scientific insights in how people perceive risks, how they make their decisions, and why 'traditional' siting does not seem to work. The most important goal in the case of the Alberta siting process was to site a facility, not to make the community better off. But as other jurisdictions started adopting this approach it became more and more community orientated and theoretically better founded. In the federal siting process this culminated in the explicit purpose to make the community better off after siting. This finding implies that the language used to motivate and explain the siting process changed somewhat. The explicit progressive pluralist reference to distributive justice in the federal siting process is one example. In the case of Alberta, the idea of distributive equity (the community benefiting from the facility) was left somewhat hanging in the air and remained at the level of economic diversification and a slight improvement of the tax base. One constant element in the discourse of the various Task Forces however, was a certain relativation of managerialism. In Alberta, the STF advocated the possibility of tailoring ('engineering') the facility to any site which was socially acceptable. In the federal process, the STF embarked on a program explicitly based on a rejection of managerialism.

There were various differing responses to the way the decision process was set up. Because the decision processes were held in so many locations, it is hard to describe the response in every single municipality. There are some general patterns however. Especially in the case of Alberta, where the approach was tried first, it can be seen that many actors involved in the ordinary decision-making institutions did not feel at ease with a community-based approach. For one thing, the decision to take away the initiative for siting from the private sector met with much resistance. One can say for instance that the Alberta government waited for around ten years in the hope that private market parties would develop proposals. And during the entire voluntary siting process the Albertan Cabinet largely remained silent except to say that it would not accept any responsibility for actually operating the facility, because that was a private sector task. Then take the Environment Council of Alberta, which was to advise on a more community-based approach and did say community acceptance was the number one criterion, but at the same time proposed that experts would select a site and would then gain community acceptance by explaining to the community why the site was best. The Alberta government actually appointed an Implementation Team that embarked on a largely technical approach and had already selected some sites. The opposition was still viable however, and the responsible minister now listened to the Siting Task Force, which started applying the ideas in the report on siting failures elsewhere.

The fact that the STFs in all three cases departed somewhat from the managerial language on siting made them vulnerable to a critique exactly from that managerial angle. To some extent, they anticipated this possibility and introduced certain - technically determined - constraints for the communities involved. But still, opposition from a managerial perspective was present in many communities, most notably in Alberta, where the Implementation Team had already selected the 'ideal' sites for a waste facility and citizens could easily ask why their community (if not on this list) was approached in the process. Certainly, the STFs left every room for a managerially-inspired interaction between participants. In fact, it does seem that their approaches should not be seen as a full departure from technically responsible siting, but rather an attempt to have ordinary citizens perform much technical study. This is indicated by the fact that the CLGs in most communities were set up to check the technical suitability of their municipality for waste disposal, nothing else. This approach had some success, most notably in Montcalm, where citizens improved the proposal, but then again still made the professional experts involved frown because they felt the citizens had no capability of studying these issues. It was furthermore important that the CLG and their STF advisors made a strong impression of neutrality, because if they did not, this would backfire greatly. Especially the presence of councilors on the CLG, and STF members blocking or influencing the free flow of information within the community or the invitation of counter-expertise could easily disturb this impression. This conclusion points to another area of discursive space for opponents. Apart from an attack on the siting process from a managerial perspective, voluntary siting was also attacked because of its compensatory element, which in many communities conveyed the message that their cooperation was being bought. This argument, part of the critique on progressive pluralism, was a powerful discursive weapon for opponents.

6.4.3 Theoretical feedback

Interests versus language

The three Canadian case studies offer some possibilities for observations on the relation between interests and language. The first important observation is that the community-based approaches to decision-making have not been inspired by decision-makers captured by the communitarian discourse. Instead, community-based approaches were tried because they offered the promise of possible success and the possibility for politicians to stay away from somewhat politically dangerous processes. That a community-based approach could work was the conclusion of a study report in Alberta and the other authorities basically followed suit when siting worked in Alberta. In this sense the choice for a certain approach for siting seems largely interest-based. On the other hand however, one can in fact observe that community-based siting does call for a certain language of explanation and secondly, invites a certain type of critique. Community-based siting, most clearly in the case of the federal siting process, tends to be defended on the basis of a rejection of the managerial approach. Such a rejection is necessary if one wants to defend the fact that practically the entire province turns into a search area, with the acceptance by the community as the dominant criterion. As I have already shown, this then invites comments from both the managerial and communitarian corners, and a critique of progressive pluralism is also often echoed in that many people resent the compensation embedded in the approach.

Feedback into some assumptions

(1) The role of experts

Like in the Dutch and UK cases, the Canadian cases shed some light on the objectivity of experts. There are indications, or at least allegations, in two of the three cases (Swan Hills, Deep River) that the experts involved were not completely neutral towards the outcomes of the process. In the Alberta process, citizens heard only one, very reassuring, opinion on the risks associated with hazardous waste incineration. This is not a matter of the members of the STF lying, because they themselves believed what they were telling the audience. One question in their selection process was whether they would feel comfortable with the facility in question in their own backyard. The expected answer to that question was yes. Still, it does seem that sending out only people who believed incineration was safe, does imply a certain bias. If one wants to know why this information was acceptable to the citizens of Alberta, then one must look at the issue of trust. The members of the STF established such good rapport with the community, and received such good help from the opinion leaders, that trust was high. In such a climate, the desire to ask very critical questions may diminish somewhat. Similar remarks can be made about the federal process. In fact, in that process the comments of some went a lot further. Especially because there was some doubt about the correctness of certain data presented by the STF, its credibility came under stress. If the allegations are correct, then one can say that the experts involved in the process were far from neutral. Especially certain people in Port Hope explicitly said that the experts had delivered the results that their principals had desired.

(2) Market parties

The role of market parties is rather limited in all three cases. In fact the failure of market parties was the reason why the governments in question stepped in and overtook the decision process.

(3) Politicians

Politicians (at the provincial and federal levels) played a role in accepting the community-based process and in determining whether their municipality would participate in the process. There was a certain degree of accountability in the process in the sense that when the siting attempts failed in 'the ordinary way', politicians from the affected areas pressed for an alternative solution. There are also indications that when politicians, especially local councilors, ignored local opinion they were held accountable by not getting re-elected. In some cases the new council then terminated the decision process. In the federal decision process, the election promise of the incoming government was an important impetus for the start-up of the voluntary siting process. However, in that case the source communities had to actively 'remind' the politicians of their promises, even to the point of somebody starting a hunger strike twice.

There are indications that politicians are accountable to special interests. There is a very wide perception among the population of the involved communities that their councilors largely favor economic development and are accountable to business interests. The case of the town of Deep River, which started participating after a hint from the largest employer in town, seems to support this idea. In most municipalities, this would have been lethal. In many places, the mere fact that councilors had decided that their municipality would participate in the decision process already made them suspect in the eyes of the public and they consequently had better not get involved. Only in some exceptional cases (Montcalm, Deep River) did the public accept strong interference from their mayors. A final example of accountability to special interests seems to be the decision of the Alberta government to (a) award an attractive operating contract to a certain private firm, which had contributed to the ruling political party's campaign, and (b) the decision to exempt certain wastes from the oil industry from the requirement to be incinerated, after the province had approved and co-financed an expansion of the Swan Hills facility on the assumption that there would be no such exemption.

(4) Voters

There was a greater role for voters, especially local voters, in these cases than there would normally be under the Canadian legislation. Particularly through the referendum, could they make their opinion known. Turnout in the places where referenda were held was large, in the range of 70 to 80 %. The fact that the referenda were inserted as an element in the decision process was a consequence of the Alberta voters actually demanding one by a threat of force. There is ample evidence of interest group formation in many towns, amongst others Montcalm, and voters did hold politicians accountable in various cases too (take the election pledge in the federal siting process, and the fact that the siting process became the subject of election debate in several municipalities. The motive of people to become active varies somewhat. In practically all communities where voluntary siting took place, citizens were mainly concerned with the health effects of the proposals. It is clear that in places where there was an attempt to override or neglect this concern in the eyes of the public, resistance quickly sprang up. Under community-based siting processes, there was a possibility to prevent this. It is good to note that voluntary siting failed (in the sense that no facility was sited) in the majority of the communities where it was tried, often because there was too much concern with health effects, either with the local council or with the local population in general[67]. This often happened even before it came to a vote, and a perception of bias in the council or STF was sometimes a contributing factor.

Most participants in the process saw their health as an 'absolute value', implying that they did not want to negotiate about it within an economic framework. Only in a limited number of cases did the decision process proceed beyond the first steps in siting decisions and were Community Liaison Groups formed. It does seem that in many of the communities where the process went further, there was a certain level of experience with hazardous substances and thereby far less intuitive fear. Take Lake Elliot, one of the towns desiring the low-level radioactive waste disposal site. '*Since we more or less live with the materials already, we don't have as*

[67] Other reasons to object include: property values, image of the area and sometimes 'fairness' (why exempt nature reserves from consideration as potential sites and not residential areas?).

much concern with it as some other communities'[68]. What also worked was the use of communicatively able experts (who translated the risks involved to well-understandable and common examples) and the use of trusted community members in the process. The selection of trusted community members, or even outright opponents to the proposals to the CLGs helped the process. The groups acted, to some extent, as a representative for the wider community, and these groups were educated to understand the technical aspects and risks involved at a somewhat deeper level than the rest of the community. Through their work, especially in Montcalm and Deep River, they communicated information within the local community towards the decision-makers and vice versa. This information could be local knowledge (Montcalm) or scientific knowledge (Deep River) and it seems that such information has enriched the decision process. It does seem that voters, after getting over their health concerns, do seem rather driven by 'self-interest', especially when their employment is at stake such as in Deep River. Especially the former chair of the CLG was very bitter about the result of the referendum in that town and said the positive result was mainly driven by concern over their employment at AECL[69]. One resident that commented on the idea of hosting a low-level radioactive waste facility said: '*The tourists we're hoping to attract here are hypothetical, but this is real. This is jobs*'[70].

(5) Parties to the court case

As none of the cases ended in court - in itself somewhat of an achievement - there is little to say about such parties on the basis of the case studies. Only in the case of the Swan Hills expansion and the start of waste importation to that facility was there a quasi-judicial hearing with the Natural Resources Conservation Board. It does seem that during these hearings, especially the one on the expansion of the facility that there was a certain desire to hinder truth finding. Actual waste arisings in Alberta were a point of contention there, but the operator of the Swan Hills facility would not openly discuss the amount of wastes they treated.

(6) Members of a community

There is strong evidence to suggest that citizens do have local knowledge that is important to the decision process. Especially in the case of Montcalm,

[68] 'Elliot Lake hoping to become nuclear waste site', The Globe and Mail, 15 January 1988, p. A-1.

[69] This concern was real. In 1988, 400 of the approximately 2,500 jobs had been lost and it was clear that more would follow. See 'Deep River citizens fight for town's future', Ottawa Citizen, 29 January 1988, page D-13.

[70] Ibid.

citizens contributed information that was crucial to the decision in the end. The ideal of coercion-free dialogue is however not strongly supported by the empirical evidence in these cases. Especially in the case of the Alberta siting process, there was an explicit recognition that the community had certain leaders and that these leaders had to be won over so that they could affect others in their turn. But also in the case of Montcalm, there were some indications of coercion. The mayor there suggested the opposition group had forced fellow citizens into signing a petition against the plant. In all three cases, council members were in a privileged position in the sense that they could decide on the continued operation of the process.

There is conflicting evidence on the question whether community members are parochial in their outlook. Deep River is an example of a municipality largely going against the will of the surrounding communities. But this does not seem to be the work of 'ordinary' community members. Minutes of one CLG meeting for instance, indicate that various citizens from the town felt other communities should participate as well. '*A member of the public expressed his opinion that even though he personally was a resident of Deep River, that he felt as though he were a member of a broader community and that he felt an expansion of the CLG's membership was justified*'[71]. The CLG acknowledged that the siting process was flawed in this respect[72], but also indicated that it did not have the authority to decide the issue and presented the matter to the council. It was the council that decided to largely exclude others. Although some members of the community apparently disagreed, there was no strong negative general response to this decision in the Deep River population. Certainly, the chair of the CLG actively worked with the municipal council to devise strategies to dominate members of other communities that opposed the waste disposal facility. In the case of Swan Hills, there are few neighboring communities, and certainly no actively opposing neighbors. This changed somewhat through the years, with some neighboring communities starting to get an increasingly negative opinion on the facility. The reaction in Swan Hills is to defend the facility, which has really become 'their facility'. In Montcalm, it was mainly the result of the work by the MHMWC that other communities were in fact informed and were asked for their opinion. It does not seem fair to say that the communities involved did not have scientific knowledge.

Especially because of their experience with hazardous substances, the communities involved were actually quite aware of the risks, and the expertise to deal with the substances was often available within the

[71] Deep River CLG Minutes, 26 November 1992. As found in the Deep River municipal archive.

[72] Deep River CLG Minutes, 29 October 1992. As found in the Deep River municipal archive.

community. The degree to which this was the case differs somewhat, with the community of Deep River having the greatest expertise in house. It has been noted that the mere fact that the decision process was not imposed, but voluntary, reduced the motivation to actively participate.

Chapter 7

Comparison of case studies

Institutions, discourses, actors, and decision quality in the nine cases

This chapter compares the outcomes of the three previous chapters. The intention here is not to make statements about the way different approaches to decision-making will work under any circumstance. Rather, the idea is to demonstrate the applicability of the theoretical framework, and to assess the potential implications of the application of certain approaches to decision-making in terms of interaction patterns and quality of the outcomes. I will discuss the institutions involved in the nine case studies, the role of various individual actors, the role of discourse, and the quality of the outcomes.

In terms of institutions, the main division is between Canada on the one hand and the UK and the Netherlands on the other. The role of the community in Canadian siting processes was at the expense of the private sector and elected representatives, institutions that played an important role in the UK and Dutch cases. On the topic of discourse: the managerial discourse is the most accepted language for debating hazardous waste issues. Structuration is 'automatic' in about half of the cases. In the other half, the role of 'enforcement measures' is important. On the actors, notable conclusions are that experts do seem somewhat biased towards the people who hire them, that the role of politicians varies with the level of conflict, and that voters do hold politicians accountable (but only for errors) and will form interest groups. Furthermore, parties in a court case do not always remain involved forever, members of a community do not always interact free from coercion. On the quality of the decisions, I conclude that emphasis on the community results in good social quality, but reduces cost-effectiveness.Finally, it is concluded that a certain level of positioning exists, meaning that people pick a certain discourse that suits their interests best. Such positioning is less frequent among people with a scientific training however.

7.1 Introduction

Introduction

In this chapter I will present the findings from my case studies and draw some conclusions. For the sake of clarity, the table below presents some data on the three cases I have studied in the three countries.

Table 1. Cases of hazardous waste siting studied

Country	Case studies (plus short description)
UK	• Seal Sands (Cleveland County) - proposal for a hazardous waste incinerator by Ocean Environmental Ltd. Rejected by local council, approved by SoS after inquiry.
	• Newport, South Wales (Newport Borough Council) - proposal for a hazardous waste treatment facility by BFES. Rejected by local council, approved by SoS after inquiry.
	• Rotherham (Rotherham Metropolitan Borough Council) - proposal for a hazardous waste landfill by VHE. Rejected by local council, rejected by SoS after inquiry
Netherlands	• North Refinery (Municipality of Delfzijl, province of Groningen) - proposals for hazardous waste treatment plant by North Refinery. Approved by authorities, but licences often quashed by the State Council.
	• Kaliwaal (Municipality of Druten, province of Gelderland) - proposal for dredging sludge disposal facility (landfill) by Delgromij. Authorities initially refused to cooperate, but province changed opinion. Much involvement of the State Council.
	• Ecoservice (Municipality of Dordrecht, province of Zuid-Holland) - proposal for hazardous waste treatment facility. Authorities approved.
Canada	• Swan Hills (province of Alberta) - proposal for an integrated hazardous waste treatment plant with incineration capability by provincial Siting Task Force. Approved by local referendum and consequently built and operated by the private sector.
	• Montcalm (province of Manitoba) - proposal for an integrated hazardous waste treatment plant without incineration capability. Approved by local referendum and consequently built and operated by the private sector.
	• Deep River (province of Ontario) - proposal for a low-level radioactive waste disposal site. Approved by local referendum but not built because of financial concerns.

Difficulty in comparing the cases

A comparison of the various case studies across the countries is problematic from a methodological perspective for various reasons. Among these reasons is the fact that the working of institutions in practice is partly conditioned by constitutional factors such as the position of local government in the administration of the three countries. This position differs. And take the difference in scale of the three countries, with practically every Canadian province being greater than the entire UK, and many times larger than the Netherlands. Also relevant might be the different degrees to which people in a local community are concerned about waste and the degree to which they trust certain institutions or not. Such differences, beyond the focus of my research, did exert a certain influence in the various case studies and influenced outcomes. Take for instance the fact that in Canada most citizens seem to have extremely limited faith in their local councilors. This has had a great impact on the practice of hazardous waste siting in that country.

Simple comparisons

The incomparability of the various cases does however not prohibit relatively 'simple' comparisons of (for instance) the discourses that were present in the decision process and thus the institutions and decision rules advocated. In all three countries one can thus study the theoretically interesting question of what happens if the legally prescribed way of decision-making is being challenged. Is the dominant discourse enforced or not? And if so, who enforces it and how? If not, why is it not being enforced? Because the three countries have been selected on the basis of the assumption that their siting processes would be different, it does seem possible to say something about the effect that different processes have on a certain population or community and the outcomes they achieve. It is not possible to predict the reaction of another population to such an approach, but the observed reactions do indicate possible reactions, identify possible problems associated with a specific approach, and help the development of discourse theory. So even though one cannot be certain that in the UK the local population would reject compensation as often happened in Canada, it is probably good for UK decision-makers to know that such a rejection is a possibility. They can perhaps learn about ways to avoid this reaction on the basis of the Canadian experiences. And although an answer to the most fundamental question, the one of the contribution of the various institutions to various conceptions of decision quality, can certainly not be definitively concluded on the basis of these nine case studies, it does seem intriguing to see whether a more community-based approach did in practice have a certain effect on the quality of the outcomes.

7.2 The institutions involved

7.2.1 Introduction: private sector versus other institutions

If one compares the roles that the various institutions played in the cases, then one notices several differences. Table 2 summarizes this. One pattern throughout the case studies is that the role of private market parties on the one hand and most other institutions on the other are inversely related. The pattern is a consequence of (a) the fact that private parties are free to do as they like, unless there is government regulation (based in law) that forbids them certain actions without approval. This in combination with (b) the fact that other parties (especially 'community members') normally only have an indirect influence on the decision, that is *via* an influence on government. If the role of the private sector increases, this implies that government exercises less control, and with it the other institutions decrease in importance.

When looking at the role of the private sector in the nine cases, one can see that this is very large in the UK and the Netherlands, where private companies were the proponent of the facility, and extremely limited in Canada, where such parties played no role in siting[1]. Especially in the UK there was often a certain pretense from the private market parties that they had selected a site on the basis of an almost scientific siting process. This pretense was intended to conceal the role of market considerations in site selection and make a more solid impression. Although often rather easy to see through, it was not really attacked during the decision process because most actors essentially saw the selection of a site as a private party responsibility. In the Netherlands, the role of the private sector varied somewhat. There was one case (North Refinery) where elected representatives and their administrators wanted to essentially forsake their duties of checking private initiative (the 'gedoog' phenomenon), thereby giving free play to a market party. In yet another they would not allow a role for private initiative in site selection for long and opted for a scientific approach to site selection, but had to concede to private initiative in the end. In Canada, the role of the private sector was limited, in part because there was no 'market' for the desired waste facilities, in part because the public would not accept any role for that institution.

[1] For the sake of argument I am confining the discussion on the role of private market parties in the text to waste disposal firms. That implies ignoring the role of landowners. These were willing to sell their land in all three countries and thus played no significant role.

Table 2. Institutions in hazardous waste siting

Country	Dominant institutions (plus short description of role)
UK	• Private market: companies made proposal and operate facilities. • Experts: help design facility, check acceptability of proposal, advise private market parties and elected representatives. • Elected representatives (in their capacity of community representative (Seal Sands, Newport) or in their capacity as head of the bureaucracy, i.e. Rotherham, Secretary of State): took decision to approve the facility or not. • Courts (mainly quasi-judicial inquiries): advised the SoS. • Community: as such not recognized. But some members sought representation to elected representatives and to an inquiry.
Netherlands	• Private market: companies made proposal and operated facilities. • Experts: helped design facility, checked acceptability of proposal, advised private market parties and elected representatives. • Elected representatives (only appointed heads of the bureaucracy): took decision to approve the facility or not, attempted to determine site in one case. • Courts (very active): reviewed government decisions, guarded the procedures. • Community: as such not recognized. But some members sought representation with experts and elected representatives, objected to higher-ranking government, and appealed decision to the court many times.
Canada	• Elected representatives (very low key): decided to install community based siting, decided to invite Siting Task Force, appointed Liaison Groups. • Experts (large visible role for 'communicatively able' ones): rough design of the facility, reassure the public, approved of the facility or not (after agreement with community - 'double veto'). • Private market parties: operated facility once sited. • Community (explicitly recognized): members could participate in the CLG or attend sessions organized by the CLG. Community as a whole had veto right.

7.2.2 Elected representatives

The different extent to which the private sector was involved in the decision process was not the only factor that distinguished Canada from the two other countries. Again inspired by public opinion, elected representatives had to remain very low-key during the Canadian case studies. Although their role was actually quite large (e.g. in deciding to apply community-based approaches and in inviting the Task Forces to the their municipalities and then selecting CLG members), they could not actually engage in the debate or sit on the CLG. Furthermore, taking decisions without holding a referendum was impossible.

The cases in the Netherlands and the UK seem to corroborate the importance of public controversy for the role of elected politicians. The 'ordinary' way of deciding proposals in both countries takes places, to a large extent, in an interaction between public officials, specialized committees of the local council, and the heads of the bureaucracy (elected representatives in the Netherlands, public officials in the UK). Local councils as a whole do not tend to get involved. In two out of the three cases in the Netherlands (Dordrecht, North Refinery) this pattern was adhered to, largely because there was no public controversy about the proposals in question. In the one case where there was a certain level of public controversy (Kaliwaal), did local councils as a whole get involved, but only *after* the case had achieved public controversy status.

In the UK, full council decided planning applications in two out of three cases (Newport, Seal Sands) because these applications had achieved the status of a public controversy, *in combination* with the fact that planning officers and elected representatives did not agree on the acceptability of the proposals. Only in the case of Rotherham did the 'ordinary' process stay intact, even though the proposal was generating quite a bit of controversy. However, here the planning committee and the public officials agreed that the proposals were not acceptable under valid planning policies. In that sense, there was no need to go to the full council.

7.2.3 The role of experts

Introduction

Experts played a role in all cases, but to a different degree, and again with an important difference between Canada and the two other countries. In Canada, or to be more precise Alberta, the ideas on community-based siting took a long time to actually be implemented. One obstacle was the Environment Council of Alberta, an advisory body consisting of experts, which advised the government of Alberta to make community acceptance the number one criterion, but at the same time saw nothing but a passive role for the public. The real innovation in the end came from social scientists with knowledge of community processes. From then on, the idea became to find a site that was acceptable for a host community and not a priori wrong from a technical perspective. The experts going into the field to gain community acceptance from then on were not so much the ones who were normally involved in licensing or siting, but were communicatively able experts, partly hired from outside the government. In Manitoba and the federal siting process, this distinction between a siting task force and the 'normal' apparatus of experts was also introduced. It caused least problems in Manitoba, and most in the federal process, where the Task Force apparently accepted a site with an uncertain status from the technical point of view and offered the community too large an amount of compensation in the eyes of the ordinary decision-making bodies.

Double veto in Canada

Despite the reduced importance of experts, the Canadian siting processes were still very much about technical issues. Especially the members of the Community Liaison Groups learned more about hazardous waste and siting and provided the link between Task Force and its experts on the one hand and the community on the other. It does seem that in various of the communities (especially Deep River), 'ordinary' members of the public often had a high degree of scientific training or experience with technical issues and could thus relatively easily participate. The 'double veto' implied that the experts who were involved in normal procedures always checked the outcomes of the decision process (although in Deep River it never got that far).

The existence of knowledgeable citizens is also visible in the UK and the Netherlands. Certainly in the Rotherham and the Kaliwaal cases, the population nearby the proposed sites contained a remarkably high level of expertise that greatly affected the decision process. The population pointed out various weaknesses in the proposals, some of which 'slipped through' the environmental impact analysis performed by experts of the proponents.

The UK and the Netherlands

In general, the role of experts in the UK and the Netherlands in the siting processes was greater than in Canada. In the UK, almost all 'statutory consultees' in the planning permission process were expert bodies. Their opinions pretty much determined the acceptability of the proposals (in combination with the planning policies of the local authority of course, but these were never a real hindrance). UK planning officers essentially gather such opinions and then weigh them up, with an important role for the 'bias in favor of the developer'. Although in two out of three cases, planning officers had to go against the councils, they did advise in favor of the proposals, thereby showing at least a great level of independence from their political masters. Their advice was ignored, but proved to be a good indicator of the outcome of the decision process in the end. The planning inspector who heard the inquiry on these two proposals, himself an expert, consented with the local authorities' planners and advised the SoS to approve the applications. The SoS had 'called in' the cases because of their sensitivity. However, as far as the SoS actually deviated from his inspectors' advice, such changes appear to have been made to guarantee uniform decision-making throughout the nation and maintain the correct interpretation of the planning system (increasing the legal rationality of the system).

In the Netherlands, the role of the Commission for EIA is remarkable. The organization effectively has a veto right on proposals because if it withholds a positive advice, the responsible authority cannot really make the decision. Especially the province of Gelderland (Kaliwaal case), which wanted to find the best locations for sludge disposal before accepting any private initiatives, had negative experiences with the Commission, which was very critical and would not accept the selection of ten locations the province made. But Delgromij and North Refinery (both private proponents) had to improve their assessment reports greatly before getting positive advice as well. It does seem in fact that this check is warranted, as the experts working for the proponent sometimes tend to make favorable assumptions from the perspective of that proponent.

More generally speaking, experts with the local and national authorities on the one hand and with the proponents on the other play the most important role during the decision process, although one must not underestimate the role of political will in this process. Especially in the North Refinery and Kaliwaal cases, political will sort of steered the experts in one direction (rejection or acceptance). The Dutch system of getting advice from experts is akin to the inquiry system in the UK as the official writing the report tends to speak to everyone, involves policy guidelines in his advice, and largely predetermines the verdict of the court.

7.2.4 The role of the courts

Introduction

The role of the courts in hazardous waste decision-making is greatest in the Netherlands, followed by the UK, and then Canada. To start with Canada, the reasons why the cases did not go to court differ somewhat. The proposals in the Deep River case never came to a point where opponents could have gone to court. I do think the potential for litigation was there in this case, specifically from the surrounding municipalities and certain environmental groups. In Alberta and Manitoba, the proposals did not end in court because there were in fact no objections. Environmental groups had been consulted beforehand and saw the need for the facilities, except Greenpeace in the Swan Hills case, but this group made only a very weak attempt at influencing the decision process. In Swan Hills, there were no surrounding communities to object initially (this changed at the time of the expansion of the facility) and in Manitoba these communities had been consulted.

Effective closure: UK inquiries

In the UK, all three cases went to an inquiry, mainly because planning permission was refused and this refusal was appealed. The inquiries were rather long-lasting events (except in Newport) and very detailed. The planning inspector hearing the case appeared very neutral to all parties involved, and allowed everyone to have a say, although he sometimes indicated that certain topics were really outside the terms of reference of the inquiry. Despite the constraints upon the inquiry related to these Terms of Reference (sometimes the SoS actually makes public statements that influence these Terms of Reference even more, e.g. by indicating there is a need for a facility), most topics of concern to the population were addressed at the inquiry. The rather litigious character of the proceedings, expressed by the multiple cross-examinations, make the inquiry clearly a pluralist beast, with each party involved making a case that is as strong as possible. The combined effect of the perception of extensive treatment of topics, unbiased planning inspector, and opportunity to make one's strongest case, does seem to satisfy the need for participation. Often the decision is taken only after a period of time (e.g. two years) which also reduces emotions somewhat and access to the courts is expensive. Together these factors ensure that the desire to go to court after the inquiry is not great.

No Closure: Dutch administrative courts

The country that really stands out is the Netherlands. All cases went to the highest administrative court in the country at least once and in two of the three cases even multiple times. The courts tend to act as agents of the legal system, meaning they tend to approach the issues before them in a rather formal way. Especially in the North Refinery case, the number of 'procedural' verdicts, without any implications for the operations of the facility, was large. Even very simple verdicts, consisting of three pages or more, sometimes remain in court for about three to four years. Although these verdicts often clarify a certain legal point and thereby help increase the predictability of the legal system in future cases, the fact that such cases take so long is defeating the purpose of achieving legal rationality for the case in question. This is because the position of licence holders (often the Chair of the State Council has already rejected requests to suspend a licence while the court ponders its position) is in the end quite uncertain because of this.

7.2.5 The role of the community

Canada as the exception?

The Canadian cases stand out as far as the role of the community is concerned. I have called the Canadian approach community-based siting and have concluded that there are subtle but important distinctions between Alberta's invitational siting on the one hand and Manitoba's and the federal government's voluntary siting and co-management on the other. Although the Siting Task Force of Alberta worked under a program that included the possibility of local vetoes of proposals, the intention was still to influence the local population into acceptance after the local council had sent an invitation to the Task Force to enter the municipality. The so-called power structure analysis, basically the identification of informal leaders, was used to get these people on board first and then influence others. The Manitoba and federal siting processes were set up differently, with especially the federal process still showing some signs that it turned into a community-influencing program, and not the voluntary approach it was portrayed to be. In any of the three cases it would be incorrect to say that the entire community actively participated. The most popular forms of participation (e.g. voting and attending hearings) involved relatively little activity from citizens. A small minority of the population however was appointed by the local council to a Community Liaison Committee to gain a somewhat better understanding of the issues at stake. To some extent these groups became community representatives, creating a certain tension with the local councils if these did not agree to the CLG's recommendation. Especially in Deep River there were problems related to the development of a right definition for 'the' community. The issue was left to the local council. In this case, being a proponent of the proposal, the Deep River council drew the lines in such a way that there would likely be approval for the proposals.

No such thing as community in the UK and the Netherlands?

The role of the community in the decision processes in the Netherlands and the UK was much less significant. The Netherlands are in fact the country where the population of two of the municipalities involved (Delfzijl, Dordrecht) was least interested, to the level of almost being completely passive. In the third case (Kaliwaal) a substantial number of members of 'the' community only became active when it was really already a bit too late. In that case, the few members of the community who did want to participate did so largely on the basis of their expertise and did not undertake very active mobilization efforts. The UK public, often described as relatively deferential, was more active than the Dutch public. In each of the cases several thousands of signatures were collected and opposition groups formed. The support for such groups seems to have been concentrated mainly in the areas around the proposed sites, with the rest of the community remaining relatively passive. It seems that the size of the towns involved (tens of thousand of inhabitants rather than hundreds as in the three Canadian communities where a proposal was accepted) contributed somewhat to this effect. As was said on siting in Winnipeg, Manitoba, in such larger cities: 'the community defines itself' but not along the lines of the local council's jurisdiction.

7.3 The actors involved

7.3.1 Introduction

In chapter 2 I have distinguished various types of actors that are potentially involved in hazardous waste siting and have formulated several assumptions about their motives and behavior. I will now compare the evidence on these assumptions for the three countries, organized per group of actors. For the purpose of easy reading, the discussion of the evidence on the assumptions begins for each group of actors with a repeat of the relevant part of table 10 from chapter 2.

7.3.2 The role of experts

Table 3. Assumptions on experts from chapter 2.

Assumptions (from which discourse, or other source?)
On the positive side:
• Are expected to be objective and neutral, will separate values from facts (managerial).
On the negative side:
• Will not be able to separate values from facts (pluralism, critique of managerialism).
• Will deliver the results that the persons paying for them want (critique of managerialism).

The neutrality of experts

The various discourses distinguished in chapter 2 do not agree on the role that experts might play in hazardous waste decisions. Especially managerialism and the critique of managerialism offer some thoughts that can be assessed - to a certain point - on the basis of the nine case studies. The first assumption expressed was that experts are objective and neutral and will separate values from facts. On that issue, it appears that in general, experts tend to function as agents of the legal framework. As far as this framework is biased (e.g. in favor of the developer) experts tend to support such biases. This is - to some extent - evidence against the assumption.

Furthermore, in the UK cases, local government experts appear to have been independent of their political masters. Planning inspectors are widely perceived as being neutral. At inquiries however, experts often argued widely diverging positions, shedding some doubt on their independence of their masters. In the case of Seal Sands, the chief planning officer, who had first argued in favor of the proposal, started arguing against it at the inquiry after his council had expressed opposition. In the Dutch cases, the Commission for EIA is widely seen as independent and neutral. Local government officials appear to be somewhat influenced by the opinions of their political masters. There is evidence that experts writing environmental assessments tend to choose favorable assumptions for the developer. In Canada, the neutrality of the experts employed by the Siting Task Forces was often contested. In two out of the three cases there was a strong desire to sway the population.

The fact and value separation

The second assumption was that experts are not able to separate values from facts. Generally speaking, I have found that the advice of experts tends to be contingent on many assumptions. Such assumptions (e.g. waste trucks will drive a certain route) are influenced by values and are not often fully discussed, although more so in the UK than in the other countries (because of the long periods of interaction between the parties). In the UK, the separation between values and facts was a story line for local public officials who expressed their opposition to proposals but at the same time indicated they could not refuse them. Planning inspectors did not accept positions unless these could be substantiated with facts (against). However, at the inquiry experts tended to substantiate the position of their masters.

In the Netherlands, experts tend to substantiate the positions of their masters as well. The advice of the Commission for the EIA however, tends to separate between 'facts' (the environmental effects predicted in the EIA reports) and the political evaluation of these facts, which is an issue for the local authorities. In Canada, especially in the federal process, the siting process was based on a rejection of fact/value separation because the objectivity of experts was questioned. Ironically, the siting processes in Swan Hills and Deep River seem to demonstrate that the separation is indeed hard to make as the desire to sway the population may have influenced the neutrality of the experts involved.

The experts as the handmaiden

The third assumption on the role of experts was that experts would deliver the results that the persons paying for them want. There is some evidence in favor and some against this assumption. In general it must be noted that parties tend to have little difficulty in finding experts to support their opinions. Note the Newport case, where most experts involved saw no reason to reject a proposal, but still a consultant was willing to argue the borough's case. In the UK, the description especially of site selection in EA reports is somewhat suspect and seems intended to make the site acceptable. Experts take widely different views at inquiries. But local planning officers seem relatively independent and at the inquiries, consensus is sometimes achieved between experts arguing for different positions. In the Netherlands, two EIA reports, written for different developers at about the same time, gave totally different conclusions on one particular waste disposal technology. In both cases, the statements substantiated the choice of the proponent. More generally: experts tended to argue in favor of their masters. In Canada, experts set the general parameters for the siting process and they did so in a neutral fashion. The Siting Task Forces seem to have had a certain tendency to hire experts who reassured the populations. There is no indication that such experts specifically changed their opinions for the Task Force, but the Task Forces picked a certain brand of experts with positive opinions about the proposals. Only in the case of Deep River were there sounds that experts had given biased opinions in the sense that they may have 'bent' information so as to achieve greater acceptability of sites.

Some general reflections

A general conclusion on the role of experts could be that experts tend to be relatively flexible creatures, unlike what one would expect under the managerial discourse. The 'trick' of the expert, so to speak, is that the assumptions they make are very important. It seems unavoidable that in the end, any decision will be fraught with assumptions that do materialize sometimes in practice, sometimes not. Especially where certain assumptions are made about the human component of facility management, these often prove incorrect. An example from each country: in the Netherlands the North Refinery plant started accepting wastes without a proper hook-up facility for trucks, which then caused a certain amount of spillage. In Canada, waste off-loading practices led to PCB leakage and at one point to an explosion. In the Newport case in the UK, waste trucks tended to ignore the route designated for hazardous vehicles. Especially the UK inquiries, and to a lesser extent the Dutch Commission for EIA and the Advisor Environmental Appeals and the Canadian CLGs, can provide good fora for discussing such assumptions, but it seems that there must be a certain level of counter expertise present to make this possible. Another general conclusion, outside the sphere of the assumptions is that experts cannot solve every dispute. Especially the North Refinery case in the Netherlands demonstrates that experts sometimes cannot determine the meaning of vague terms such as Best Available Technology. In that particular case, the consequence was to leave a facility in operation, despite the emission of 'black list' substances.

7.3.3 The role of market parties

Table 4. Assumptions on market parties from chapter 2.

Assumptions (from which discourse, or other source?)
On the positive side:
• Know their own interest best and will defend it if need be in court or in politics (pluralism).
• Can be expected to define their own interest so that it includes the interests of others (pluralism - the harm principle).
On the negative side:
• Will have difficulty in defining their interest in complex issues (managerialism, critique of pluralism).
• Will not think of anybody but themselves (critique of pluralism).

Know your own interest

Market parties do play an important role in the decision processes in the UK and in the Netherlands. It is hard to say whether the first assumption, that private parties know their own interest best and will defend it, if need be, in court or in politics, holds true. Perhaps a comparison of the Canadian cases, where private market parties played a limited role, with the cases in the other countries is the best way to actually gauge the knowledge of their interest.

I have already said that market parties tend to have a great interest in correctly assessing the market for their facility and in getting approval for their proposals as quickly as possible. This interest is driven by the fact that enormous investments are at stake and that the decision process itself tends to be quite expensive. It appears that especially in the Swan Hills case, it was problematic to assess the market for waste treatment at the facility. The province of Alberta took up a certain responsibility for the commercial success and paid a hefty fine. On the other hand, the far less pretentious facility in Manitoba seems to provide an example that such losses are not inherently embedded in a community-based approach to siting. What is often said to be lacking in community-based siting, where private developers are absent, is an actor that can effectively play the role of proponent of the facility. It does seem that in each of the Canadian cases, Siting Task Forces, although not always designed for that task, did take up a certain advocacy role. This role needs to be played by at least one actor, because the decision process runs a large risk of 'going nowhere' without a proponent.

In the Netherlands and the UK all companies involved had sunk large sums of money in the decision process and the design of the facility. They did undertake very strong attempts to advocate their proposals both in politics and in court. Especially the Kaliwaal case in the Netherlands is a good example of a very persistent proponent, which brought the issues before court countless times and continued to lobby for its plan for about 20 years. Note that if the role of proponent is picked up by a private party, then other actors can take up different roles. The province of Gelderland for instance, after ten years of resistance against the involvement of a private market party in sludge disposal, decided to leave the initiative to Delgromij. This opened the possibility for the province to sit back and start playing an arbitrating role instead of the role of proponent, which it had attempted to play for some time.

The harm principle

The second assumption, that private parties can be expected to define their own interest so that it includes the interests of others (the harm principle) does not receive much support from the six UK and Dutch cases studied. In fact, one can observe a limited level of sophistication among private developers when it comes to understanding the public and making their proposals more attractive to a local population. In the UK, all private developers effectively bypassed the population and did not try to win over the local public. In fact, they seem to have had an eye only for councilors and public officials and for the question of whether the technical quality of their proposals was good enough to aquire the approval of the regulators.

In the Netherlands the situation is somewhat similar with the exception of the Kaliwaal, where objections to sludge disposal from the provincial authorities were taken away by accepting 'voluntarily' an obligation to develop a natural area at the site after completion of the landfill. Interestingly, this move gave the developer a certain leeway in the negotiations with the province on licence conditions. Because the nature development was not legally required, the proponent could threaten to take that part away from the proposals. Because of the weak support for the assumption that parties take care of the interest of other parties, the fourth assumption, which is the opposite and which holds that parties think only about themselves, is perhaps more credible.

Complexity

The third assumption was that private market parties would have difficulty in defining their interest in complex issues. This assumption does not appear to hold true. Although hazardous waste siting does involve certain difficult issues in a technological sense, most parties knew quite well what their interest was. Waste disposal companies tend to have in-house expertise on waste disposal technologies, and if not, they hire the expertise from elsewhere. They have greater difficulty with the economic side of the issue than with the technical side. Particularly because the waste market was in flux near the end of the 1980s, it was difficult to predict the actual demand for waste services. Such flux was caused especially by the development of new technologies (from landfill to incineration to treatment) and by rapidly changing government regulations. Such developments (note the Seal Sands, the Kaliwaal, and the Swan Hills cases) can make investments (that seemed worthwhile initially) unprofitable rather quickly (or vice versa). This is somewhat ironic as the debates about the acceptability of the proposals often rage on for months, if not years.

7.3.4 Politicians

Table 5. Assumptions on the role of politicians.

Assumptions (from which discourse, or other source?)
On the positive side:
• Will be accountable to the people because they need their support (pluralism).
On the negative side:
• Are accountable only to special interests (managerialism, critique of pluralism).
• Will only use hazardous waste decisions for political gain, even if it is strictly a technical issue (managerialism).

Accountability to the people

The first assumption on the role of politicians in hazardous waste decision-making was that politicians will be accountable to the people because they need support to stay in office. Accountability to the electorate was an important factor in Canada and the UK, but less so in the Netherlands, mainly because of the differences in the level of controversy[2].

[2] The explanation for such differences was not studied here. However, in two out of three cases in the UK, the local public had negative experiences with existing proposal or had already been 'sensitized' by an earlier waste proposal. Such sensitization was less present in the Netherlands. In Canada, it seemed that communities became sensitive in part because *other* communities had reacted in a negative fashion. The first Albertan community where resistance had occurred had a very active local doctor with a strong concern over health

In Canada, the community-based siting process was installed largely as a consequence of the fact that politicians feared for their jobs if the decision process were allowed to continue under the normal regulations. When the voluntary siting process was started, it turned out, paradoxically, that they were perceived as most accountable if they stayed out of the decision process. In the UK, local politicians play an important role in the course of a decision process in the sense that if they approve of a planning application, they effectively cut off the population from the appeal system. In all three UK-cases however, politicians appeared very sensitive to the opposition (always a minority of the population), and rejected planning applications. Siting was never an election issue in the UK, but that was mainly because none (or hardly any) of the elected members of councils and Parliament were in favor of the proposals. To be in favor of the proposals would have potentially implied important electoral consequences. The timing of the SoS's decisions especially is remarkable in this respect: the unfavorable decisions were announced just *after* the general elections. The system in the UK is however such that the national level takes over in the appeal process. Although in all three cases it was the SoS himself who decided the appeals, he did so mainly as an agent of the planning system, and thus saw to it that the decision would be proper in legal terms. To this end, the SoS generally followed the advice of his inspectors, except where certain conditions (e.g. local monitoring of the Seal Sands incinerator) did not fit within the legal framework.

In the Netherlands, local elected politicians, in two out of the three cases, had no problems with the proposals, probably just like most inhabitants of their municipality. In the one case where there was some public opposition - the Kaliwaal case - the most relevant local council was against the proposals practically from the start of the decision process. The negative opinion of this council was largely based on the findings of experts that the Kaliwaal was an important ecological reserve, an argument that opposition groups used as well. Public attention to the proposals in this case dropped for an extensive period of time, but the resistance of the local council remained, instilled in a zoning ordinance. The Kaliwaal case is quite unique in that the community nearest to the proposed site did not live in the municipality that exercised control over land use. The nearest community fell under a municipality where the council never wholeheartedly resisted the proposals, as its mayor preached faith in the higher authorities and experts. Only when public opposition in that specific community erupted, did their council join the fight against the proposal. This happened on the basis of an explicit

---

risks associated with chemicals. I have often heard that such trusted figures can swing the mood in a community.

reference to the amount of local opposition and to local democracy. Thus, this council was responsive to the local public.

Accountability to special interests

The second assumption contends that politicians are only accountable to special interests. Although the term 'special interest' is somewhat derogatory and perhaps not always applicable, I do think that one could make a case that both the waste disposal firms and environmental (and opposition) groups are cases of special interests. In Canada, where local politicians are widely seen as pro-development and pro-industry, the, often indeed very inviting, attitude of councilors towards the waste industry or task force was clearly not a reflection of broader consent within the community. In fact, for community-based siting to be successful, local politicians had to refrain from actually participating in the process. In some cases they lost their jobs if they violated this principle, in some others (Deep River, Montcalm) they successfully interfered in the siting process and rescued it from going under. In the case of Alberta, the decision of the Alberta Cabinet to exempt certain oil industry wastes from hazardous waste regulations seems to have reflected a certain responsiveness to that type of industry. Environmental groups played no role in the Canadian siting processes, except in the federal siting process where one action group gave counter-expertise to the communities involved in voluntary siting.

In the UK, there is no indication whatsoever that politicians were accountable to the waste industry. In fact, in all three cases they rejected the proposals from the start, although they could not publicly state so. The accountability of politicians to action groups was limited. In all three UK cases, very active action groups were founded to oppose the proposals. But politicians tended to refrain, at least publicly, from mingling with these action groups until they had formally decided to reject the proposals. From then on, they tended to cooperate with the action groups (in Newport they did so from the very beginning).

In the Netherlands, the case of North Refinery demonstrates a clear-cut case of responsiveness to special interests. The municipality of Delfzijl wanted to accommodate industry as much as possible and did so by largely circumventing legal procedures. This fitted the national agenda to some extent, as the concern of the two national ministries involved was to develop waste treatment capacity. The other two cases do not show a certain responsiveness of elected representatives to special interests, although one may wonder whether in the Kaliwaal case the province could still refuse the proposals after having had to admit that Delgromij's plans were the only viable solution to the sludge disposal problem.

Political gain? What gain?

The third assumption, which predicts that politicians will use hazardous waste decisions for political gain receives only weak support. The fact of the matter is that if the proposals are uncontroversial, then they do not offer much possibility for political gain at all. This appeared to have been the case in two Dutch case studies. In such situations, the 'normal' rule that economic development is good for the municipality can be easily applied and local politicians can only use the proposals in the sense that they can seek to be associated with the new development, e.g. by opening the facility. This happened indeed. It also happened, to a degree, in Swan Hills (Alberta), where the provincial decision to site a facility in their community was celebrated in the presence of politicians. This was however *after* the proposals had proven to be accepted by the public. The people of Swan Hills considered the role of their local member of the Provincial Legislature in acquiring the facility quite great and in fact staunchly supported him from that time. In other communities, where politicians spoke out on behalf of hazardous waste facilities *before* the public had embraced the idea, they then came under enormous pressure and had to back-off. In the UK, all local politicians, in all three cases, spoke out against the proposals that were being made in their areas, but only after the council had formally taken a decision to reject the proposals. Saying that politicians sought to gain from the decision process does not properly explain this phenomenon. Instead, councilors strongly feel the obligation to decide applications neutrally. Only when the decision process enters a more litigious stage (the inquiry and the period leading up to it) can they freely defend their opinions. Members of Parliament do not live under such a constraint, and here one can see that almost immediately after the various proposals became public, MPs spoke out against them, irrespective of their party affiliation. It seems that as in Canada, politicians better avoid expressing their support for the proposals because it may cost them votes.

7.3.5 Voters

The active voter

Especially in the UK and Canada, voters were active participants in the decision process. This was less so in the Netherlands. Again, this is related to the level of controversy associated with the proposals in the constituency. The first assumption on voter activism was that voters would hold politicians accountable if they do not take the right decisions. This certainly happened, and especially in the Canadian cases, where voters practically took elected politicians hostage in one case. The answer came in the form of community-based siting, which was in many ways also a way to de-emphasize the role of elected representatives in the decision process. In the Netherlands and the UK, as far as I know, no political victims fell in connection with the proposals. This is at least partly related to anticipation of the public reaction by elected representatives.

Table 6. Assumptions on voters.

Assumptions (from which discourse, or other source?)
On the positive side:
• Will hold politicians accountable if they do not deliver the right decisions (pluralism).
• Will form interest groups to influence decisions in the right direction (pluralism).
On the negative side:
• Do not have an interest in decision-making and are not sufficiently educated to form an opinion (pluralism and managerialism).
• Will behave selfishly: favor economic development over the environment and will express 'absolute values' (critique of pluralism).
• Will form interest groups to further their own particular interest and prevent achievement of the public interest (managerialism).

Interest group activism

The second assumption was that there would be interest group formation in connection with the proposals so that voters can influence decisions. There are some remarkable differences between the cases in the three countries in this respect. In Canada, we can see that already established interest groups (environmental groups) were either committed to the siting process by a long process of consultation (Alberta, Manitoba) or were engaged in the siting process as counter-experts (federal process). And, if we look at the three communities that volunteered, we see that only in one community (Montcalm) was there formation of an interest group. This result is logical in a sense, because the formation of opposition groups was a reason for the Manitoban task force to break off the decision process. The fact that the process in Montcalm was not terminated is exceptional and related to strong interference by an elected politician. The fact that virtually no opposition groups formed in these three communities is probably partly related to the fact that the CLGs took up part of the critical role normally performed by such groups. Especially in the case of Swan Hills, the CLG was organized in such a fashion that the staunchest opponents were invited on board, giving the CLG credibility in the public eye.

If we compare the Canadian pattern with that in the UK, we can see that in every UK case an interest group comprised of local citizens formed. Here too, the already existing environmental groups were inactive (although Friends of the Earth played an assisting role in Seal Sands). The local action groups were all successful to a degree in organizing a petition, in informing the public and in challenging the proposals at the inquiry. It does seem however that the level of activism required for forming the interest group and keeping it running was only delivered by a relatively small group of people. The amount of time and energy required to sit through an inquiry is enormous and was only invested by a small group of people, with the rest of the population mainly looking on.

The situation in the Netherlands differs from that in the two other countries in that established interest groups (mainly environmental groups) play a dominant role in the decision process. In two out of three cases (North Refiner, Ecoservice) no local groups formed whatsoever but environmental groups were active (very active in the North Refinery case). In the third case (Kaliwaal), established environmental groups did participate, but seem to have had difficulty in orientating themselves towards the project, which did contain a 'nature creation' package and was therefore supported by the WWF. This led to a certain level of infighting between action groups, with some opposing the proposals, others supporting them. Meanwhile, a local action group did form, but this was relatively late. This was a consequence of the fact that the (initially) few opponents thought they could win the debate on the proposals by engaging in a technical debate and had a limited eye for the mobilization of the population.

The careless voter

The third assumption was that voters would have no interest in decision-making and would not be sufficiently educated to form an opinion. There is evidence of limited interest for the decision process in all three countries, albeit that the Dutch communities were least interested. The general pattern does seems to be that those who live near the proposed site, say in a two kilometers radius, tend to be most active and are the ones (if at all) that form interest groups, organize petitions, or participate in another fashion. The rest of the voters tend to be onlookers, although people with a specific interest in this type of proposal (e.g. in the Seal Sands case: somebody who had operated an incinerator) may link up with the opponents. Armour (1996) has indicated that in the Canadian communities where community-based siting processes occurred there was a certain fatigue among the population after a period time, related to the process being too participatory and seemingly offering little threat to the population.

The second part of the assumption, the low degree of education in the population, should probably be rejected. It must be observed that in most cases, the local voters experience the proposal, at least to some extent, as an issue involving (absolute) values. The decision whether or not to accept the proposals is therefore one that people feel they can make themselves quite well. But even in the realm of the technical aspect of the decision process, voters tend to do quite well. Of course it is true that most citizens have little experience with hazardous waste and will never be able to grasp the technical issues in great detail. However, especially the Canadian cases demonstrate that most voters are able to understand the issues, and if asked to become a member of a CLG, can follow the discussions. What does seem crucial in making this 'work', ironically, is a certain level of distrust that motivates people to keep asking critical questions. Especially in the Swan Hills case, there are some doubts as to whether this happened sufficiently. Furthermore, in each of the three countries, there is at least one case where knowledge or expertise of the ordinary public was an important contribution to the decision process. The case of Rotherham in the UK is probably the most spectacular example of local citizens having an extremely high level of expertise, but the cases of Montcalm (flood risk) and the Kaliwaal (river soil) were also good examples. If anything, such examples demonstrate that from a managerial perspective, it is quite a good idea to have citizen participation in the decision process.

Absolutism from the voter

The fourth assumption was that voters would behave selfishly, opt for economic development and formulate absolute values. These assumptions do seem to hold true to some extent. The fact that the most active voters are located nearby the proposed sites is an indication to that effect. If we look at the concerns that people tend to have, then we can observe that hazardous waste facilities often lead to questions about health effects, risks of accidents, and about the implications for the local environment and landscape. Health concerns are often great and are used by opposition groups to mobilize resistance. Symbols associated with death (such as skulls) are therefore a logical mode of expression for opposition groups. Figure 1 is an example and represents the letterhead of an invitation to a meeting of the action group in the Rotherham case.

Especially the concern for health effects tends to be formulated in absolute terms, as their health is something that most people do not wish to negotiate about at all - it is an absolute demand. The 'standard' way of regulating hazardous waste decisions is very much geared towards 'optimizing' risks to an extent that is cost-effective. Given the fact that the 'absolute value' of health is implied in the decision process, this standard way of operating is often unacceptable to those with health concerns. Opponents demand absolute certainty that the facilities will be safe and tend to 'invent' scenarios that prove that absolute safety is not guaranteed. The demand of absolute safety is however not one that can be satisfied in the real world. The fact of the matter is that such absolute certainty cannot be guaranteed for any type of development, and in that sense the controversial nature of hazardous waste disposal facilities (in the UK, in Canada) is a function of risk perception and sensitivities in the local population. Apart from the aforementioned concerns, the perception that people have already suffered quite a bit from industry (the UK) or that there is a potential to become the dustbin of the country (UK, Canada) seems to play a role here as well. Ironically, the two arguments are somewhat at odds with each other. Exactly the fact that the three UK regions involved were industrial regions implied that they produced large amounts of hazardous wastes. These had in fact been exported to other regions until then.

Figure 1. Use of symbols by the action group in Rotherham.

So whereas initially it seems that people are more concerned with their health than with economic development, this is only partly true. It must be observed that hazardous waste facilities do not offer such a great opportunity for economic development. The fact of the matter is that most facilities offer very small numbers of jobs. Still, in a small community (Montcalm, Swan Hills) or in community with high unemployment (Delfzijl), dozens of jobs are already quite a few. In Montcalm and Swan Hills (and Deep River too), people overcame their concerns about the health effects of the proposals under the guidance of the respective task forces. This was, in part, possible because a certain degree of experience with hazardous chemicals was present. After overcoming this hurdle, they in fact demonstrated a certain interest in the possibility of increasing employment in the community or in the strengthening of the tax base. In the two cases in the Netherlands where the community was generally speaking inactive (North Refinery, Ecoservice) it also appears that there was a certain degree of experience with heavy industry that considerably reduced the perception of a grave risk. Employment was the dominant theme in the local press in these two cases, in the sense that the angle chosen in reporting on the proposals in the local press spoke mainly of the number of jobs involved and the size of the investment.

Do interest groups undermine the general interest?

The final assumption on voters, that they form interest groups that consequently prevent the attainment of the public interest, is difficult to answer, as it is not exactly clear what the public interest is. To some extent this discussion borders the topic on the quality of the decisions in the case studies, a discussion that follows later. The term 'public interest' derives from the managerial discourse however, and in that discourse decisions must be made that increase the aggregate utility for a relevant constituency. The problem with siting, it also appears from my cases, is that facilities do tend to concentrate the costs on certain local populations in the interest of a greater population, usually regionally (county, region) or nationally defined. One argument in two out of the three UK cases (Seal Sands and Newport) was that there was a need for waste disposal facilities in a certain region. The SoS in public addresses confirmed this need. Viewed from that perspective, the resistance of local groups was indeed against the public interest.

However, need is not the only determining factor of the public interest in the managerial discourse. Facilities can only be in the public interest if they are technically safe and cost-effective as well. It is interesting to note that in both of the UK cases just mentioned, there were clear indications that the facilities were technically in order. Still, the local council would not approve them. Like the issue of need, this points to 'unsubstantiated' (in managerial terms) resistance, and one should perhaps conclude that the local communities indeed tried to block the attainment of the public interest. Looking at the case of Rotherham however, both the need and the technical quality of the proposals were suspect. Resistance against proposals thus helped further the public interest there.

Suppose we look at the other case studies, with the two criteria of need and technical quality hand in hand. Then one can see that the need for the facilities in the Dutch cases was pretty much established in most cases (e.g. North Refinery received government subsidies to build its facility, there was a severe sludge problem). However, the technical quality of the Kaliwaal proposals, and to some extent North Refinery's proposals was somewhat suspect. Especially in the Kaliwaal it was the opposition from ordinary citizens that played an important part in identifying weaknesses in the proposals, thereby helping achievement of the 'public interest'. In Canada, the outcomes of the decision processes seem to suggest a certain tension between the local level and the larger scale. In pure managerial terms, the need for the facilities was pretty clear (although the low-level radioactive waste problem was considered a non-problem by some). Public resistance against proposals that were made along the 'ordinary' route (Alberta, federal process) or an anticipation of such resistance (Manitoba) led to a greater emphasis on the community. It does seem that this has had 'negative' consequences from the perspective of the managerial quality criterion of cost-effectiveness. The facility in Swan Hills was expensive to operate in part because transportation distances were quite high, and in part because certain additional (extraordinary in the literal sense of the word) were incorporated into the facility. In Deep River, the local community desired quite expensive guarantees from the federal government that did not seem to meet the cost-effectiveness criterion either. But in Canada it was (in the federal process at least) an explicit choice to depart from the 'traditional approach' and pay more attention to the interest of the local community.

7.3.6 Parties to a court case

Table 7. Assumptions on parties to a court case.

Assumptions (from which discourse, or other source?)
On the positive side:
• Will make the strongest possible case for them selves and thereby support truth finding (pluralism).
On the negative side:
• Will make the strongest possible case for them selves and thereby hinder truth finding (critique of pluralism).
• Will not rest until the last legal resort has been tried (critique of pluralism).

Introduction

Given the fact that the courts were not involved in the Canadian cases, they offer little insight in the behavior and motives of parties with a conflict before the court. The Dutch and UK cases offer a lot in this respect however, as each Dutch case went to an administrative court at least once but often far more often, and the UK cases all went to an inquiry.

Truth finding

The first assumption about the behavior of parties in a court case is that they will make the strongest possible case for themselves and thereby support truth-finding. The second assumption states in fact the opposite: that parties hinder truth finding by their desire to present their case as well as possible. The difference between the two assumptions stems from diverging expectations about the way in which court parties 'duel' in the courtroom. Those who are pessimistic about the truth-finding capability tend to stress differences in resources between the actors involved and are afraid that because of their partiality actors will withhold information from the court. Those who are more optimistic tend to assume that parties are able to make coherent arguments and attack weaknesses in the opponent's case, which in turn implies that all relevant information will come out.

There is a striking similarity between the Dutch court cases and the UK inquiries in that their set-up does not look like the image that one often reads about courts. In both cases, the 'judge' (in the UK a planning inspector) is not a passive third party to whom the parties present their case. Instead, the judge is active, for instance by gathering information himself or by raising issues that were not brought forward by either party. This difference is caused by the fact the court hearings tend to seen as part of an administrative decision-making process (this is no longer formally the case in the Netherlands but the tradition still exists). The goal of the court case is therefore not the solution to a conflict between two parties who themselves identify what their conflict is about. Instead, the goal is truth finding per se. This factor is very important because it de-emphasizes the role of the parties in the court case, a consequence that is most clearly observable in the Dutch court cases, where the Advisor Environmental Appeals is often involved. This (technical) advisor to the court does much of the preparation work, including conversations with the parties, a checking of the relevant documentation, and a review of the relevant legislation. The conformity of court decisions with this advice is so overwhelming (in the 70% range; deviations from the advice often stem from procedural errors) that there can be serious doubts about the usefulness of court hearings per se: the case made by the parties does not matter so much. But one could also argue on the other hand that the Advisor then becomes the most important actor that the parties in the dispute must seek to convince. Indeed, there is some of that going on because parties tend to speak to the Advisor, but most of the information presented to the Advisor, however, was already on the table during the licensing process (i.e. the environmental assessment report and the objections from opponents). This information does indeed often make the strongest case possible for the parties.

However, the strengthening of one's own case sometimes happens to such an extent (see my remarks on the role of experts and their bias) that truth-finding is potentially hindered. Take the environmental assessment report in the Kaliwaal- or the Rotherham cases. In both situations, the proponent had prepared a report that suggested that all be well with the proposals. In each case it does seem however that there was a certain level of wishful thinking embedded in these reports and this was detected at the licensing stage. Because of the different aggregation rules in the two legal systems, the suspicion of weak parts in the assessment report had different consequences, implying that the court cases had different functions. In the

Netherlands, there is a general prohibition of disposal activity unless the proponent can demonstrate it to be safe. As there were clear gaps in the information presented by the proponent, the licensing authorities could relatively easily demand that additional information be supplied. In the UK however (bias in favor of the developer) it was largely the authorities' responsibility to make a credible case proving that the information in the assessment report was not valid. Especially the Environment Agency focused on the risk of water pollution and posited certain scenarios that were meant to 'prove' that such pollution was a credible possibility. The company in question however could simply posit that this scenario was not credible and then appeal. At the inquiry, the burden of proof continued to be upon the opponents of the proposals. The Environment Agency failed miserably in this respect, but the action group instead succeeded in making water pollution credible. In this sense, this particular inquiry did help in truth-finding.

More generally, it seems that the UK inquiries are far more geared towards truth-finding by party interaction than the Dutch court cases. Inquiries last much longer than Dutch court cases (usually a limited number of days) and offer much opportunity for cross-examination and 'grilling' of opponents. The effectiveness of this process depends to some degree on the resources of the parties. The local authorities involved were all able to field well-paid and highly regarded experts, but for opposition groups this was much more difficult, if not impossible. Considering this, the latter conclusion is somewhat discomforting because opposition groups sometimes do contribute to the decision process (Rotherham case, Newport case). In the Newport case, the information on hazardous goods vehicles movement from the opposition group was simply not believed, whereas it probably was correct. One wonders what would have happened if the information had been brought forward by a certain well-respected expert and not by some citizens on the basis of press clippings. Remarkably, in some cases (especially Newport) the inspectors were able to forge a certain consensus among the parties, or rather the experts working for the parties, on some issues such as risk assessment.

I rest my case

This immediately follows to the third assumption, which was that parties in a court case will not rest until every legal resort has been tried. This assumption seems to hold true in the Netherlands, especially in the Kaliwaal- and North Refinery cases. The reasons why there are so many court cases in the Netherlands are multiple. First, the threshold for going to court is relatively low, certainly financially. Secondly every single licence that is needed for the development in question can become the subject of a separate court proceeding. Thirdly, the court cases themselves tend to be rather unsatisfying for the parties involved as the courts have a tendency to avoid the substance of the conflict if they can find procedural errors. Furthermore, the actual interaction between the parties during the court cases is rather limited and the verdicts often take years. These factors imply that the conflict never ends and continues to hang in the air. By contrast, UK inquiries allow for much interaction between the parties, last long and cover just about all the topics in one proceeding, and are not about procedure but about substance. And finally, access to the courts is expensive and is subject to approval from the court itself, which can thus relatively easily 'decide' cases by saying that the party that wishes to challenge a certain decision does not stand a chance.

7.3.7 Members of the community

Table 8. Assumptions on community members.

Assumptions (from which discourse, or other source?)
On the positive side:
• Are engaged in coercion-free relationships and are in dialogue with each other (communitarianism).
• Have local knowledge, which is crucial for the decision process (communitarianism).
On the negative side:
• Will be parochial in their outlook and not look beyond their own community (critique of communitarianism).
• Do not have scientific knowledge, which is crucial for the decision process (managerialism).
• Do not want to participate in the decision process (pluralism).

Introduction

The role of members of the community was not great in the Dutch cases, whereas in the UK some members were very active and others more passively involved. Only in the Canadian cases was there greater involvement of community members, although the emphasis was, again, on a smaller group of citizens who manned the CLGs.

Coercion-free dialogue

The first assumption was that members of a community are engaged in coercion-free relationships and are in dialogue with each other. The evidence is rather negative on this assumption, although the size of the community in question does seem to matter to some extent in respect of the possibilities for dialogue. Apart from the fact that landowners sometimes forced their tenants from the land they were using (often without much consultation), there is no evidence of coercion between community members in the UK cases and the majority of the Dutch cases. To some extent the issue of coercion between community members was somewhat less relevant per se in the Dutch and UK cases because the emphasis was not on this institution and those citizens who wanted to participate did so mainly in interaction with the local authorities. In the one Dutch case where there was some coercion (North Refinery), this was mainly in the form of negative responses to the participation of a concerned citizen in licensing procedures.

In Canada however, the emphasis was in fact on the community, and community processes were thus much more important. If one looks at the experiences in Manitoba and Alberta, one cannot escape the impression that a certain level of coercion was present. In the case of Montcalm for instance, the opposition group collected signatures and was said to have done so by pressuring other members of the community. In Swan Hills, siting was described as 'manipulation with integrity'. Some of that manipulation came from within the community ('the community leaders'). The mere existence of community leaders, also confirmed by the Montcalm case, implies that the implicit assumption behind the ideal of coercion-free dialogue, that the community members are equal, is not fulfilled within the communities. The members of the communities are not equal in standing and are certainly not equal in their knowledge of hazardous chemicals. Those that have a higher standing, perhaps in combination with a more intimate knowledge of chemicals, tend to more often speak at public gatherings and attempt to influence the decision process more often than others who may remain silent throughout much of the decision process. Most of the 'dialogue' thus took place within smaller circles of the community, or was in fact a one way process, with many community members listening. One should however not just focus on the formal interaction moments. I have been told that the informal network of relations within the communities was much more important in achieving community opinion than the formal moments of interaction. I have not studied this informal interaction process, but would be inclined to speculate that community leaders still play an important role in it, be it with probably more possibilities of interaction between them and 'ordinary' members of the community. The case of Deep River stands out somewhat in the sense that the Chair of the CLG told me that dialogue was difficult in the town because of the economic interests at stake there. He was convinced that the outcomes of the decision process were mainly a reflection of the inhabitants' desire to keep their jobs. There had been 'real' dialogue, but then within the CLG. The CLG had advised against the proposals, but was consequently overridden, in his opinion because of this economic motive.

The importance of having local information

The second assumption on members of the community was that they have local knowledge, which is crucial to the decision process. The fourth assumption on the other hand was that members of the community would lack scientific knowledge, which is crucial to the decision process.

My conclusion is that both types of information have appeared to be quite crucial, but that it is quite hard to always distinguish which information is local and which information is scientific. The problem, so to speak, is that many members of the community happen to have a high level of education, sometimes in the sciences, sometimes not. The implication is that much 'scientific' knowledge is in fact present locally, which makes it local and scientific knowledge at the same time. The same applies the other way around. There is little reason why such knowledge that is present in the local community should not be called scientific knowledge, except that a scientist has not yet gathered this knowledge. Take the knowledge on flood risks in the Montcalm case. Whereas the 'scientific' constraint maps of the province of Manitoba indicated that a certain spot was flood-prone, local farmers knew it was not. The farmers derived their knowledge from their own senses and memory, the scientists probably from maps demonstrating the alleviation of the terrain near the river. The maps were not exactly correct however. After comments from the local farmers, the maps were 'improved' and this enabled the selection of a site that had indeed not been flooded since then, even though floods had occurred in Montcalm. Or take the Rotherham case. A mining engineer with dozens of years of firsthand experience with the soil under the preferred site knew more of the situation than the expensive experts hired by the proponents with their maps and calculations. At least that is what the planning inspector hearing the case believed on the basis of the cross-examination he saw. Scientific or local, it does seem that the local community is quite a source of information and can in fact contribute to the quality of the decision, 'even' in managerial terms.

Parochial communities

The third and fifth assumptions about the members of the local community suggest that members of such a community will be parochial in their outlook and will not want to participate in the decision process. I have already commented on the latter assumption when discussing the role of voters and there is little to add as voters and members of the community are analytically distinct categories, but much less so in practice.

That leaves the issue of the presence of a parochial outlook or not. It does seem that this assumption holds a certain level of truth. People tend to get involved in hazardous waste decisions if proposals are made for sites that are relatively close to them. Otherwise they tend to remain passive. That in itself generates a certain parochial element, because members of the community tend to think up more arguments against a certain proposal than arguments in favor of a different proposal. Most communities which are so afraid they become a dustbin for their region tend to be shocked when they find out that their community is actually a net exporter of hazardous wastes, and often remains one even if the facility were to be accepted. Certainly in the UK, the dustbin argument was often used, but did not have much substance to it. Most opposition groups in the UK cases mainly fought the one proposal they were facing, without suggesting an alternative. One reason why these groups may have felt it was justified to do so was that they tended to feel that there was no need for the proposed facility. Interestingly, this argument proved largely correct in the case of Seal Sands.

The Canadian cases, exactly because of their emphasis on the local community, are once again an interesting source of information on this assumption. In each of the three cases, the local population was not overly concerned about the neighboring communities. This is logical to some extent in Swan Hills, with other communities tens of kilometers removed, but think of its 'competitor' Ryley, which was completely surrounded by a county where the rest of the population was absolutely against the facility. A similar situation existed in Deep River, where the opinion of the neighboring communities was also quite clearly negative and largely ignored. In Deep River, this result was achieved despite the protests of a limited number of members of the community, and largely because the municipal council and CLG were against the involvement of other communities. In Montcalm in the end, other communities were informed and asked for their opinion, which resulted in favorable responses. Yet, this was not so much the initiative of community members, but of the Hazardous Waste Management Corporation.

7.4 The role of discourse

7.4.1 Introduction

In this section I will discuss the presence (or absence) of the various discourses in the nine case studies. Broadly speaking, I will focus on two questions. The first is which discourses were used and by whom. The second is whether the use of certain discourses had the desired effect those using it desired. Important from this perspective is whether discourse structuration took place at all and why. One remarkable conclusion that can be drawn from the case studies is that the combination of conservative pluralist and managerial discourse dominates the discussion in all nine cases, despite the fact that the Canadian cases were community-based. This outcome is achieved in different ways however. In the UK, awards of costs punish deviations from the discourses embedded in the law and inquiries are clearly structured according to the legal framework. In the Netherlands, the courts play an important role in upholding the law. In Canada, finally, community-based siting processes stimulate a technical debate at the local level (or sometimes try to direct such a debate to a certain conclusion).

7.4.2 The presence of various discourses

Introduction

The fact that the mixture of managerial and conservative pluralist discourses embedded in the law in the three nations is dominant in the end does not imply that there are no attempts to change this. It does seem that opponents often use progressive pluralism. The table on the next page presents an overview of the various discourses and their presence in the different case studies.

On the basis of table 9, one can see that all discourses (and the critique they have drawn) are all present in at least some of the cases I have studied. I will explain the content of the table by discussing the presence of various discourses in the various cases. Before doing that however, I wish to make two remarks. The first is that there is a certain affinity between various discourses that makes it somewhat difficult to actually separate them from each other. The managerial and the conservative pluralist discourses, for instance, have a certain affinity, as discussed in chapter 2, in that both assign a major role to experts and both consider utility an important criterion. The second is that there is a certain degree of 'positioning' involved in most cases, meaning that actors use a certain language to defend a position and select a language that serves them best. This topic will receive separate treatment under the topic of the relation between discourse and interests. For now, I just wish to posit that actors do seem to use various different discourses during the decision process, sometimes tailored to the arena in which they are discussing their opinion.

Managerialism

In a relatively 'pure' form, managerialism seems to be carried only by expert institutions such as the Environment Council of Alberta and the Atomic Energy Control Board in Canada. The Dutch province of Gelderland (the Kaliwaal case) was also inspired by the idea that an ideal site would be found for sludge disposal sites as they stressed the identification of proper criteria for identifying sites. These actors were relatively averse towards the involvement of private parties in the decision process and tended to advise that resistance against site selection should be overridden.

Table 9. Discourses present in hazardous waste siting.

Discourse	Present in which case studies (by which actor, which story line)?
Managerial	• Canadian cases (Environment Council of Alberta, Alberta Hazardous Waste Implementation Team, Atomic Energy Control Board, Deep River municipality, and several opponents to siting proposals: 'siting on the basis of proper siting criteria', 'override local resistance if need be', 'overly expensive solutions'). • UK cases (proponents, planning officers, planning inspectors, opponents, SoS: 'proposal must be allowed unless substantiated reasons to reject can be formulated', 'although the proposal may increase the risks locally, this must be compared to decreases elsewhere')*. • Dutch cases (proponents, Commission for EIA, local officers, opponents and elected representatives). Province of Gelderland: 'siting on the basis of proper criteria', 'avoidance of haphazard and ad hoc development')*.
Critique of man. discourse	• Canadian cases (all Siting Task Forces: 'engineering around social acceptability').
Conservative pluralist	• Canadian cases (Alberta government, Manitoba government: 'no involvement in the operation of the facility'). • All UK cases (planning inspectors: 'credence to the projections of the private market'). • Dutch cases of North Refinery (Municipality allows start up of plant without relevant licences) and Kaliwaal (State Council: private market party has the right to make proposals).
Critique of conservative pluralist discourse	• UK cases (opponents: 'profit at the expense of our health'). • Dutch case of the Kaliwaal (opponents: 'profits at the expense of our health').
Progressive pluralism	• All Canadian cases (Siting Task Forces: assistance to the community). Cases of Swan Hills (proponent of expansion paid for intervention from opponents), Montcalm (opponents: 'fairness' if plants are protected, why aren't we?), Deep River (Siting Process Task Force, compensation of damage). • UK cases (opponents: 'dustbin', 'fairness', we have already suffered from industry). • Dutch cases (Ministry of the Environment and competing firms: controls on the functioning of the waste market). Case of the Kaliwaal, (opponents: 'fairness', we have already suffered from excavations).
Critique of progressive pluralism	• Canadian cases (opponents: 'bribery').
Communitarian discourse	• Canadian cases (opponents: 'Siting Task Force must not mingle in the community'). Low level radioactive waste process (Siting Process Task Force: 'the community should be better off'). • UK cases of Seal Sands (opponents: 'sufficient support from the community before starting protest') and Newport (Borough Council: 'proposal is against the community's interest').
Critique of communitarian discourse	• Canadian cases (opponents from other villages: 'the affected community is larger than one municipality'; managerial thinkers: 'the result will not be technically optimal'). • The UK case of Newport (proponent of facility, planning department, planning inspectors, SoS: 'reasons for refusal of planning permission must be substantiated' and 'local risks are offset by reductions of risks elsewhere').

* Most of the actors mentioned here propagate a mixture of conservative pluralism and managerialism.

The purely managerial position is not taken very often however. Especially in the Netherlands and the UK, most actors that pleaded, one way or another, for a certain degree of managerialism in the decision process, did so from the idea that a mixture of managerialism and conservative pluralism should be applied. This implied that most of these actors generally accepted a certain role for the private sector in selecting sites and waste treatment technologies. Especially the selection of a site, 'wrapped' in managerial language by the proponent, was not vigorously questioned. Rather, the debate focused on the suitability of the proposal for the proposed site. Although such debates could be extremely technical, profitability of operation was generally accepted as an argument for or against certain positions in the debate. The actors that took such positions include the proponents in the UK and Dutch cases, the Commission for EIA in the Netherlands and most administrators plus most elected representatives involved in the Dutch and UK cases.

The Canadian community-based processes, to some extent, were a reflection of the critique of managerialism (communitarian and progressive pluralist tenets were also present). Especially the fact that community acceptance was such an important criterion necessitated the Siting Task Forces to relativize the role of experts in the decision process. On the other hand, and somewhat ironically, the task forces sometimes indicated that the engineering possibilities for tailoring a facility to a certain spot were great, which then again seems indicative of quite a high level of trust in experts. Also, the Task Forces could not be permitted to completely ignore technical requirements in the siting process. The process was not to be seen as bribing local communities and not as a process that would just about accept any site, because communities would not feel protected enough entering the discussions. Hence the constraint maps. Because of the same reason of credibility, the issue of compensation was avoided during the CLG-deliberations. During these deliberations, the focus was, again, on technical suitability of the various sites. Thus managerialism was rejected to some extent, but then again replaced with a new form that allowed local citizens to be 'managers' themselves within the confines of pre-set limits.

Conservative pluralism

The idea that private market parties should determine just about every aspect of the decision (pure conservative pluralism) was not present in any of the decision processes. However, it was clear that various actors were very keen to guard a major role for the private sector, in a mixture with managerialism. In the Canadian cases, especially the Alberta one, the Siting Task Force had a tendency to aim for reduction of the role of the private sector after a site had been found. This tendency was resisted by the provincial government, which clearly considered its interference in the siting process exceptional. In the Kaliwaal case in the Netherlands, the province of Gelderland 'drifted off' from the legal framework and would not accept a role for the private sector. The Commission for EIA went along with the provincial approach and forced it to go 'all the way' with its scientific approach of finding a site. This was impossible because of various reasons. Among them the fact that Delgromij (the proponent in this case) kept asking the court for interference so that the role of private parties could be restored, which the courts subsequently did.

As was to be expected the critique of conservative pluralism had a certain presence in the cases where the private sector had the greatest role (the UK, Netherlands, Canada before community-based siting started). Especially opponents in the UK often used this critique, which can be roughly summarized to 'we should not trust private companies with these types of wastes since they are only out there to make a profit'. In Canada, this feeling was so strong that it made siting efforts by a private company impossible. In the Netherlands the feeling was not very strong, except in the Kaliwaal case, but there it was secondary to the managerial agenda of the opponents.

Communitarianism

Communitarian thought was largely absent from the decision processes, although some opponents in the UK were apparently inspired by this line of thinking. Furthermore, opponents of the community-based siting processes used the discourse sometimes in order to argue against 'outside' involvement in their deliberation processes. For these opponents, the strong presence of a Siting Task Force and a Task Force monopoly on expertise was unacceptable. In some cases, they also argued against the dominant, 'managerial', way of speaking during the CLG deliberations. Those using the communitarian discourse deplored the fact that a 'philosophical no' 'or a local, however uninformed, no' was not acceptable (Manitoba siting process). These actors thus effectively asserted that community-based siting was not really community-based at all.

The critique of communitarian discourse is of course present only where the discourse itself was used. Especially the cases (Deep River, Ryley) where delineation of the relevant community was done in a relatively partial way (to enhance siting success), this resulted in a certain negative response from other communities. They then pointed out that the affected community was really greater than just the population of the community that was allowed to vote. Also, the 'established' licensing communities of experts did not favor community-based decision-making because they feared it would not result in the right decision. This feeling was particularly strong among the radioactive waste licensing community, which was somewhat aroused by the Siting Process Task Force, which had proposed a change of disposal techniques. In Newport (UK) finally, the Borough Council had decided to use 'the interest of the community' as a reason to refuse planning permission. This led to a wholly negative reaction from just about everyone in the 'planning chain' as most other actors felt that this was not a reason for refusal in itself, it had to be 'substantiated'. As far as local opponents of the proposals in Newport used the increase of risks as an argument against the facility, this was resisted by referring to the fact that risks would be reduced elsewhere.

7.4.3 The consequences of the use of discourses: structuration or not?

Two questions

There are various questions related to the use of discourses. First one may wonder whether the use of a discourse has a particular effect upon other parties or not. Secondly and relatedly, one may ask how, in situations of diversity, a certain discourse achieves structuration (or reverse, in situations where structuration is not achieved, how this is possible).

The effect of discourse

To start with the effect of various discourses: my impression is that in each of the three countries, one needs to use managerial discourse to be considered credible. Take for example the fact that most proponents in the UK used managerial language to explain why they had selected a certain site, whereas their choice was essentially commercially motivated (and to some extent by land use plans). Although one would expect that in situations of 'community-based siting' this would be different, this is not the case, because even during this type of process, the focus is on the technical suitability of a certain site and one is actually forced to use managerial language. The Siting Task Forces quickly learned that to do otherwise made them vulnerable to accusations of bribery and irresponsibility. The effectiveness of the use of other discourses throughout the various cases is limited, although the criterion for effectiveness I use: that the arguments affected the outcomes of the decision process; may influence this judgement. It does seem that especially progressive pluralist discourse, with its emphasis on fairness (e.g. in the form of the 'dustbin argument' or the 'we have already suffered' argument), is something that opposition groups use for mobilizing opposition. The same goes for the communitarian discourse ('the proposals are counter to the interest of the community') to some extent. However, after the mobilization of resistance, these discourses tend to lose their value. In the UK cases, the argument of the 'dustbin' tended to be countered with figures about waste production in the community in question. Usually, this pretty much silences the argument because the communities in question tend to be waste exporters. Only in the Seal Sands case were various proposals in the end 'linked' to study the substance of the dustbin argument. The progressive pluralist argument of 'we have already suffered' was present in all UK cases and in the Dutch Kaliwaal case. In the Kaliwaal-case, it was countered by a reference to the future, which would see a much more enjoyable site. More generally, it seems that hazardous waste is seen as part and parcel of the industrial society, and is therefore everyone's responsibility. The argument that something is 'against the interest of the community' tends to be 'rewarded' with a call for explanation and substantiation of statement in managerial terms. This goes even for the Canadian cases. As one critic of the siting process in Manitoba said: 'a philosophical no won't do'.

Structuration

On the issue of structuration, there are cases where structuration worked, and where it was problematic. Table 10 provides an overview.

Failed structuration

Among the cases where structuration failed are the Canadian ones. There, despite a legal framework that emphasizes private sector initiative and that assigns a relatively inconsequential role, the actual siting took place on a community-orientated basis. But also in the Netherlands and the UK, there were attempts to challenge the legal framework. In the Netherlands, in one case (North Refinery) there was an attempt to bypass formal procedures in the interest of a private market party. In yet another case (Kaliwaal), there was an attempt to bypass private market parties instead. In the UK, the Newport case was the most obvious attempt to escape from the legally validated structure for the decision process.

In pondering the reasons why structuration failed, one needs to take into account that the Canadian cases are unique examples of a perfect match between the level of government that decides on the proper procedures and the level of government that decides to not apply this framework. In addition, there was little pressure from a party that wanted to continue under the 'normal' procedures. Take the case of Alberta: the province itself was responsible for devising legislation on land use planning and environmental licences. Its decision to apply this legislation only after a community-based decision process had been completed was authoritative in the sense that the determination of procedures was in the provincial realm of the authorities. The province was helped by the fact that private developers were not overly interested in developing waste disposal capacity and did not therefore pressure the province into application of the ordinary framework of decision-making. If we compare this situation with the one in the Kaliwaal case or the Newport case, we can see that the proposed deviations from the legal framework there (in a managerial and communitarian direction respectively) were not taken by the body in control of decision procedures (the province and borough respectively). This made their approach susceptible to (quasi-) judicial review, also because there was in fact a private party interested in siting a facility, which demanded that the 'ordinary' framework for decision-making be applied. It should be noted that such parties (Delgromij and BFES) had great difficulty in actually getting things to go their way. The reason for that is that the public bodies involved had quite a central position in the decision process (both in the sphere of pollution control and land use planning) and secondly that the courts are a rather slow decision-making institution.

Table 10. Structuration of discussions on hazardous waste.

Case	Structuration by the legal framework or not?
Seal Sands (UK)	Yes. 'Dustbin' argument in various communities led to linked inquiry. Attempts from local government to further change the focus to the national level and establish positive planning rejected. Request for award of costs against the county.
Newport (UK)	Difficult to achieve. Community interest argument used by the local planning authority. Then rejected by planning inspector and SoS. Led to award of costs as requested by the proponent, but later quashed by the court.
Rotherham (UK)	Yes. Initially, the opposition group stressed the interest of the community, but was told to advance a policy argument in the interest of their case.
North Refinery (NL)	Difficult to achieve. Local authority bypassed formal procedures, with permission from the Ministry of the Environment. Structuration in the end achieved through the courts.
Kaliwaal (NL)	Difficult to achieve. Provincial authority refused private party involvement in waste disposal. Various court cases needed to restructure the debate. Came at a price: the proponent offered a certain amount of compensation.
Ecoservice (NL)	Yes. No challenges whatsoever. Little interest from the community.
Swan Hills (CAN)	No. Proposals extremely controversial, but not very commercially attractive either, leading to little pressure from private parties. One county council was physically threatened and various Members of the Provincial Legislature were scared of their position. Struggle between 'managerial' and 'invitational' siting advocates.
Montcalm (CAN)	No. In anticipation of controversy, the provincial Cabinet decided to follow the example of Alberta.
Deep River (CAN)	No. After much controversy, the federal Cabinet decided to follow Manitoba and Alberta. There was then an explicit design for the decision process, but this was not followed either.

Finally, the fact that the 'real' argument for refusal of permission was sometimes concealed. Especially in the case of Newport, the borough cited four reasons for refusal of planning permission, three of which appear to be mainly intended to voice arguments that would fit within the system, whereas the interest of the community was of far greater concern to the council. Note that proponents are not the only ones having great difficulty in restoring the application of the legal framework. The case of North Refinery demonstrates that local authorities that bypass the legal framework are difficult to influence by the opposition. In this case, the local municipality did not say much about its acts in public. The few words that were stated indicate that approval for the facility would occur without doubt because there were no technical problems associated with the facility. Apart from the fact that this can be doubted (the Inspectorate did not agree at all), the real reason for the very cooperative attitude of the municipality was economic development.

Structuration success

When looking at the cases where structuration did in fact take place (in the Netherlands: Ecoservice; in the UK: Seal Sands, Rotherham), one can see that there was either little interest in the decision process (Ecoservice) or that all the actors involved simply adhered to the discourses embedded in the legal framework. In the Ecoservice case, opposition was close to nil, but the few opponents present were using managerial language. In the Rotherham-case, opponents had initially stressed the moral unacceptability of the proposals, but were informed after some time by a member of the planners' organization (RTPI) that they should start developing a case on the basis of policy arguments if they were to stand a chance. In Seal Sands, opponents had from the start been focusing on a range of issues, including the need for the facility, the health effects, and the effects on the image of the county. Throughout the process, these arguments were motivated with increasingly sophisticated evidence, climaxing in a health report to the inquiry.

In addition to the cases where structuration was achieved with very little effort, there are some where structuration was only achieved after much difficulty. I would count two Dutch cases (North Refinery, Kaliwaal) and one UK case (Newport) among this type of case. If we then look at the way structuration was in the end achieved in these cases, one immediately notices that the UK system of achieving structuration is more effective and leads to fewer continuing problems. In the Dutch cases, especially those of the Kaliwaal and North Refinery, the courts had to decide conflicts countless times, without apparently really satisfying anyone. The UK inquiries appear to largely satisfy most parties (apart from the fact that no one likes to invest so much time in a decision). The difference between the UK and Dutch systems is that in the UK, the inquiry is a long and intense battle that allows for much personal interaction between the parties. Even though they are planning inquiries, they do touch on a broader range of issues (including pollution control) and are overseen by someone who is regarded as very impartial. Although the Dutch courts are also generally regarded as impartial and have a tradition of being seen as part of the administrative decision-making process just like inquiries, there are great differences. Dutch court cases are of relatively brief duration and therefore allow for little interaction, are prepared largely outside the view of the parties involved (by the Advisor Environmental Appeals), and are fragmented in the sense that for each licence, there is a different possibility of appeal and the discussion of related issues is not allowed. It is no wonder that most parties in the Dutch system do not feel very satisfied after judicial proceedings and look for ways to have things go their way anyway. The tendency of the courts to first look at procedural issues does not enhance the legal rationality (in the sense of the predictability) of the outcomes at all. In the Netherlands, court verdicts often surprise the parties involved and are received as an opportunity for renewed disagreement or a challenge to try yet again. In the UK, emotions may fly high during inquiries, but then cool down so that a 'rational' discussion can take place. Several years then tend to pass before a decision is taken, which often explicitly refers to the report of the respected planning inspectors. This, in combination with relatively inaccessible courts, shields the decision process from eternal continuation.

Interestingly, the Canadian cases provide some evidence on legal rationality as well. Especially in the Alberta and Manitoba processes, where Siting Task Forces were really 'learning on the job', there was relatively little guidance on how the process would be conducted. There was no explicit guidance on the issue of compensation for instance, which implies that such compensation was assumed by the community but not achieved in a binding way. Especially in Manitoba, local populations learned from the Alberta experience and demanded a contract with the province, which would contain binding conditions to protect the local community. Yet another step towards legal rationality was the federal siting process which is in many ways a culmination of the experience gathered in Manitoba and Alberta. However, it appears that certain factors that were present in Manitoba and Alberta (especially the desire to actually get siting approval) did prohibit, to some extent, the achievement of the community-based agenda. In community-based siting, the fact in particular that competition between communities may arise and that elected politicians consequently must decide a winner, does not enhance legal rationality. Local communities that decide to volunteer have by no means the certainty that they will actually be selected.

7.5 The quality of the decisions

7.5.1 Introduction

Having discussed to some extent the legal rationality (in the sense of clarity of decision procedures ex ante) of the decision processes, I will now move towards a discussion of the indications for the substantive quality of the outcomes. As discussed in chapter 2, there are various different rationalities, each associated with a certain conception of decision quality. The table below sums them up.

Table 11. Rationalities and criteria for decision quality.

Type of rationality	Hinges on; more concrete definition of decision quality
Technical	Efficient and effective production; is the decision acceptable on the basis of proven scientific facts (e.g. environmentally safe) and does it deliver the greatest good for the greatest number (not too expensive given the benefits)?
Economic	Purification and extension of the economy; was agreement between free parties achieved and compensation given, does the decision enhance economic growth?
Social	Integration of members of the society and avoidance of conflict; was the decision acceptable to members of the local community, was it perceived to be in their interest, was there consensus on the decision?
Political	Support for the political system and solution of problems; did the decision leave the support for the political system (and individual politicians) intact and was the problem solved?

7.5.2 Technical rationality

Introduction

On the issue of technical rationality, the case studies in the three countries do seem to reflect certain developments, which suggest that technical rationality itself is not something constant. Especially the shift from incineration technology, typically 1980s, to treatment of hazardous wastes and waste prevention reflects the development of new technologies that were initially not available. The fact that the Seal Sands incinerator was not constructed in the end attests to this change. To some extent, the problems of the Swan Hills facility could also be seen in this light. In the Netherlands, sludge disposal practices evolved to become a separate arm of the waste regulation tree in a period of about 15 years, resulting in changed ideas about sludge disposal facilities. The broader issue is that a facility may have a fantastic design, yet not be constructed because technologies have advanced further. The differences in perception of the practice of co-disposal would seem even more baffling. Whereas the UK expert community considered the combined disposal of hazardous and 'ordinary' wastes completely responsible, this was not at all the case in the Netherlands and Canada. In order not to complicate matters too much, I will stick to a narrow conception of technical quality for now. The matter of interest is therefore whether the proposal in question was technically acceptable from the perspective of knowledge that existed at the time of the siting debate in the country of interest. Another question is whether the proposal improved in the just stated sense during the decision process and for which reasons.

Technically suitable solutions

Generally speaking, most of the proposals I have studied appear to have been technically suitable solutions to waste disposal. In practically every case a serious look was taken at technical quality. Take for instance the Newport case, where the Borough's technical advisors warned that an air treatment system was necessary and further information needed on the operability of the plant. To some extent, the discovery of such issues was quite normal. However, there were various proposals where doubts about the technical quality of the proponent's proposals went somewhat deeper than in others. These cases included Rotherham, North Refinery, Kaliwaal, and Deep River, where the level of safety of the operations at the plant was rejected. In Swan Hills, there was mainly doubt about the cost-effectiveness of the facility, which was also the case in Deep River.

Taking a brief look at the problems in each of these cases we see that in Rotherham, the proposal of the proponent was in the end rejected because of various reasons, one being that that the quality of the groundwater would be threatened. In the case of North Refinery, it does seem that the Inspectorate had the greatest concerns about the quality of the proposals, but that these were not widely shared and perhaps largely based on a certain level of distrust. In the case of the Kaliwaal, the technical quality of what the proponent was proposing was suspect too, and the proposals had to be improved because of this. Whether or not the solutions, an improvement of the soil, was in the end sufficient was somewhat uncertain, but the province itself guaranteed emergency measures in case something would go wrong. In Deep River, the quality of the AECL location as a site for low level radioactive wastes was controversial, but would have been subject to additional testing before any decisions were made definitive. The proposals in this case, along with the Swan Hills proposals, have been lamented for their lack of cost-effectiveness. Especially in the Deep River case, the federal government in the end said that the costs associated with the proposal could not be borne. In the case of Swan Hills, certain measures to increase safety of operations (e.g. tucks with double wall tanks) and the relatively great distance to waste generators was said to have negatively influenced its operations.

The identification of weak spots

Who was responsible for identifying weak spots in the design of the facility? The proposals in Rotherham were rejected in the end as a result of the combined effort of the local authority, the NRA and an opposition group. The fact that the opposition group had so much expertise in the field of geology was particularly remarkable and important for the decision. In the case of North Refinery, facility design changed only on paper, and the private developer in practice largely ignored the advice of the Inspectorate, including the advice on floating roofs in storage tanks, thereby leaving a residue of doubt about the technical quality of the facility. In the Kaliwaal case it remains rather speculative whether the problem with the proposals, i.e. the incomplete layer of clay on the soil of the pit, would have been revealed without the contribution of some of the citizens. It does seem however that they did indeed play a role in identifying this problem. In the Deep River case in the end, the CLG pointed out certain problems with the site, but the Task Force's experts denied these existed. Testing by the government regulator AECB may have shed some light on the 'real' situation at the site, but it never got to that point because of economic considerations. The cases just mentioned demonstrate that members of a community do not just hold 'ordinary information' but also scientific information, and that if they do not have scientific information, they can acquire it. This does not reduce the importance of the distinction between ordinary and scientific information, but it does indicate that the source of such information need not always be paid experts.

Some qualifications

It should be noted that even in cases where the proposals were relatively good, there is a difference between the quality of their general design on the one hand and their situational and comparative quality on the other. Especially the Manitoba and Alberta cases were examples of Siting Task Forces 'selling' a completed design to various communities. This design may have been good in itself, but the siting decision was still improved by the contribution of the CLG about flood-prone areas in Montcalm. And in Seal Sands, the judgement on the quality of the proposals made by Ocean Environmental was greatly helped by the linked inquiry, where two other proposals were also discussed. 'Even' for ordinary citizens it was clear that Ocean's proposals were relatively better in technical terms than ITE's.

7.5.3 The economic quality of the proposals

Free market?

The economic quality of the proposals is, of course, in part related to the cost-effectiveness of the facility, which has already been discussed. The economic aspect of decision quality has three dimensions: agreement between free parties, compensation, and economic growth. It should be noted that at the core of regulation in each of the three countries, there is the reduction of free transactions between market parties. In this sense, there is little free market for hazardous waste siting out there. One exception to that general conclusion is the ground market. The proponent of the facility needs ownership of the land to actually build the facility, and the exchange of ownership is generally a transaction that market parties are free to make. It is important to realize that the costs of land are really only a minor part of the total costs associated with the projects I have described. In none of the cases was it a problem to find a landowner who was willing to sell the land, be they municipalities with industrial land, or simply private owners. Especially when agricultural land is transformed into a site for a hazardous waste facility, such a change of use does indicate a rising land value, and the owners receive an interesting price for their land. If and only if, the hazardous waste facility consequently still operates with a profit, then the economics of the proposal must have made sense and economic growth is achieved. In itself however, this growth is rather 'artificial' in the sense that government regulations drive the need for hazardous waste treatment, not so much the 'free' choices of market parties. Take for instance the Rotherham case. In that case, the proponent saw a need for the facility, in part because of the company's own activities in the general region. The authority in question however, indicated that another proposal was being developed and supported by the relevant policies. The local need for the waste facility was thus not said to exist. The planning inspector who heard the case accepted this point. Because of this, it is rather difficult to say much more about the economic quality of the proposals other than that they are operating profitably or not.

Profitability

In the UK, the Seal Sands facility was in the end approved, but by the time this happened, the market for incineration had evaporated and the facility was not built because of economic reasons. In Newport, the facility does still operate and in fact operates with a profit. On the other hand, BFES relatively quickly decided to sell the facility (because of strategic choices), which in this case implied that the company took quite a loss on the plant because its investments never fully paid out dividends[3].

In the Netherlands, the North Refinery facility was a marginal operation throughout much of its existence. At some point the facility even stopped treating hazardous wastes. The ambitious RUN project that was approved in the late 1990s and was intended greatly to expand the facility is uncertain, mainly because the ownership of North Refinery changed and the plans of the new owner had yet to be announced. The facility at the Kaliwaal has yet to operate, but can be expected to become a major cash cow for Delgromij. The Ecoservice plant in Dordrecht was commercially a limited success until taken over by a larger company and its operations were coordinated with other facilities.

Compensation

Like the issue of economic growth, with regard to the various proposals the issue of compensation is hard to answer outside the realm of the ground market. However, in the Canadian cases, compensation of the community was, to a varying degree, part of the community-based siting process. The community of Swan Hills has fared well with its incinerator, at least until recently. The number of jobs generated by the facility, at some point circling around the 160, has greatly declined recently, but the facility did help the development of the town by economic diversification. The community of Montcalm fared less well, with the plant being a relatively marginal operation. Deep River, finally, demanded too much from the federal government, leaving it empty-handed in the end. Ironically, that was the case where compensation had been most explicit on the agenda. The Canadian cases do shed an interesting light on the UK and Dutch cases in the sense that 'compensation' of the community was not part and parcel of the strategy of the proponents in these countries. There were some feeble attempts in Rotherham through setting up a visitor's center. In the Netherlands, only Delgromij offered compensation in the sense that the firm decided to add 'nature creation' to its proposals at the request of the province. In both cases, it cannot be said that 'the' community had asked for such compensation. Neither was it consulted in any way on these plans. This in itself is an indication of the fundamental philosophical differences underlying the decision processes.

3 Interview with Mr. T. Butterfield (formerly with BFES), Coventry, 5 January 2000.

7.5.4 The social quality of the decisions

Integration and conflict avoidance

There are various aspects to the social quality of the decisions, including the integration of members of the society and the avoidance of conflict. It should be noted that in the UK cases, there was a certain increase of integration of members of the local community because opposition groups were formed and relatively a lot of members of the community took an interest in the discussions (the most contentious local political issue in years). At the same time however, the opposition groups (except perhaps in the case of the Seal Sands proposal) felt cut off from the rest of the community which remained largely passive in the decision process. This feeling of division was then somewhat compensated by the fact that the local councils were unanimous in their rejection of the proposals, signaling unity in the community. In the UK cases, there was quite an intense level of conflict, not so much within the local communities, but rather between local councils and opposition groups on the one hand and an external developer on the other. It is my impression that this conflict had a positive effect on the decision process in the sense that many people felt a strong motivation to actively scrutinize the proposals before them.

In the Dutch cases, the integration of the community was not really an issue in two out of the three cases, as there was no involvement of the community at all (North Refinery, Ecoservice). In those cases, there was little conflict, although in the North Refinery case, the proponent and the Werkgroep squared off at each other quite intensely. This was however a conflict between two parties outside the local community. The North Refinery case, particularly during later stages, does demonstrate that in a community that strongly supports the chemical industry and the creation of employment, those asking critical questions may receive negative feedback from their fellow citizens. In the Kaliwaal case, it seems that the local opponents did little to mobilize the community until late in the decision process. When a certain level of interest did develop in the end, this did not create division within the community as the conflict was defined as one between the community and outsiders.

In Canada finally, the community-based siting processes were a result of a threat of rampant conflict. Looking only at the three communities that volunteered in the end, one can see that the integration of the community was still under stress in Montcalm and to some extent also in Deep River. Generally speaking however, the communities as a whole chose to volunteer with a few individual citizens as the exception. I have been told that in Swan Hills, one family left the town after the decision process out of fear of the facility and because they obviously did not share the local agenda of economic development. With reference to this agenda, I have been told by several interviewees that the communities standing the greatest chance of actually getting the facility are the ones that have a coherent and shared vision about their economic future. When looking at Swan Hills, Montcalm and Deep River, one can indeed see that economic diversification (Swan Hills, Montcalm) or maintenance of a certain industry (Deep River) was already on the agenda before the siting process started. However, I have not studied the degree to which this separates these communities from others. In Montcalm and Deep River there was a certain level of conflict between opponents and proponents. Especially in Montcalm, this energy was used effectively to create a critical discussion about the proposals by inviting opponents to the CLG.

So far I have restricted the analysis of integration and conflict avoidance to local communities. It is clear however that the issue can perhaps also be discussed on a somewhat greater geographical scale. This is because the interest of one local community may be at odds with that of another. It is interesting to note that frictions with other communities did arise in two out of the three Canadian cases (Swan Hills, Deep River). Especially in the Deep River case, the local municipal council essentially tried to override the concerns of the surrounding communities. Although one should not exaggerate this, such acts do have a rather negative effect on the standing of the municipality with other communities, which may in the long term work against the volunteer community. In the other two countries, especially the UK, one can see that the interest of other communities is taken up by the national government. Especially in the Newport case, the planning inspector made clear that Newport might want to avoid any risks to its citizens, but that the unacceptable implication would be that other communities would have to bear these, or even greater, risks. And indeed, the surrounding communities did not protest when Newport was forced to accept the facility. In the Netherlands, the mechanism for protecting others than the local community involved is potentially strong, as boundary rules pose virtually no hindrance to broad participation. However, the relatively strong powers of ordinary citizens tend to be used solely in a negative fashion, that is, mainly to object. Note that the attempts of the province of Gelderland (Kaliwaal case) were inhibited in part because its selection of multiple smaller sites for sludge disposal met with resistance in each single case, whereas the wider population remained silent. The political fall-out associated with imposition is considerable and there is little possibility of forcing local municipalities to change their zoning ordinances if these do not permit the facility.

Acceptability, the community's interest, and consensus

There are three other issues that need to be addressed under the heading of the social quality of the decisions. They relate to the acceptability of the decision to the members of the community, the existence of a perception that a decision was taken that was in their interest, and the existence of consensus or not.

To determine community acceptance, one should see whether the decision that was finally made was in accordance with the will of the local community. Especially in the UK and the Netherlands there is little reliable data on community acceptance, apart from the opinion of the local councils perhaps. Having said that, it does seem that in the Netherlands, the local communities were relatively unconcerned about the proposals, with the exception of the Kaliwaal case. This suggests that the decisions to approve the facility in two out of the three cases were socially beneficial. In the UK, Rotherham MBC and the SoS would not allow the proposals, which is a decision that most members of the community probably approved of. In the Seal Sands and Newport cases, there are signs of disapproval, although even the opponents had to admit in the Seal Sands case that the proposals were technically in order.

In Canada, the referenda do give a more reliable impression of the local opinion, and the decision to volunteer was carried by a strong majority of the local vote. However, in the case of Deep River, the federal government in the end decided against the proposal, a decision that was then deplored by the local community. In a sense, this case is the reverse of the two UK cases where the local public deplored the decision to allow the facility. In the UK, the acceptability of the decision was somewhat enhanced through the use of inquiries. To suggest that the passivity of the Dutch populations involved signifies agreement with the proposals gives rise to serious thoughts. Yet it does seem that the local council agenda of economic development was widely shared in the populations of Delfzijl and Dordrecht and was thus more likely perceived as being in their interest than against it. The UK cases, to some extent, exhibit the same pattern as the Dutch cases, in the sense that the local population was largely passive. It is difficult to interpret the meaning of this, but given the large amount of signatures on the various petitions against the plant and the unanimous opposition the local councils, one could perhaps assume that the proposals were not seen as in the community's interest. The central government's decision to subsequently approve the facility in two cases was thus deplored and seen as against being the local interest, whereas the decision to reject one application for planning permission was then probably regarded as being in the community's interest.

Finally the issue of consensus. It does appear that in very few communities there was true consensus on the proposals at stake. It is again difficult to judge the passivity of the greater parts of the populations in the UK and the Netherlands, but in the Canadian cases, even in communities with massive support for the plans (Swan Hills: 78% of the voters) there is still a significant part of the community that opposes the facility. It does seem therefore that a certain degree of imposition is inherent to hazardous waste facility siting, no matter which decision method is used.

7.5.5 The political quality of the decisions

Canada: important role for political rationality

From the perspective of political quality, it is important that decisions do not destroy the support for the political system or individual politicians and that waste problems are solved in the end. It seems that in the Canadian cases, political rationality played a major role in the decision to structure the debate differently rather than normally. Politicians were under severe pressure to take no decisions which would be against the opposition in the various local communities. They feared for their seats, and in many instances where they ignored the resistance, they did lose support. But at the same time, they had committed themselves to responsible waste treatment, either as part of the general industrial development agenda (Alberta) or by election pledges (federal process). The rabbit that came out of the hat in the end, community-based siting, seemed to offer a possibility for satisfying both requirements. This worked in Alberta and Manitoba, but not in the federal siting process, where the problem was not solved in the end. In that federal siting process, technical and political rationality were at odds with each other in the sense that the solution of low level radioactive waste problems in the source communities along the path of the agreement between STF and local community would have required investments beyond what was considered to be cost-effective. Perhaps in part because of transitions in government, the election pledge to the source communities became relatively less urgent throughout the voluntary siting process. In the nuclear industry, many saw the issue of low-level radioactive waste as a non-issue.

UK: a division between the local and the national level

In the UK, it seems that there is a clear divide between local politicians and other politicians. Local councilors are constrained in the expressing their opinion by the planning system, which demanded their neutrality. This restriction is clearly present, witness the fact that councilors in all cases took care not to be associated with the opposition groups. Even in the case of Newport, where councilors themselves were among those who initiated the opposition group, this was true as councilors withdrew from the action group after the first meeting. This restriction was not as great for local MPs, who tended to speak out against the proposals rather quickly regardless of their political affiliation. In two out of the three cases, the judgement about the technical acceptability of the proposals by local officials resulted in an advice to approve the facility. It was at that moment when it became clear that the two councils involved (Cleveland and Newport) were not neutral and actually opposed the plans. It is remarkable that in both cases, the issue was consequently redefined by these councilors changing from a technical issue to an issue involving local democracy. In both cases, the council leader started playing an important role, and this leader drafted motions for the full council to reject the proposals. Although opposition parties could have had an interesting opportunity for developing a political profile that separated them from the majority party here, this did not occur and the councils unanimously voted against the application. At the central government level, the pressures were consequently somewhat different. Even in cases where a significant number of MPs spoke out against the proposal (Newport), the SoS resisted the political pressure and acted as an agent of the planning system, thereby resisting an opportunity (perhaps) for political gain. Although there was a great degree of disappointment in the communities where the SoS decided to allow the facility despite local objections, this did not lead to much reduction of support in the political system. Most people I spoke to tended to accept this decision, essentially because they felt that the inquiry had offered them a fair process at influencing the decision.

The Netherlands: lack of opposition (mostly)

In the Netherlands, the lack of opposition in two out of the three cases allowed local politicians to claim a part of the success (employment) associated with the sites. This claim was less evident in the case of North Refinery, where the public's enthusiasm of politicians had to remain under wraps because they had effectively bypassed the legal framework. Only in the case of the Kaliwaal were there initially fewer opportunities for political showcasing, but the proponent's gesture of adding nature development changed this somewhat. It offered provincial politicians an opportunity to safe face (the province had been opposed to the facility for a long time) and to actually be associated with a positive development in the area. In terms of problem-solving, it does seem that in the North Refinery case, the Minister of the environment had the increase of waste treatment capability high on his priority list. So high in fact, that he was willing to allow start up of a plant that was technically suspect in the eyes of his own officials and was operated by a company with an ambiguous reputation. Because the Dutch hazardous waste regulation system revolves in part around the issue of capacity regulation, this was an important part of the discussion in the Ecoservice case as well. The proposal there was ruled to contribute to the solution of hazardous waste problems in the Netherlands.

In the case of the Kaliwaal, the provincial executive decided in the early 1990s that the solution of the dredging sludge problem would become a high priority, and as there was only one viable solution (Delgromij's proposal) they decided to go along with it, as long as it was technically acceptable. Only in this case local politicians experienced a certain level of pressure from the local public. It is remarkable that none of the parties in the local council was in favor of the proposals. There was a threat from opponents that they would turn the issue into an election campaign if any of the parties expressed support for the proposals. This did not miss having its effect. In fact, one prominent opponent was invited to join one of the local parties and became a council member. Locally, political opportunity thus demanded resistance, and this is what the politicians did. This then pitted the municipal council (and its officers) against the province and the private developer. In turn, this led to a considerable delay in solving the sludge disposal problem.

7.5.6 Overall assessment of quality: patterns?

Introduction

If, on the basis of the quality parameters, we take a look at the decisions that were made by the relevant authorities, we arrive at the table on the next page.

Patterns of quality? Some assumptions

The question quickly becomes whether there are any patterns in these scores that, no matter how tentative, could perhaps connect certain approaches to decision-making to certain outcomes. The prime candidates for such patterns would revolve around the different roles the community plays in the various countries. Especially the fact that the community plays an important role leads to the expectation that more socially acceptable decision are taken in Canada. At the same time, one would expect that the UK method of decision-making leads to less socially acceptable decisions. If we were to look for 'suspects' that can be expected to carry the burden of increased social responsibility in Canada, one easily arrives at technical and economic quality of the decisions. Again, the UK and Canadian cases would then be extremes on a scale, with the UK system then generating technically and economically better decisions and the Canadian system worse decisions.

Table 12. The quality of hazardous waste decisions in the cases.

Quality Criterion / Case	Technical quality of proposal	Economic quality of licence decision	Social quality of licence decision	Political
Seal Sands (UK)	+ (no changes throughout decision process)	+/- (not built)	- (proposal disliked)	+/- (conflict not resolved)
Newport (UK)	+ (limited changes)	+ (commercially successful)	- (facility disliked)	- (conflict not resolved)
Rotherham (UK)	- (not built because of risks)	+/- (overlap with competing proposal)	- (proposal disliked)	+ (support for rejection, waste problem still solved)
North Refinery (NL)	+/- (doubts from Inspectorate, leakage)	+/- (commercially limited success)	+ (indifference)	+ (helps solve waste problem, little conflict)
Kaliwaal (NL)	+/- (improved through decision process)	+ (will be profit maker)	- (resistance)	+/- (helps solve sludge problem, but results in conflict)
Ecoservice (NL)	+ (surpasses standards)	+ (is commercial success)	+ (indifference)	+ (helps solve waste problem)
Swan Hills (CAN)*	+/- (relatively expensive)	+/-	+ (greatly desired)	+ (waste problem solved)
Montcalm (CAN)*	+ (improved through decision process)	+/- (only profitable parts opened)	+ (greatly desired)	+ (waste problem solved)
Deep River (CAN)*	+/- (suspect from the perspective of containment and too expensive)	- (too much compensation demanded; not built)	+ (greatly desired)	- (rejection by federal govt. deplored, waste problem not solved)

Explanation: + = good, +/- = different opinions, - = bad. * For the Canadian cases, only the communities that volunteered are considered in this table.

Community-based decisions: high social quality, low technical quality?

It would seem that these patterns indeed exist to an extent, although the 'double veto' character of the Canadian decision processes obscures the relation somewhat. If we look at the Canadian cases, we can see that community-based siting tends to lead the decision process towards locations that are not near the waste generation points. This implied higher transportation costs. Also, it does seem that the guarantee of certain tax payments (Manitoba), or the introduction of extra safety measures at the request of citizens' committees (Swan Hills), or the promise of job guarantees (Deep River), reduces the cost-effectiveness of waste disposal. Extra costs are added and these consequently reduce the technical quality of the decision somewhat. Community-based siting does have its own dynamic to some extent, and especially the change from community-based discussions to the realm of private market parties or representative democracy is difficult. To put it more simply: in Swan Hills, certain assumptions were made about the waste market, a facility was sited on that basis, and then suddenly a private operator had to be found, which was difficult and resulted in a reality check of waste projections. The province however was willing to fork millions of dollars out to keep private operation of the facility attractive. This in itself reduced the incentive for the private operator to be very critical of his expenses, including for instance a large donation to opponents of the expansion of the facility (intervenor funding). What did keep costs in check was the fact that, until the provincial government decided it would be Swan Hills, the town was competing with another town and was therefore relatively modest in its demands. Not so in Deep River, where the only community still in the race had quite high demands in the end. But there, the double veto was consequently used, meaning that the elected representatives would not accept the deal between town and STF. Only in Manitoba was the connection between community-based siting and waste projections really solved in a technically (and economically) acceptable way. The government kept feeding waste projections into the decision process (incrementalism) and refused to invest in the facility. The downside, which to an extent corroborates the inverse relation between social and technical quality, was that the community obtained much less employment than it had expected initially, thereby reducing the social quality somewhat. Note that these remarks only pertain to the cost-effectiveness part of the technical quality criterion. Community-based siting does not appear to result in less adequate facilities from a risk perspective. In fact, the local communities were capable of increasing the technical quality of the decisions.

Top-down decisions: low social and high technical quality?

The opposite expectation of social unacceptability of decisions in the UK also came out, except for the Rotherham case, where technical and social acceptability considerations pointed in the same direction. It is clear however that if proposals are technically of good quality, the UK government will impose them on the local community, which results in poor social quality of the decision (but then only from the local perspective as explained). This is however considered practically irrelevant judging by the effect the explicit focus on technical quality of the proposals in the decision process. Community opposition is indeed a valid and material planning consideration, but only if there is substance to the local opposition. The fact that commercial considerations (the need for the facility) is not very productively debated during the UK inquiries is reflected in the outcomes to some extent. Note that the need for the facility in Seal Sands dissipated in the end and the facility was not constructed. In a commercial sense, therefore, the decision to allow the facility was not a very good one, but this is rather unfair as risks associated with a lack of economic quality are laid squarely in the laps of private market parties.

7.6 The interest-discourse divide

As stated, in all three countries, the managerial discourse (within certain confines) is the most credible way of speaking. This applies even to the community-based cases in Canada. This implies that parties, in order to convince others, at least in the regulatory arenas involved, need to use this discourse. Outside the regulatory arena parties may have other considerations for their opinions - these may be communitarian in character for instance - but they serve little function than that of mobilization. In the regulatory arena, it does not really matter so much what the reason for the position of the actors is, they are still forced to defend it in managerial terms. To some extent this creates an uncomfortable situation in the sense that the 'real' arguments are not always in the open.

The conclusion that actors are forced to use one discourse implies that, as Hajer said, interests are indeed constituted through discourse. However, this should not be taken to imply that people derive their interests through discourse (that is, perceive only interests that are embedded in that discourse) nor that interests have no existence outside discourse. To clarify further, these interests are partly 'objective'. Take for instance the health aspect of the debate. This is such a fundamental concern for most people that it is felt before it is even communicated. Surely, when one formulates an argument on the issue, one may need a certain discourse to explain one's position (e.g. it is (un)fair to impose health risks on the local population in the interests of a larger community), but this does not imply that the interest did not exist outside discourse. This then explains the phenomenon of positioning. Legal rules allow the expression of interests only in a certain way. People who previously spoke in completely different terms, may realize that strategically they had better change their tune in order to be more effective in defending their interest (see the Rotherham case).

7.7 Conclusions

7.7.1 Introduction

As this chapter is largely summarizing and conclusive in nature itself, it is less necessary to summarize its content in a concluding section. Instead I will present some of the more general conclusions that I draw from the nine cases. These can be grouped in two categories. Most of them will be within the realm of the theoretical framework that I have outlined in chapter 2. Here I will address what I consider the most relevant findings from the case studies. Some other conclusions are related to observations about the theoretical framework. Specific use can be made of the applicability and completeness of the framework to analyze hazardous waste siting.

7.7.2 Conclusions within the theoretical framework

The discourse-institution connection

The most remarkable cases from the perspective of discourse and institutions are the Canadian ones. One thing that is unique about the Canadian case studies is that they combine an emphasis on the community with a largely technical public discussion of the proposals. These cases are unexpected from the perspective of my theoretical framework in the sense that I had hypothesized that discussion along managerial lines would be solely among experts. In a time of increasing levels of education among the general public, a trend that is corroborated by various case-studies where 'ordinary citizens' were experts in relevant fields, this experiment is interesting. It does seem that the outcomes of the Canadian debates are supportive of greater community involvement in the sense that volunteer communities did come forward (solution to gridlock) and in the sense that the technical quality of the decisions was good. On the other hand, there are some concerns about what happened in these cases. The cost-effectiveness of the solutions was not great, implying that sites were chosen that are expensive to operate. The quality of the decisions in terms of pollution containment was however good and in fact improved during the decision process, be it not much.

A 'mismatch' between institution and discourse also seems to exist in the UK and the Netherlands, where the courts and inquiries present an opportunity for managerial debate. As already explained, this is a consequence of the view that these procedures are part of the administrative decision-making process. The role of experts, the independent role of the 'judge', the emphasis on truth-finding are all indications that these courts do not meet the description for decision-making in courts that I had derived earlier (see the previous sections).

The dominance of the managerial discourse

It should be interesting that, despite the fact that the decision process took place in clearly different institutions in the three countries, the dominant language was still managerial. The other languages that are available, most notably the progressive pluralist and the communitarian languages met with critique from most parties involved. Whereas they may use these languages in calling for the mobilization of resistance, it does seem that most people would not want to depart from the managerial language. Certainly, an overly overt critique of the managerial language makes proponents of facilities vulnerable to the criticism that they are being opportunistic. Alternatively, a reference to compensation (progressive pluralism) quickly leads to the accusation that the proponent is corrupting the local decision-making process. A claim of being in the interests of the community tends not to be very credible due to the fact that 'the' community is divided.

The need for a proponent

Although the development of Siting Task Forces towards proponents of siting in Canada must be disapproved of in the context of 'voluntary siting' (it can perhaps be approved of in the context of invitational siting), it does seem that this development is inspired by an 'objective' need to have a proponent in the decision process. This decision process does not automatically progress in the direction of a site. Instead, there must be a plan designing the facility to be sited and the people participating in the decision process need an image of who is proposing the facility in question. Without this, people will be confused and become suspicious as demonstrated in the Manitoba case.

The need for disagreement

In addition to this, I feel that the discussions of the proposals in the UK and the Netherlands were more complete than the discussions in Canada. To some extent this is a function of motivation. The role of trust in the Canadian cases is quite large, and when members of the community develop a certain level of trust in those proposing the facility it does seem that they become less critical. The idea that the facility may enhance local employment adds to this, even to the point where people feel they are competing for the facility and must thus be moderate in their demands. The UK cases and the two Dutch cases where there was opposition demonstrate that a critical public (be it environmental groups, ordinary people, or the experts from the Commission for EIA) can often detect errors in quite crucial parts of the proposal. What drives these people is a certain hostility towards the proposal that encourages them to check just about every bit of information they can find on the proposal. These people are extremely motivated to participate and tend to bring out the weaker points in the decision process.

Decision quality

It does seem that there is a certain pay-off between technical and social decision quality. The Canadian cases, if anything, demonstrate that political rationality necessitates acceptance of this pay-off. It is clear that this is not the case in the Netherlands and the UK. In the Netherlands, the local communities involved were not very concerned with the proposals, whereas in the UK, the decision process effectively buffered the resistance. The advantage of adhering to the legal framework is of course that legal rationality is enhanced: the predictability of the decision process increases. Apart from the cost-effectiveness concern, which some will describe as the only reasonable way a just way of incorporating social costs into waste treatment, the legal rationality of community-based siting was not of much use. Especially in the case of Alberta, where the process was being invented while it was being implemented, this led to questions about how the decision process would progress and what means could be used by the Siting Task Force and what not. Especially the fact that local communities started demanding contracts with the provincial government indicates that there was a need for predictability, even after the decision process has been completed. Especially because the connection from provincial decision process to private market parties was difficult in Alberta, Manitobans and later the people in Ontario started demanding control over the facility after it was sited.

7.7.3 Some conclusions about the theoretical framework

Structuration: enforcement methods

What is rather remarkable from the perspective of the theoretical framework in chapter 2 - much less so from the perspective of someone with knowledge of hazardous waste decision-making - is that structuration is not merely a matter of story lines. I have already called the idea of story lines and the affinity that people might feel towards them somewhat simple. It does seem that the case studies show that structuration is achieved by non-lingual means. The courts are in fact explicitly designed as an enforcement structure to supervise adherence to the legislation. Where local authorities or others use arguments that are not relevant under the legislation - such as the interest of the community - these bodies tend to determine that the decision should be quashed. In addition, the UK system recognises the system of awarding costs, which puts those actors that do not adhere to the relevant discourse at serious risk of losing money. This implies that discussions are more than an intelligent conversation between independent people (to paraphraze Hajer), but that even their positioning in terms of discourse is not just a matter of choosing a strategically smart 'argumentative position' but also reflects power. Especially local authorities run the risk of judicial problems if they deviate from the legally-prescribed discourse, and generally speaking they will want to avoid them. The possibilities that existed in Canada for community-based siting must be understood in part from the fact that the authorities making collective choice level decisions about decision procedures were the ones deciding to postpone the application of ordinary procedures. Furthermore, private market parties were not very interested in hazardous waste siting and did not push for upholding the law either. This would seem a relatively unique combination.

The concept of discourse coalition

In addition to the fact that story lines and affinity are not the only factors responsible for positioning or discursive interaction more generally, it would seem that the concept of discourse coalition is somewhat difficult to use at the concrete choice level. The coalitions that are present in the case studies seem not so much based on affinity, but much more on a shared interest - for instance resistance against a certain proposal. This generates a rather uneasy feeling in using the concept to describe the way local opposition groups unite and then argue a position in the decision process. It does seem that the concept is better suited to analyze debates at the collective choice level than at the concrete choice level, where it is somewhat 'over the top'. Therefore, I would suggest that it is a 'middle-range concept'.

Critique of discourses as an influencing factor

It is quite remarkable to note that decision procedures - especially in the Alberta case - seem not so much motivated by a certain discourse but by scientific research of what works in siting and what does not. The province of Alberta is hard to place in terms of the three discourses that I have distinguished. The choice for a community-based approach was not driven by communitarian thought as I had anticipated, but much more by a pragmatic anticipation of the local reaction towards siting proposals. The case of the federal siting process is similarly difficult to place as it seems mainly based on a rejection of managerialism controlled by experts. There one could apply to some extent the argument of progressive pluralism as there was much attention to distributive justice, but the siting process that was advocated was also largely based on the idea of a managerial discussion at the local level instead.

Chapter 8

Summary and conclusions

This chapter summarizes the main findings of the book and presents some conclusions. The previous chapters have demonstrated the important role of the conservative pluralist and managerial discourses. This is reflected in the large role that the private sector and experts are supposed to play in these decisions. Technical and economic quality are deemed most important in the legislation in all three countries. Structuration, the shaping of actual decision processes by the law, took place in about half of the case studies, whereas in the other half structuration was only achieved after much difficulty or not at all. Although discourse analysis claims structuration will be achieved through story lines, it seems more accurate to say that enforcement mechanisms such as the courts secure structuration. In the Canadian cases structuration was not achieved in the sense that different institutional arrangements played the lead role rather than the ones envisioned by law. The role of the community and arguments related to the community's interest were larger than expected on the basis of the law. This came at the expense of the private sector parties, elected representatives, and technical argumentations. The qualities of the outcomes in the nine cases I have studied seem to reflect the prominence of the different institutional arrangements. In Canada, the social quality of the decisions was good, but the economic and technical quality somewhat dubious. In the UK, where the national government overrode local resistance in some cases, social quality was poor, but technical quality good. I criticize discourse analysis for a number of reasons, although the 'milder' form that recognizes interests seems attractive. I conclude this chapter with recommendations for the practice of environmental decision-making, among them the recommendation to continue to use managerial language to discuss hazardous waste decisions.

8.1 Introduction

My initial interest with this study was the quality of environmental decisions. In the course of my work I have come to focus on the role that different types of institutions may play in attaining good decisions. Also, I have come to assume that various languages, which contain different conceptions of decision quality, exist. Such languages are, together with ideas about the proper institutions for taking decisions, embedded in discourses. Discourses are shared ways of comprehending the world and they are lingual in nature. As shared ways of comprehending the world, they offer 'ready-made' ways of understanding the world and directives how to tackle problems. Discourses can have a strong effect on the practice of environmental decision-making if they 'institutionalize' in government legislation guiding such decision-making processes. If this occurs, then the aggregation, authority, boundary, information, and pay-off rules in such legislation express the thinking of the discourse in question.

The three discourses that can potentially have such an institutionalizing effect are the communitarian, pluralist and managerial discourses. Each has its own outlook on decision-making, which emphasizes different types of institutions and a different language (i.e. arguments, type of information) for decision-making. Broadly speaking the managerial discourse stresses the role of experts and science, the pluralist discourse the role of private parties and their interests, and the communitarian discourse local communities and dialogue on the basis of local knowledge. Choices on legislation that will guide environmental decision-making are made at what can be called the collective choice level. There, lawmakers, in interaction with others, decide what the rules will be. I have assumed that the three discourses have a presence at the collective choice level and compete for 'hegemony', which would imply that they start to dominate the thinking behind the legislation, which is in the end adopted by Parliament. The means for achieving hegemony at the collective choice level, according to Hajer (1995), include 'story lines', which are simple emblems for understanding the choices at hand. Behind such story lines, 'discourse coalitions' form, which seek to influence legislation. It can be expected that collective choice debates are also influenced by decisions made at the constitutional level, but that factor was outside the scope of my study.

Discourse analysis is not blind to the existence of interests, but those advocating discourse analysis do not always agree on its importance. In the most extreme version of discourse analysis, interests are assumed to be driven by discourse alone as people cannot think outside discourse. A milder version, advocated here, suggests that interests are 'constituted' through discourse, which means that whereas interests exist separately from language, they must always be communicated through language, which therefore defines them. By contrast, the standard political science view of decision-making assumes that interests drive decision processes. If this were true, then discourse - if relevant at all - would be inspired by interests, with the discourses providing nothing but a menu from which to pick an argumentation that suits the interest at hand.

Not just the collective choice level is of importance in environmental decision-making processes. At the concrete choice level, the institutions assigned with the task of making the decision and using the language that was considered the most relevant, actually have to work with the rules that were devised. If the legislation that was developed at the collective choice level succeeds in guiding their actions and argumentation, I speak of discourse structuration. It can be expected that the non-dominant discourses at the collective choice level 'return' at the concrete choice level and try to challenge the discursive order established at the collective choice level. The 'weapons' are then the same story lines that were assumed to be of relevance at the collective choice level. It is interesting therefore if having occurred, how the discursive order that was established at the collective choice level is maintained and the debate structured the way it was meant to be.

The rest of this chapter

In this chapter, I will summarize my analysis of the debates at the collective and concrete choice level. I will discuss whether institutionalization has occurred along the path described by discourse theory, whether discourse coalitions and story lines are actually observable in practice. Similar topics will be covered in the section that summarizes the concrete choice level debates. There, I will also discuss whether concrete decision processes are indeed structured by certain discourses and if so, how structuration is achieved or broken down. I will also discuss the implications of different ways of structuring the debates for the quality of the decision-making process itself and the outcomes. After having discussed these two issues, I shall move on to a discussion of the implications of my findings for discourse analysis. After that, I shall conclude with feedback into the discussion on environmental decision-making.

8.2 The collective choice level: institutionalization

The presence of the three discourses and discourses coalitions

Within regard to institutionalization the countries I have studied follow a largely similar pattern. The basis discourse is conservative pluralism, which was at its strongest in the 19th century. Then the countries became industrialized and the need for regulation increased. Initially the proponents of new regulations used the progressive pluralist discourse, with an emphasis on fairness. This remained without much result, but the subsequent reversal of the reformers to managerialism was attractive to more parties than just the reformers and took hold, be it in combination with conservative pluralist ideas. Together these two discourses have dominated the collective choice debate since then, with some periods (especially after the Second World War) showing a stronger managerial pretension than others (e.g. since the 1980s the emphasis is again much more on the market). The communitarian discourse was largely lacking from the collective choice debate in all three countries.

It does appear that the conservative pluralist discourse is the 'basic' discourse in all three countries in the sense that the right to initiate development for private parties is held in high esteem. The discourse has a certain affinity with managerialism in the sense that adherents to the discourse find that government interference is warranted if experts rule that such interference would increase utility and therefore be in the public interest. It has less in common with progressive pluralism, which stresses distributive equity. This explains the limited success that the reformers initially had.

The climate for regulation changed as a consequence of various developments, which cannot all be explained by discourses alone. Especially the fact that certain diseases at the time were contagious for everyone (not just the working class), that the value of labor greatly increased, and that workers united in labor unions, made it easier to realize that regulation could be in everyone's interest. Businessmen and politicians alike started to see the added value of regulation. The 19th century was one of great technological progress, and it therefore saw an increase in the trust in science. A new breed of professionals (planners, doctors, and engineers) came up that saw a twofold task for itself. The first part of their agenda was the advancement of the profession itself by setting entry standards and developing norms on the curriculum. The second part of their agenda was to increase their role in society, by obtaining roles in the private and public sectors. Regulation was an attractive field for them as they considered conflicts resolvable by science and they simply needed employment that the private sector would not offer. They linked up with those seeking government interference on the basis of the progressive pluralist discourse and started pleading for an increased role in government. Some of the new professionals considered themselves the only ones capable of implementing the regulations that they demanded from government.

In none of the three countries did the new professions become the ones behind the guiding wheel of society. In fact, the desire to develop in this direction was relatively weak. The professionals who took up important positions in both the private and the public sectors de-emphasized the - progressive pluralist - reform agenda more and more, by using a combination of their own choice and outside pressure. The new professionals quickly found out that in order to find employment, they should not just stress the technical quality of their ideas, but pay attention to the economic quality as well. Especially the engineering professions started inserting business economics in the curricula they controlled. This shift towards the private sector's desire to make profits increased their attractiveness for private employers. In the public sector, regulators were forced to make a similar turn, if they had not already made it themselves. The initial planning legislation, especially the one in the UK, was quite managerial in tone in that it allowed infringement upon private property and intended to isolate planners from private sector influences by denying opportunities for representations. This legislation was literally imported into Canada, and the legislation in the Netherlands showed similar tendencies.

The actual implementation of such legislation was shaped, to a large extent, by intervention from the courts. In the UK, the courts initially cracked down on regulators in the interest of private property owners, and instituted the requirement for consultation with property owners. They later made a creative turn towards acceptance of activities by administrators that were overseen by their political masters. This implied a great level of discretion for UK land use planners, which did not exist in the two other countries. One must observe however that this discretion was structured by political guidance, which emphasized a bias in favor of the developer and de-emphasized the importance of binding plans in favor of case-by-case decision-making. This bias reflects the desire for economic (industrial) development that politicians had. A similar desire existed among politicians in Canada, but the pre-occupation of the Canadian courts appears to have been with the protection of the interests of landowners of land adjacent to proposed developments. They often attempted to prohibit unfavorable development, only to consequently see their decisions rejected by elected representatives. The resolve to interfere with the private sector on the part of Canadian planners was not great. In fact, they were placed under the supervision of planning boards, controlled by private sector membership. The constellation led to a situation wherein planners became 'administrators' of private sector development. Once such development had occurred, planners were to make zoning ordinances that protected it from infringement from other proposals. In the Netherlands, regulators were not ambitious in their attempts to regulate private development. And when they did want to exercise their powers, they often found the courts put heavy emphasis on legal certainty for landowners across their paths. As a consequence, few land use plans were actually developed and ownership continued to largely determine land development.

Given the limited success of land use regulation before the Second World War, it is somewhat of a surprise that planning became so popular after 1945. Military planning had strengthened the faith in planning however, and there was a general feeling that reconstruction under central guidance was necessary as the war had destroyed much of the resources of all three countries. The land use planning system expanded under the influence of a new managerial discourse coalition of elected politicians and planners, with the silent support of the general public. Especially in the UK and the Netherlands, development rights were expropriated and all land was placed under government control. However, in neither country was 'positive planning', complete control over land use on part of the government, ever achieved because the dominant discourse of conservative pluralism was too strong. There were some attempts at the introduction of positive planning, motivated by progressive pluralist thought, but these never gained sufficient political support. In the 1980s, the conservative pluralist discourse reasserted itself quite strongly and land use planning was robbed of its pretensions, which had grown considerably in the 1960s and 1970s on the basis of procedural planning theory. The fact that land use planning was not at all delivering the superior results it had promised was an important explanation for this, along with the broader - and international - shift in ideological climate towards neo-liberalism.

One of the areas where the land use planning system failed was the environment. From the 1950s on, a new set of regulations developed in this sphere. Although the extent of the environmental regulation system has only increased since then, the language operating the system has not really changed much and developed in a similar fashion like the language of the land use planning system. Again, the emphasis was put on regulation in the interest of the public and an important role of experts. However, governments only wanted to interfere in cases where intervention had an added value. Pollution was not to be minimized, it was to be 'optimized'.

The presence of story lines

The managerial discourse was the only discourse that could make inroads into a landscape dominated by conservative pluralism. The reason for its relatively strong power was that it offered the promise of avoidance of political conflict (which was threatening at the end of the 19th century) and corruption (a real problem in Canada) and the suggestion that the economic pie could be made much larger so that all could benefit. The founders of reform movements and the early members of the new professions advanced a progressive pluralist and a sometimes strongly managerial line of thinking, but became more 'realistic' once responsible positions were in sight. In fact, to get into such responsible positions they had to stress the fact that regulation, if done properly, would also be in the interest of businesses. The story line of increased utility, the public interest, was both the entrance ticket to government and business but at the same time implied a constraint on the managerial agenda. Instead of replacing elected representatives and business leaders, the new professionals became their assistants. The consequent mixture of managerial and conservative pluralist discourse, supported by elected representatives, is extremely hard to resist as it combines the story lines of managerialism (the public interest, science, neutrality), with those of conservative pluralism (especially freedom) under the sanctioning of democracy. Only *within* this mix have some story lines changed in importance through time. There is an affinity between conservative pluralism and managerialism in that utility is considered extremely important by both discourses, and that both attach great weight to an involvement of experts. However, the conservative pluralist discourse starts from a different point: freedom of enterprise, whereas the managerial discourse tends to plead for greater involvement of the state. It does seem that there are clear shifts in the emphasis on either freedom or state control in all three societies I have studied. There are times when the emphasis is on state planning and there are times when there is more emphasis on freedom. This points to the fact that 'the public interest', itself an enormously powerful story line, is not constantly defined in the same way. The fact that various groups, with often widely different interests, claim to support the general interest alludes to this fact. Environmental groups are often said to be public interest groups. Yet they want something completely different to developers who also may refer to the public interest.

The role of the various institutions

It does seem that the hegemony of the conservative pluralist and the managerial discourses translates into the selection of relevant institutions for decision-making along the expected lines. The role of elected representatives, experts, and private market parties is the greatest, with the courts having a more distant role and the community as such none. The limited role for the community does not imply that there is no citizen participation in the decision-process, however. In fact, all three countries do recognize citizen participation rights. Indicative of the strength of the dominant discourses, the main reasons for such participation are that landowners must have a say in the decisions that are made regarding their land, or that the population can contribute to the quality of the decision in managerial terms.

The notion of decision quality

Given the mixture of managerialism and conservative pluralism embedded in the legislation, the dominant concerns from a decision quality perspective were with technical and economic quality. As stated in chapter 2, political rationality is at the meta level and thus the question is raised whether the choice for technical and economic quality was rational from the perspective of support for the political system. There are few indications that actors at the collective choice level felt that these quality criteria were irrelevant or should be replaced, even though it does seem that some would have stressed the economic quality of decisions somewhat more than the technical quality. It appears that the social quality of environmental decisions was considered less of a concern at the collective choice levels in all three countries. The fact that a community could potentially oppose a proposal was and is largely irrelevant under the decision-making laws. Overseeing the system of decision-making are the courts. The courts have the responsibility for upholding the predictability and clarity of the legal system and this can be called a meta-rationality, just as the selection of decision procedures by politicians needs to reflect a meta-rationality. However, note that the quasi-judicial tribunals in the UK and Canada, and the administrative courts in the Netherlands, also have a role in supervising the technical and economic quality of the decisions that were taken. This is largely due to their (in the Netherlands: past) association with the administration. Especially in Canada and the UK, a certain amount of contention before the tribunals (or inquiries) is considered helpful in testing these types of quality. In the Netherlands, the parties in the dispute play a much smaller role and the Advisor Environmental Appeals largely determines decisions.

8.3 The concrete choice level: siting debates

Introduction

Some of the questions I have tried to address at the collective choice level were also studied for the concrete choice level. Nine cases were studied, three in each country. I studied which institutions were most important in the case studies, which discourses were used, and which discourse coalitions were present. In addition to that, the case studies offered an opportunity for assessing the degree to which structuration of the debate took place. In about half of the case studies, the legal framework did structure the debate, but in the other half structuration failed to some degree. This section addresses the questions why structuration was achieved in some cases and not in others. Finally, I studied the quality of the decisions that were taken and attempted to link these to the institutions that were involved.

The institutions involved

The greatest difference in terms of the institutions that were involved in the decision processes was that the Canadian cases showed an emphasis on the community and a reduced role for the private sector and elected representatives, whereas the Dutch and UK cases showed an important place for the private sector and a far less important role for the community. The choice of institutions for this type of decision processes appears to be influenced by considerations of political and legal rationality. The set-up of the Canadian siting processes that I have studied was distinctly different from what the law prescribed and must be explained by the fact that the private sector was not making proposals and thus had no interest in maintaining an important role, whereas decision-making involving private market parties and elected representatives was too politically sensitive. In the UK and the Netherlands there were some occasions where the role of the private sector in siting was contested, and one case where elected politicians and government experts decided to refrain from directing private sector initiative, but these deviations met with resistance from the courts. The courts acted as agents of the legal framework and rejected deviations because they would endanger the predictability of the decision process.

The discourses involved

Although all discourses (and their critiques) are found in siting debates, it does seem that the mixture of conservative pluralism and managerialism embedded in the law is the pre-eminent way of talking. The progressive pluralist and communitarian languages are also used, but then mainly as a means to mobilize resistance. Interestingly, this finding even applies to the Canadian cases, which were 'community based'. These cases demonstrate that many people in these countries are not at ease with a siting process that would only emphasize community acceptance and nothing else. Too obvious mentioning of the possibility of compensation leads to accusations of bribery and therefore fails too. Instead the emphasis must be on the technical part of the siting proposals, which is discussed in a managerial way. It does seem that various members of the community are able to engage in such a technical debate, but there is quite a strong temptation on the part of those running the siting process to sway the population. It is also interesting to note that 'the' community, which has a veto right, is perfectly capable of ignoring the negative advice of some of its members who formed so-called Citizen Liaison Groups. In the case of Deep River, the CLG advised negatively on the technical acceptability of the proposals that were being made to the community, but the population voted in favor anyway.

Discourse coalitions?

My conclusion is that the term discourse coalition is a middle-level theoretical concept, which is therefore of limited use at the concrete choice level. Although people do tend to associate and form groups, such groups seem not so much glued together on the basis of discourse, but rather on the basis of a shared purpose (defeating the proposal for instance). As already discussed, the managerial discourse is the most dominant mode of speaking during this type of decision processes. This implies that even opponents tend to use managerial discourse, just like private market parties, politicians and experts. To call this a consequence of discourse coalition formation is a mistake however. Rather, it is a consequence of the enforcement mechanisms that guard the structure of the debate.

Discourse and discipline

The case studies confirm that discourse and discipline are related to an extent. Waste facility siting did not involve physical force in most instances. In all three countries, arguments and reasons must be given for positions, and only certain arguments are accepted, whereas others are rejected. This is not a matter of comprehensibility or language per se, but a matter of selective speaking and listening. Some terms mean little in the debates, whereas others mean a lot. The proponent of a facility, so much is clear, is forced to present his proposal in a managerial way (e.g. the site was compared to various others on the basis of multiple criteria and was the best) and others are expected to reply in that same way. To some degree, this is almost 'automatic'. There is a large degree of anticipation among the participating actors, which implies that they 'automatically' choose the managerial discourse as the language to express themselves, even if they oppose. People 'sense' that if they are willing to make the claim that the proposal will be detrimental to their health, that they will need to substantiate this claim with evidence. The fact that they are afraid carries hardly any weight in itself.

Structuration achieved

I have considered the reason why people do automatically adjust to the relevant discourse. It would seem that the increased level of education among the public, together with dominant social values play an important role in this. The level of education among the public, in all three countries, was remarkable high. In at least one of the three case studies in each country, there was a member of the public who had as much expertise in the relevant domains as the experts hired by the proponents. People with such expertise seem drawn to this type of decision process, certainly when it is in the area where they live. The Dutch case of the Kaliwaal and the UK cases of

Newport and Rotherham are clear examples of cases wherein local people with scientific training entered the decision process. The arguments they bring are never communitarian or pluralist, but strictly managerial as they try to win the decision process on the basis of technical arguments alone. This can go so far (see the Kaliwaal case) that they forget about the possibility of mobilizing the local public, I would say because community resistance is not a very relevant criterion in their opinion.

However, the opinion that scientific training and managerial argumentation is not just shared by the people who have such training, it is very widely held in the three societies. This can be drawn from the fact that opposition groups tend to elect members with academic training as their heads or as their spokespeople. Take the fact that the representatives of the Werkgroep Eemsmond (North Refinery case) were academically trained, or take the fact that all chairs of the CLGs in Canada had some degree of scientific training. In other words, 'ordinary' members of opposition groups, but also members of local councils, consider a certain degree of scientific training relevant too. There are however cases where not all participants immediately recognize the importance of scientific training. Especially the Rotherham case comes to mind. In that case, a relatively 'soft' (in the sense of non-violent) correction method was activated, in particular a lecture by a planner who told the opposition to start making policy arguments or simply lose the case.

Non-structuration

I have described how in cases of non-structuration, it is important that a decision to remove siting from the hands of the private sector be taken by the government level that determines the legislation on decision-making. If this lesson is not heeded, then actors that propagate a different approach to decision-making are vulnerable. Especially in the UK, local authorities have only so much leeway, and stand to be corrected by relatively tough mechanisms such as an award of costs if they do not adhere to the legislation. It is clear that this instrument has a certain effect on decision procedures because the proponents of facilities do refer to this possibility and actually use it, whereas planning officers warn the councilors against this possibility in cases where such councilors intend to use non-relevant planning arguments.

The inquiry too, is an instrument of structuration, just like administrative appeals in the Netherlands. Deviations from the legal system in the Netherlands and the UK tend to end before such (quasi-) judicial bodies, and the presentation of a dispute by a private developer tends to lead to the reconfirmation of the proper legal framework. Because such court cases are expensive and take quite some time, I have called the cases where they were needed (Kaliwaal, North Refinery, Newport) cases where structuration was problematic. It should be noted that inquiries and administrative courts are 'structure-enforcement institutions' in part because they hardly distinguish between cases that are brought to them because structuration has failed and cases that are brought to them because there is a conflict *within* the relevant terms. There were fundamental differences between the resistance of the local council in Newport on the one hand and Rotherham and Cleveland on the other. In the first case the stated reason for local resistance was communitarian, in the others it was managerial, or presented as managerial. But during the inquiries only the technical part of the argumentation was discussed. Especially in Newport, the inquiry was relatively brief because the opposition was much more philosophical in nature than technical, and the philosophical arguments were basically not discussed. It does seem that inquiries 'normally' do work in reducing public opposition to proposals, in part because of their fairness and thoroughness in technical details. This dynamic was not able to work in the Newport case however, as the technical argumentation of the council was so weak. In a sense, the fact that 'closure' of the debate was not achieved there should thus not be a surprise.

Still, the council or the public in this case did not go as far as the public in Canada, which demonstrated a tendency to react aggressively, even to the point of physical violence. The cases where this happened (Alberta and the federal siting process), demonstrate the fact that power can be exercised through discourse, but only up to a certain point. If physical force is used, this is a clear indication that the structure embedded with the legal framework is not acceptable to the public. Power can then not be exercised through the discourse in question. Fundamentally then, the public authorities have the choice of using counter-force or making a discursive turn. It is interesting that none of the Canadian authorities involved was willing to use counter-force, but opted for a discursive turn. The reason for that would seem to be related to political support. To allow a private developer to make proposals for facilities in various communities would have had negative effects on numerous political careers. The discursive turn was made easier by a fortunate combination of circumstances. A body (the province) that also controlled the ordinary decision procedures introduced the new decision approach and there was a relatively weak desire from the private sector to site a facility under the normal procedures. It should be noted that the discursive turn of the Canadian authorities was not complete. 'Ordinary' decision procedures stayed in place, but with a local veto added. And even though the emphasis was on the local level, the language of discussion continued to be managerial. This is in part because of popular demand. A more communitarian or progressive pluralist-based decision process was difficult for the public to accept. There were various instances where the emphasis on social acceptance and compensation was used to discredit the decision process when it started being implemented. This is, again, an indication of the need for science based decision-making. The crucial difference with the UK and Dutch approach was then the institution where this debate was to take place: the community. This unlikely combination seems to have worked to some extent. Certainly, the technical quality of the Canadian decisions (except for the cost-effectiveness part) seems uncontroversial.

8.4 Remarks about discourse analysis

Introduction

So far, I have largely remained within the realm of discourse analysis as a theoretical framework for analyzing the hazardous waste debates. In this section I shift my attention to some remarks about discourse analysis itself. I will discuss some of the difficulties associated with discourse analysis and some of its shortcomings, all from the perspective of a relatively positive view of what discourse analysis, the way it has been applied here, can bring. Among the reasons to be positive about discourse analysis is the fact that its core - power is exercised through language as institutions function on the basis of this language - does seem to hold true in all three societies I have studied. Hazardous waste debates are not won by physical force, they are won by arguments. Which arguments can be used, how they are used, and what their effects are is thus very important.

Discourse analysis is demanding

I feel that discourse analysis offers an interesting perspective on environmental decision-making. The attention to ideas and language on the one hand and the effects they have on legislation and concrete siting debates on the other offers an interesting perspective and helps understand what went on in such debates. As explained in chapter 3, discourse analysis is quite demanding on the analyst however, and is not a perspective often applied, at least in the political and policy sciences. Discourse analysis is demanding because the analyst is driven towards a deep analysis of events, the involvement of various actors, the actions they undertook, their interests and their motives, and the way they motivated their actions. Chapter 3 is based on secondary analysis of existing literature and I have found this literature often wanting in the sense that little detail was provided of the thoughts and discussion processes that went on in the three countries since the late 19th century. Very often, descriptions remain just that, or give interpretations with much hindsight. The theoretical perspective from which the descriptions are made is often not made clear, but tends to be interest-based, which implies that the descriptions remain at the level of 'who was in favor and who was against, and who won the debate'. Similar difficulties exist in analyzing actual siting debates. Fortunately, these debates tend to be well documented and much of this documentation is public so that it can be used for the analyst.

Discourse analysis: not the extreme form

Apart from the fact that discourse is demanding, I saw the rejection of a purely discursive understanding of reality confirmed. My discussions on the topic of the relationship between interests and language lead me to conclude that it would be incorrect that people are 'captured' by a single discourse and simply cannot think outside that discourse. The milder version of discourse analysis appears to be supported to a great extent by the empirical evidence I have gathered. Note for instance the 'positioning' of actors in the concrete level siting debates suggests that actors do make an assessment of the language they will use in reply to the proposals they have been confronted with. In fact, their use of language may differ per phase of the decision process, or even per audience. It does seem that the three discourses have different functions, with the managerial and conservative pluralist languages providing the required way of speaking, and the communitarian and progressive pluralist languages providing ways of mobilizing actors. It seems that the actors involved in hazardous waste debates are often quite aware of these different functions and tend to shift from one way of speaking to another, depending on the public and the forum, as they see fit. The least flexibility in this respect can be found among those with a scientific training. It does seem that for practically all actors with scientific training, a managerial approach to decision-making is the only way to proceed. These actors often despise political influence, citizen involvement, the application of the criterion of social acceptability. The fact that *their* suggested approach to decision-making benefits especially them seems to escape these experts. At a fundamental level, modernization has taken place in many corners of the three societies. This development surfaces in the cases that I have covered in various ways. The first is that various members of the public have scientific training themselves and can thus comprehend the decisions at stake quite well. It is clear that managerial discussion is the norm in all three societies. We have come to expect that such important discussions are made in a scientific fashion and most members of society are willing to speak in this mode.

The role of non-linguistic factors

However, it should be observed that interests are not always constituted through discourse. Whereas it does seem that power is exercised through discourse, the fact remains that power of the state and other actors is also based on other factors, such as the possibility of using force. Especially the focus on the collective choice debates in the 19th and early 20th century made clear that the new professions had to fight quite hard in order to gain their present important positions at the decision-making table. The fact that the ruling elite at the time was willing to listen to them in the first place was in part a consequence of the fact that they promised an apolitical solution to the tensions that threatened to topple society. It is of interest that community-based siting was in part also a reply to the readiness of the Canadian population to use force.

Discourse analysis: the way to productive decisions

One often hears the criticism that discourse analysis is far better suited to analyzing differences than in actually contributing to solutions to resolve these differences. Take the fact that in certain situations, different actors may advance different lines of reasoning that are not well connected. Whereas one actor may stress the acceptability of proposals for the community, yet another may stress the scientific quality of the proposals. This criticism would seem particularly relevant if indeed people cannot think outside the borders of 'their' discourse, which makes a discussion which falls on deaf ears. Fortunately, my study demonstrates that 'facts' can sometimes close debates. A significant example from this perspective is the way the UK authorities countered the 'fairness argument' used by some of the opposing localities. These figures demonstrated that the localities were exporters of wastes and were thus actually bothering other localities with their waste. Indicative of the political nature of the debate however, the source of this data is relevant. For them to have a 'closing' effect, it helps that they are derived from statements by an opponent (as happened in the UK, the data came from the local waste plan). However, facts and data will not always close debates and discourse analysis offers little thoughts on how to structure debates where people do not share a common language. My study indicates however that for most people in hazardous waste siting it is much less the language that is used as it is trust in the institutions that are using them that explains their resistance. The Canadian cases show that institutional redesign, while maintaining the managerial language, is possible to some extent.

Discourse coalitions: a middle range concept

As explained in the previous chapter, the discourse coalition concept seems less suitable for use at the concrete choice level. Furthermore, the assembly of 'discourse coalitions' does not take place on the basis of story lines, but much more on the basis of shared goals.

Enforcement structures:

I concluded on the basis of my case studies that it is not just story lines that attract people to a certain discourse. Especially for the discourses embedded in the law, the courts act as 'enforcement agents', implying that they oversee whether the decision processes were run the way they should according to the law. Resistance against court verdicts is difficult and the verdicts tend to be final. In the UK, the award of costs was another such enforcement structure, which had a serious impact on the decision-making process.

Cultural factor

Although I have focused only on formal decision procedures in the three countries, there is another aspect that influences decision-making, which becomes quite clear if one carries out research in the three countries in question. This factor is that of administrative culture, and specifically openness. Although this is just a small indication of openness it is in a sense telling that visitors to local council offices in the UK are kept away from the offices where administrators work. Having had to study local council archives quite often, I have experienced that a researcher never gets access to file storage rooms and needs to ask for certain documents in order to be able to see them. Files are checked for confidential material and this is generally removed before the researcher gets to study the documents. Making photocopies tends to be extremely expensive at local council offices (up to £2, at time of research). Such hurdles are often crossed by the personal kindness of administrators, but one does not get the impression of a warm welcome initially. In Canada and the Netherlands on the other hand, access is often given to file rooms directly, photocopying tends to be free of charge and is sometimes done for a researcher. If there is 'coffee-round' one tends to be served as well. Again, these are just small incidents, but they do indicate that in the UK there is a greater hesitation towards sharing information.

8.5 Implications for environmental decision-making

8.5.1 Introduction

In this final paragraph I will present some recommendations from my study in the practice of environmental decision-making. An important source for my advice is the empirical evidence I have collected in the various case studies. I am aware that there are multiple perspectives from which to give advice - including the private sector's perspective, the opponents' perspective and the perspective of an elected politician. Here I will only give some reflections that could be useful for those engaged in the collective choice debates and that could assist them in their choices on the set-up of institutional arrangements.

8.5.2 The practice of hazardous waste siting

Introduction

Taking into consideration of the limited possibility of generalizing the findings from a case study in one country to another, I will discuss some of my thoughts on hazardous waste siting in each of the three countries, in the form of a discussion of their stronger and weaker points of the way decision-making was advanced, and then present some ideas for reform.

Hazardous waste siting in the UK

I feel that the UK approach to siting has various strong elements. Among those I would count the fact that the private sector is responsible for making proposals and thereby bearing the risk that from an economic perspective the proposals make little sense. It is in many ways laudable that waste issues are (or were) being translated into land use plans (however the fact that local authorities were at the same time responsible for such planning and had ownership in waste disposal companies made less sense). Furthermore, the professional independence of planning officers appeared to have been great in the cases I have studied, which helps their reputation as impartial and neutral decision-makers. Third, I found the system of an inquiry, with its (within the system) lack of bias, its comprehensiveness and its opportunity for direct and antagonistic interaction between opponents a strong element. The inquiry is an instrument that is capable of denying an appeal. Fourth, the possibility of a linked inquiry, which introduces a comparative perspective on proposals, seems to have strengthened the decision process in the one case where it was applied. Finally, the strong control from the central government level ensures that parochialism in hazardous waste issues stands no chance.

There are also weaker parts to the UK approach to decision-making. Particularly the fact that the local population has no right to appeal an approved waste facility seems somewhat odd. Furthermore, it seems that much is made of scientific site selection, whereas in reality sites are chosen mainly because of commercial reasons. Third, waste companies are not very advanced in terms of informing the public in a timely manner. Fourth, the possibility of filing two similar planning applications for the same development seems somewhat odd and is used only to pressure the local authority. Fifth, ordinary citizens participating in an inquiry are not well protected in the sense that deadlines tend to be violated by the proponent without little reaction from the planning inspector. Sixth, the almost omnipresent perception was that hazardous waste facilities would be incompatible with the classification term 'industrial use'. It was surprising to many authorities that this was not the case, and that their land use plans therefore offered ample suitable locations for waste facilities. Often this was not the intention of the local authority when designating a certain site as suitable for industrial use. Finally, sometimes no independent check of Environmental Assessment, outside the responsible authority and the statutory consultees, is carried out. Both tend to have limited knowledge in many relevant aspects pertaining to the proposal, and sometimes do not seem to have the capability of responding in time.

In conclusion, I feel that the decision-making process in the UK is in many ways well thought out, although participation by the local public may be enhanced somewhat more. Especially the inquiry adds to the satisfaction of the UK public with the planning system. I would propose that the UK government introduce a form of intervenor funding to assist the main parties in the inquiry (opposition groups) in preparing themselves. Furthermore, measures to ensure the 'objectivity' of environmental assessment might be introduced.

Hazardous waste siting in the Netherlands

The Dutch system of decision-making had several strong points as well. Here too, the responsibility for making proposals is in the hands of the private sector, which therefore also bears the risks associated with project failure. In two out of the three cases these companies chose municipalities that were happy to host them. In the one case where this was not so, the proponent offered compensation in a relatively advanced way. The system of decision-making is relatively open to ordinary citizens, who are invited to participate even in the setting of the Terms of Reference for environmental impact assessments. In addition, the independent expertise of the Commission for the EIA seems to have a positive effect on the quality of assessment documents.

The downsides of the Dutch system are also evident. First, the 'gedoog' phenomenon reared its head in two out of the three cases. This implies that (strictly speaking) illegal activities are allowed. The negative result is uncertainty for those involved. Furthermore, in two out of the three cases, the valid land use plans were so old as to render them practically meaningless for the decision process. This implies that a test of the sites from a planning perspective was not possible. Other comments are mostly related to the fact that in case of opposition it is extremely difficult to get closure. Closure is hard to achieve because the procedures lack a clear focal point, in which the proposal is discussed in its entirety. Furthermore there is little opportunity for interaction between opponent and proponent before the courts. The courts themselves are extremely slow in decision-making, even if they reach relatively simple verdicts, and do not assist the attainment of closure to decision processes by giving 'procedural' verdicts quite often. Access to the courts is so easy that it becomes attractive for 'busybodies' to seek recourse for each and every licence or other decision. This does not serve the decision process, nor the busybodies either.

It seems that the Dutch system of decision-making could perhaps be served by the introduction of an inquiry-like institution. This means a quasi-judicial approach that is experienced as fair and unbiased and that offers more opportunity for interaction between the parties opposing each other. It would be relatively easy to expand the current practice of advice by the Advisor Environmental Appeal to an inquiry-like procedure. Furthermore, the horrendous and artificial divide between the various aspects of the proposal could be made somewhat less rigid; for instance by allowing a discussion of the proposal in its entirety at the inquiry.

Hazardous waste decisions in Canada

Obviously, hazardous waste decision-making using the community-based approach has great advantages in terms of the social acceptance of the proposals, which was lacking in the UK. The innovative idea that the local community obtains a certain level of control over the management of the facility guarantees that social support remains intact. In addition, community-based siting has the potential to use the locally available knowledge and seems to meet the fundamental idea of fairness as it is partly about compensation. The awareness of hazardous waste problems among the population increases and people develop a better understanding of the issues at stake. Finally, conflict can be avoided by the community-based approach, certainly if the process is terminated when signs of opposition develop.

Despite these advantages, community-based siting can only be used for proposals that are extremely controversial. This is because in most other cases it would be difficult to find citizens who are willing to play the active role that is demanded of them in community-based siting. Other conditions limiting its usability are that there must be no private sector actors that seek to locate a facility under the normal legislation, plus that the body implementing community-based siting must be the one that controls siting legislation. A second drawback relates to the cost-effectiveness of the solutions that are selected in the end. The location of sites that emerge out of voluntary siting were not ideal from an operating perspective and safety measures beyond industrial norms are required, which also increase the costs of operation. A third (but correctable and not unique to community-based siting) problem is that if data on waste arisings is not continuously updated, one runs the risk of establishing a facility that is too big from the perspective of demand. Fourth, the emphasis on one community and the delineation of this relevant community sometimes excludes others who have a passionate opinion about the proposals. Whereas such actors can probably respond by using the ordinary forms of participation in licence procedures, one must doubt the likelihood that this has any impact after the whole 'circus' of community-based siting has been completed. Fifth, being an extra-legal process, community-based siting introduces several uncertainties for the communities that participate. To some extent, such uncertainties hamper the success of the approach. Especially the fact that a pre-emptive approach may follow a failed community-based siting process reduced the attractiveness of participation. This was because the identification of suitable sites during voluntary siting could have led to their selection later on, even if communities dropped out. Furthermore, the fact that a community cannot be certain of being selected as the host after volunteering to do so makes participation less attractive (an effect that is somewhat compensated by the fact that some communities will reconsider their refusal if another is volunteering). Finally, the real possibility that the siting process evolves through manipulation of the public would seem dubious from a moral perspective. Such manipulation results if attempts are made to informally sway opponents, if an impression of economic compensation is allowed to continue when incorrect, and if experts start producing one-sided stories about the risks involved. In such situations, one can wonder whether members of a community are deciding in their own best interest (see Bressers (1984) for a discussion on this topic).

Community-based hazardous waste siting poses some dilemmas. Although one would like a certain level of predictability to take away some of the downsides related to uncertainty, the process is probably better off not being formally legislated. It might help to have some documents outlining the approach and stating certain minimum guarantees (one suggestion: guarantee that communities that find a suitable location yet decide not to go ahead, will not be confronted with a top-down approach later). It is my impression that the Manitoba approach was the best practice as it avoided some of the difficulties experienced in the other two processes. The process was 'incremental' for a long time in the sense that waste generation data kept being fed into the process, which implied that communities were not approving a facility that would never be built in the end. Also, the topic of compensation probably needs to be avoided more than the jobs associated with the facility. Furthermore, the surrounding communities need to be made aware of what is going on and need to be asked for their approval. Finally, the idea of writing a contract between province and community, detailing how the facility will be controlled afterwards makes sense.

8.5.3 The practice of environmental decision-making

The difference between hazardous waste decisions and other issues

It is clear that hazardous waste decision-making is not representative for all the decision processes that one can classify under the heading of 'environmental decision-making'. Hazardous waste treatment does not involve by any means the level of complexity and technical uncertainty that other problems possess. Certainly, hazardous waste siting is not a problem on a world scale and can be addressed locally or regionally at most. This makes hazardous waste decision-making relatively easy in comparison with other environmental decision-making processes. On the other hand, there are many 'environmental' decision processes that resemble hazardous waste siting in that they are local, involve relatively limited scientific uncertainty, but do raise a lot of opposition from the local public. The findings of my study may have some relevance for that type of development.

The concept of sustainable development

Sustainable development is the dominant goal in environmental policy. The concept is clearly multifaceted, as it has not just an ecological, but also social and economic dimensions. Welfare of human societies, defined in the broadest possible sense, must be maintained and, if possible, be increased from generation to generation. It would seem that sustainable development, with its several facets requires the involvement of various differing institutions so that all dimensions of the concepts are addressed. Decisions must not only be right from an ecological perspective, but also from economic and social perspectives. My study seems to suggest that managing the attainment of these three goals at the same time is difficult. Especially solutions that make sense from the technical and economic perspective seem to meet with social concerns. If decision procedures then emphasize social concerns, it is more difficult to satisfy the demand of technical and economic rationality. Furthermore, the legal frameworks of the three countries I have studied are much more geared toward the technical and economic aspects of decisions than towards the social aspects. This makes decision processes that are based on social acceptability less predictable and therefore less 'legally rational' for those participating.

There are different possible perspectives on balance that need to be struck between social quality on the one hand and technical and economic quality on the other. Some argue that it is normatively desirable to start emphasizing social quality more. As far as there are negative consequences in terms of consequences for the cost-effectiveness of decisions, these people feel that this must be seen as an internalization process, meaning that previously unrecognized costs are now incorporated. On the other hand, there are those who feel that a loss of cost-effectiveness must not be accepted, because it could for instance hamper economic development. My personal opinion, on the basis of the case studies I have done, is that the argument of cost internalization does not make a lot of sense. I would subscribe to a theory of complicity, meaning that people, as members of a modern and industrialized society and benefiting from the profits that it brings, should sometimes accept developments they personally do not like. In addition, I am greatly concerned about the effect that talk of compensation has on the decision process. It does seem that the possibility of compensation encourages people to accept certain risks and accept technically unsuitable solutions. This is by no means a sustainable solution.

The division of responsibilities between institutions: presupposition

My study has demonstrated that it is possible to structure hazardous waste decision processes in various differing manners from an institutional perspective and that this tends to have quality implications. Especially on the basis of the empirical information on the way several actors behave in this type of decision process, I think I can present some of the stronger points and weaker characteristics of every institution and on the basis of that develop some - rather generic - institutional advice. Before doing that, I wish to express my own personal opinion that the managerial discourse is the only discourse that is suited for this type of decision process. This is mainly because I feel that the communitarian and progressive pluralist alternatives (stressing the community or fairness) will result in an unproductive exchange of arguments. The concept of a harmonious community, free of coercion and full of dialogue, appears an attractive ideal but is highly fictional in practice. Furthermore, the discussion of whether it is fair or not to site a facility in a certain location always leads to the question whether it is fair to then shift the burden on to somebody else. In both cases, productive discussions are only possible if arguments are substantiated with evidence (for instance somebody else is producing all the waste or local health will suffer), which implies that one automatically enters the domain of managerialism. This conclusion, however, pertains especially to the implications a facility will have once a site has been chosen. I personally believe that site selection should be on the basis of explicit discussion in representative institutions beforehand (where fairness and community interests can be addressed), or lacking such plans, by the private sector. The idea that sites can be chosen managerially, that is to say, in a completely rational fashion, is not credible at all.

The role of the private sector

Involvement of the private sector in hazardous waste decision processes seems necessary in order to guarantee that the proposals make sense economically (private market parties are probably better in collecting market information because of their experience), because it will be clear who is the proponent, and because he will pick up the costs if the facility is an economic failure. On the downside it seems that private market parties are not very sophisticated in their dealings with local audiences and sometimes do not even care about the local response. A great number of such private companies has a dubious reputation in terms of their environmental record, which of course raises eyebrows among the local public.

In terms of institutional advice, the requirement that somebody who wishes to operate a waste disposal facility must be a fit and proper person seems a very good idea. This requirement is law in the UK and the Netherlands and will in the long run increase trust. In addition, guidance may be developed that instructs waste disposal firms to liaise with the local public and to add non-financial forms of compensation to the proposals they make.

Elected bodies

The positive side of the involvement of representative institutions in this type of decision processes is that proposals can be considered from the perspective of an entire community. Representative institutions, no matter what their failings, can be held accountable and are elected for a constituency in its entirety. Especially if they develop land use plans that demonstrate a certain political consensus on where waste facilities must go, they can steer the siting debate in the direction that is politically most acceptable. On the other hand, elected representatives are partly responsible for the low level of trust that exists in the waste regulation system. That is because they have at times been sensitive to industrial demands for lenient waste regulations. Also, local councils often fail to actually plan for waste disposal and are somewhat parochial in their outlook when they judge proposals.

The main institutional recommendation would be that elected bodies provide for a balanced (in the sense of not just reflecting industrial interests) and planned approach to hazardous waste (especially in the sense that locations are chosen beforehand). In doing so, representative institutions can play a responsible role in developing answers to hazardous waste problems.

Experts

In a managerial discussion, expertise is very important. There is ample evidence that experts can be unbiased and that they can engage in a productive debate, even if hired by different parties. It is important that experts contribute to the development of norms that describe the best practice so that the discussions on risks and health effects have a certain foundation. Experts often develop a certain consensus on such norms and are the ones with intimate knowledge about such norms and possible developments in them, which makes them central actors. On the downside it does seem that experts often say what they are paid to say and are thus biased in their statements. They are certainly not always neutral and often not very open about the uncertainties and assumptions embedded in their statements. Also, the norms that experts develop are, like every human construct, open to criticism and are often based on cost-considerations for industry.

Biases of experts can be detected best by other experts. Expertise in the waste sector is relatively scarce. Care must be made that expertise is embedded in the process and has no ties to any of the parties in the conflict.

Courts

Decision-making through the courts - if done the Dutch way - has relatively few advantages. Decision-making through inquiries appears to be an effective way of working on expert consensus (if attainable at all) and creating interaction between the parties involved. These inquiries appear to be capable of closing the discussion quite well, and are generally accepted by the parties as the final verdict. The Dutch courts (more than the UK inquiries) offer an opportunity for independent review of the issues before them, which in itself can have an added value. The downside of decision-making through the courts is that judges are not accountable. Furthermore, in the Dutch situation their verdicts are rather procedural in nature, which prohibits (just like the relative lack of interaction opportunities) real closure of the conflict.

My advice would be to change the Dutch system of administrative courts to one that involves inquiries. Certainly, the opportunities for interaction between parties must increase.

Communities

One of the positive conclusions about communities from my research is that they often contain much relevant expertise and local knowledge (the difference is not that clear in practice). This implies that at least certain members can participate in a debate on a managerial footing and can contribute to the quality of the debate. On the downside, the concept of 'community' often seems relatively useless to describe the connection between inhabitants of certain municipalities, certainly in the larger industrial towns of the UK and the Netherlands, as there is little interconnection between these inhabitants. Added to that - most strongly in the Netherlands - there was a tendency to remain relatively inactive during the decision process. In the cases where community-based decision processes were more possible - the Canadian ones - it appeared that there was a certain level of parochialism among the members of the community. In addition, relations between the members of the community were certainly not free from hierarchy, which implied that interactions between members of the community did not always look like 'dialogue'.

The advice on the role of the community would be to tap the expertise within the community as much as possible. Ideally, this should be done by a private proponent or by the local council at the waste planning stage. It is seriously doubtful that hearings are a good method for this, because opponents easily overtake them. The use of workshops in an informal atmosphere may be a better idea.

The central role for the managerial language

Despite the differences between hazardous waste siting and other environmental issues, I feel it is safe to assume that the managerial discourse is widely considered the most suitable language for discussing environmental disputes in all three countries. At least in hazardous waste siting, opposition to proposals reflects not so much a rejection of this language, but a lack of faith in the institutions that are in control. The progressive pluralist and communitarian discourses seem to be used mainly for mobilizing opposition, not so much for actually engaging in the debate. Progressive pluralism, which offers compensation as the way to get community consensus, is easily perceived as an attempt at bribery and is therefore only useful in a low-key fashion and in the form of non-financial compensation. How a community debate should be structured (e.g. which arguments count and which ones not) in a debate set up along communitarian lines remains somewhat too vague for most people. The 'philosophical or ill-informed no' as one adherent of communitarianism called it, holds little attraction for most members of the three societies in question. Add to this discovery that 'ordinary' people have certain unexpected capabilities in the sense that they often possess relevant expertise or the ability to master the technical aspects of decision-making. Thus the conclusion may be that the Canadian approach to decision-making made a lot of sense.

However, the embrace of the Canadian approach to siting and its potentially wider application merits some qualification: it does seem that a similar set-up of decision processes for more types of developments would be quite demanding upon the administration and citizens. It should therefore only be instigated if a certain level of controversy arises within the ordinary system. By 'localizing' a debate and holding it in managerial terms (while avoiding the issue of compensation), one could counter much opposition. The UK cases show something important as well however: by allowing a (non-local) party that is perceived as neutral, by allowing contentions but fair hearings, and by allowing extensive personal interaction between opponents on all aspects of the development, one can apparently also achieve a certain degree of satisfaction with the outcomes. However, this satisfaction would seem to depend on various socially-shaped conditions and perceptions, for instance on what constitutes 'fair play' in a decision procedure. What constitutes as fair play in the UK may not at all count as such in the Netherlands or in Canada. This then implies that approaches to hazardous waste siting must be tailored to the country or region in question. This insight, I feel, is often missed in the North American literature on the problem of siting, where the failure (in the sense of gridlock) of the traditional method of siting has led to the advice to introduce 'new' methods (voluntary siting or compensation). My findings suggest that in some situations 'traditional' methods of siting can actually work. Secondly, that new siting approaches come at a certain cost in terms of decision quality and must therefore only be used in exceptional situations where trust in the ordinary institutional arrangements (broadly similar in all three countries) is limited. Finally, the Canadian experience does suggest that, even from the narrow perspective of gaining community acceptance for proposals, voluntary siting often fails.

And what of the limits to the managerial language? I think some of these were brought out by the case studies and the discussion in chapter 3 rather well. There are strong indications that science cannot be the new 'Archimedean point' in politics (see chapter 2). To be called an Archimedean point, science would have to remain in one position and be free from conflicts about where the point is located. None of these conditions are met. Take the Dutch approach towards dredging sludge (Kaliwaal case). If one compares the 'scientific' underpinnings to the practice of sludge disposal as they were twenty years ago with the underpinnings as they currently are, one can see great differences. What was irresponsible then, seems responsible now. Therefore, science changes its position. Also, conflicts about science are omnipresent. Take the clearly different perception of co-disposal in the UK on the one hand and Canada and the Netherlands on the other. Whereas 'the' scientific community in the UK looked favorably upon the practice, this was certainly not the case for their counterparts in the Netherlands and Canada. This difference had all sorts of systemic consequences, for instance the need for defining hazardous wastes in the Netherlands and Canada, a need that did not exist in the UK. International cooperation between countries, such as in the European Union and the

OECD, is bringing about the convergence of practices here. It is ironic that exactly at the same time as the UK is introducing a similar terminology and approach to hazardous waste as the other two countries, scientific studies in the Netherlands are beginning to question the need for separating hazardous wastes from other waste streams.

The exact process that causes international convergence or even national scientific consensus has not been explicitly studied here. Yet I would feel safe in saying that for citizens and action groups, the practices in other countries are increasingly easy to retrieve. Concretely speaking, this enables action groups to attack a co-disposal facility in the UK, by indicating that the Netherlands decided to stop 'co-disposing' of their wastes. It would seem interesting to find out how action groups or ordinary citizens exactly go about picking an option they like best. It does seem that certain intuitive feelings play a great role in explaining these preferences. But apparently, this is not much different in the scientific community. It so happens that officials in the Netherlands laugh about the ridiculous idea of co-disposal, whereas their UK counterparts consider waste separation equally ridiculous. And if a Best Available Technology for treating emissions from a waste treatment plant can not be devised (North Refinery case), the solution is to leave the plant open. The only solution to the problem of lack of stability and disagreements in science would seem to be openness about these issues. I have seen very clear and frank admittance of uncertainties in the Kaliwaal case and in the Canadian low-level radioactive waste process. Such admittance is somewhat tricky in the sense that the great discursive power of science is abandoned, in return for uncertainty. However, the increasing scientific literacy of the population in combination with the fact that science can really not pretend to be able to predict every consequence of a project for an extensive period of time (such a pretence is false), necessitate it.

References

Alberta Environment, *Environmental impact assessment guidelines*, Edmonton, 1985.
Alberta Environment, *Review of Alberta's environmental impact assessment system*, Edmonton, 1980.
Alberta Environment, *Environmental impact assessment guidelines*, Edmonton, 1977.
Alberta Municipal Affairs, *A guide to land use planning in Alberta*, Edmonton, 1987.
Alder, J. and D. Wilkinson, *Environmental law and ethics*, London, 1999.
Allen, R., *Waste not, want not. The production and dumping of toxic waste*, London,1992.
Andrews, R.N.L., Hazardous waste facility siting. State approaches, in: C.E. Davis, and J.P. Lester (eds.), *Dimensions of hazardous waste politics and policy*, New York, 1988, pp. 118-128.
Arblaster, A., *Democracy*, Buckingham, 1994 (2nd ed).
Armour, A.M. *Modernizing democratic decision-making processes. From conflict to cooperation in facility siting*, paper for the conference 'The environment in the 21st century: Environment, Long-term governance and Democracy', Fontevraud, 1996.
Armour, A.M., *Socially responsive facility siting*, Waterloo, 1990.
Arnstein, S.R., A ladder of citizen participation, in: *The Journal of American Institute of Planners*, 1969 (vol. 35), pp. 216-224.
Ball, S. and S. Bell, *Environmental law*, London, 1995.
Baren, N.G.E. van, Planhiërarchische oplossingen: een bron voor maatschappelijk verzet, Amsterdam, 2001.
Barker, A. and M. Couper, The art of quasi-judicial administration. The planning appeal and inquiry systems in England, in: *Urban Law and Policy*, 1984 (vol. 6), pp. 363-476.
Barringer, K., *An assessment of public participation under the Manitoba Environment Act*, Winnipeg, 1990.
Beck, U., From industrial society to the risk society: question of survival, social structure and ecological enlightenment, in: *Theory, Culture and Society*, 1992 (vol. 9), pp. 97-123.
Benidickson, J., *Environmental law*, Concord, 1997.
Blowers, A., *The limits of power. The politics of local planning policy*, Oxford, 1980.
Boardman, R. (ed.), *Canadian environmental policy. Ecosystems, politics, and process*, Toronto, 1992.
Bosma, K. and C. Wagenaar, *Een geruisloze doorbraak. De geschiedenis van de architectuur en stedebouw tijdens de bezetting en de wederopbouw van Nederland*, Rotterdam, 1995.
Bressers, J.Th.A, Analyse en evaulatie van de beleidsinhoud, in: Blommestein, H.J., J.Th.A. Bressers, and A. Hoogerwerf, *Handboek beleidsevaluatie. Een multidisciplinaire benadering*, Alphen aan den Rijn, 1984, pp. 53-75.
Brouwer, H. and A. Tompot, Milieu-effectrapportage en participatie, in; *Beleidswetenschap*, 1995, no. 4, pp. 301-322.
Brubaker, E., *Property rights in the defence of nature*, Toronto, 1995.
Burnheim, J., Power-trading and the environment, in: *Environmental Politics*, 1995 (vol. 4), no. 4, pp. 49-65.
Burns, T.R. and R. Ueberhorst, *Creative democracy. Systematic conflict resolution and policy-making in a world of high science and technology*, New York, 1988.
Castle, G., *Hazardous waste facility siting in Manitoba. Case study of a success*, Vancouver, 1992 (unpublished draft article).

CEAA (Canadian Environmental Assessment Agency), *Canadian environmental assessment process. The citizens guide*, Ottawa, 1994.
CEC (Clean Environment Commission), *Report on public hearings. Phase I - Stage II Manitoba hazardous waste management program.* November 12, 1986- January 20, 1987 (Volume I), Winnipeg, 1987.
CEC (Clean Environment Commission), *Report on public hearings.* December 5, 1983- February 2, 1984, Winnipeg, 1984.
Champion, J.B., *Yes in my backyard. The successful siting of a hazardous waste treatment centre near Swan Hills*, Alberta, place unknown, 1989.
Chem-Security, *Alberta special waste treatment centre environmental protection plan*, Calgary, 1987.
Coase, R.H., The problem of social cost, in: *Journal of Law Economics*, 1960 (vol. 30), pp.1-44.
Connoly, J. and G. Smith, *Politics and the environment. From theory to practice*, London, 1999.
Cooke, R.J., *The development of a hazardous waste management system in Manitoba*, Winnipeg, 1988.
Cramer,. J., *De groene golf. Geschiedenis en toekomst van de milieubeweging*, Utrecht, 1989.
Cullingworth, J.B., *Urban and regional planning in Canada*, New Brunswick (NJ), 1987.
Currier, B., Northern exposure. Giving radioactive wastes the cold shoulder; an activist's notes, in: *This Magazine*, March/April, 1993, pp. 35-40.
Dahl, R.A., *Modern political analysis*, Englewood Cliffs, 1991 (5th edition).
Dahl, R.A., *Democracy and its critics*, New Haven, 1989.
Davoudi, S., Planning for waste management. Changing discourses and institutional relationships, in: *Progress in Planning*, 2000 (vol. 53), pp. 165-216.
Department of the Environment, *Making waste work. A strategy for sustainable waste management in England and Wales*, London, 1995 (a) (Cm 3040).
Department of the Environment, *The effectiveness of planning policy guidance notes.* A study undertaken by Land Use Consultants for the Department of the Environment, London, 1995 (b).
Department of the Environment, *Waste management paper no. 4. Licensing of waste management facilities.* Guidance on the drafting, supervision and surrender of waste management licenses, London 1994 (3rd edition).
Department of the Environment, *Planning, pollution and waste management* (prepared by Environmental Resources Limited and the Oxford Polytechnic School of Planning), London, 1992.
Department of the Environment, *Speeding planning appeals.* A review of the handling of transferred written representation planning appeals. Report of an efficiency scrutiny, London, 1985.
Diesing, P., *Reason in society. Five types of decisions and their social conditions*, Urbana, 1962.
Dietz, Th., Democracy and science, in: O. Renn, Th. Webler, and P. Wiedemann, (eds.), *Fairness and competence in citizen participation. Evaluating models for environmental discourse*, Boston, pp. xvii-xix, 1995.
Disco, C., Class collision or class collusion. Dutch engineers, capitalist enterprise, and the state 1885-1920, in: *Sociologisch Tijdschrift*, 1981, no. 8, pp 714-755.
Doern, G., *The Atomic Energy Control Board. An evaluation of regulatory and administrative processes and procedures*, Ottawa, 1976.
Doern, G. and T. Conway, *The greening of Canada. Federal institutions and decisions*, Toronto, 1994.

Doyle, D. and B. Sadler, *Environmental assessment in Canada. Frameworks, procedures and attributes of effectiveness. A report in support of the international study of the effectiveness of environmental assessment*, Ottawa, 1996.

Driessen, P.P.J., A.P.C. van der Heijden and P. Leroy, *Locatiekeuze van afvalverwerkingsinrichtingen: een beleidsevaluatief onderzoek*, Nijmegen, 1990.

Dryzek, J.S., *The politics of the earth. Environmental discourses*, Oxford, 1997.

Dryzek, J.S., *Ecological rationality. Environment and political economy*, Oxford, 1987.

Dudley, G. and J.J. Richardson, Arenas without rules and the policy change process. Outsider groups and British road policy, in: *Political Studies*, 1998 (vol. XLVI), pp. 727-747.

ECA (Environment Council of Alberta), *Hazardous waste management in Alberta. Report and recommendations*, Edmonton, 1980.

Elder, P.S. and J.M. Keeping, The charter of rights and community planning, in: *Plan Canada*, 1986 (vol. 26), no. 8, pp. 206-214.

Evers, F.W.R., Beleidsvisie op de toekomst van milieu-effectrapportage, in: *Milieu*, 1986, no. 5, pp. 176-177.

Etzioni, A., *The new golden rule. Community and morality in a democratic society*, New York, 1996.

Etzioni, A., *The spirit of community. Rights, responsibilities and the communitarian agenda*, New York, 1993.

FEARO (Federal Environmental Assessment Review Office), *The federal environmental assessment and review process*, Ottawa, 1987.

FEARO (Federal Environmental Assessment Review Office), *Environmental assessment panels. Procedures and rules for public meetings*, Ottawa, 1985.

Fischer, F., *Citizens, experts and the environment. The politics of local knowledge*, Durham, 2000.

Fischer, F., Citizen participation and the democratization of policy expertise. From theoretical inquiry to practical cases, in: *Policy Sciences*, 1993 (vol. 26), pp. 165-187.

Ganz, G., *Understanding public law*, London, 1987.

Geerts, R., *Slibberging in de Kaliwaal. Een onderzoek naar de risicobeoordeling door omwonenden*, Nijmegen, 1999.

Gerrard, S. *Hazard, risk and waste management. The constant struggle for tolerability*, Norwich, 1994.

Gier, A.A.J. de, J. Robbe, and C.W. Bakkes, *De actio-popularis in het ruimtelijke ordenings- en het milieurecht*, Utrecht, 1999.

Gray, T.S. (ed.), *UK environmental policy in the 1990s*, Houndsmills, 1995.

Grant, J., From 'human values' to 'human resources'. Planners' perceptions of public role and public interest, in: *Plan Canada*, 1989 (vol. 29), no. 6, pp. 11-18.

Greenpeace, *Burning questions. Incineration in the UK*, London, 1991.

Grove-White, R., Land use law and the environment, in: Churchill, R., J. Gibson and L.M. Warren (eds.), *Law, policy and the environment*, London, 1991, pp. 32-47.

Gunsteren, H.R. van, *The quest for control. A critique of the rational-central-rule approach in public affairs*, London, 1976.

Haan, P. de, T.G. Drupsteen, and R. Fernhout, *Bestuursrecht in de sociale rechtsstaat. Deel 2: bestuurshandelingen en waarborgen*, Deventer, 1986 (3rd ed.).

Hajer, M.A., *The politics of environmental discourse. Ecological modernization and the policy process*, Oxford, 1995.

Haigh, N. (ed.), *Manual of environmental policy. The EC and Britain*, London, 1992 (with regular updates).

Haigh, N. and C. Lanigan, Impact of the European Union on UK environmental policy making, in: Gray, T.S. (ed.), *UK environmental policy in the 1990s*, Houndsmills, 1995, pp. 18-37.

Hall, P.A., Policy paradigms, social learning and the state. The case of economic policy-making in Britain, in: *Comparative Politics*, 1993, pp. 275-296.

Hardin, G., The tragedy of the commons, in: *Science*, 1968 (vol. 162), pp. 1243-1248.

Harrison, K. and G. Hoberg, *Risk, science and politics. Regulating toxic substances in Canada and the United States*, Montreal, 1994.

Hazardous Waste Management Committee, *Hazardous waste management in Alberta*, Edmonton, 1981.

Healey, P., Sites, jobs and portfolios. Economic development discourses in the planning system, in: *Urban Studies*, 1999 (vol. 36), pp. 27-42.

Heijden, A.C.P. van der, *De locatie gekozen. Milieuaspecten bij de locatiekeuze van afvalstortplaatsen*, Nijmegen, 1998.

Heilman, J.G., Some thoughts on a rule based conceptual framework for institutional analysis in water policy research, in: Bressers, J.Th.A., and L.J. O'Toole (eds.), *International comparative policy research: preparing a four country study on water quality management*, Enschede, 1992, pp. 59-83.

Heldeweg, M., Deskundigen-advisering in het milieurecht, in: *Milieu en Recht*, 1995, no. 1, pp. 9-14.

Hessing, M. and M. Howlett, *Canadian resource and environmental policy. Political economy and public policy*, Vancouver, 1997.

Hisschemöller, M., *De democratie van problemen. De relatie tussen de inhoud van beleidsproblemen en methoden van politieke besluitvorming*, Amsterdam, 1993.

Hodge, G., *Planning Canadian communities. An introduction to the principles, practice, and participants*, Scarborough, 1998 (3rd edition).

Hoeven, M. van der, *De drie dimensies van het bestuursrecht. Ontstaan en vorming van het Nederlandse bestuursrecht*, Alphen aan den Rijn, 1989.

Hough, B., Standing in planning permission appeals, in: *Journal of Planning and Environmental Law*, 1992, pp. 319-329.

House of Commons Environment Committee, *The environmental impact of cement manufacture*. Report together with the proceedings of the Committee. Third report Session 1996-1997, London, 1997 (124-I).

House of Commons Environment Committee, *Toxic Waste. Volume I.* Report with appendices together with the proceedings of the Committee relating to the report, Second Report Session 1988-1989, London, 1989 (a).

House of Commons Environment Committee, *Toxic Waste. Volume III.* Appendices, Second Report Session 1988-1989, London, 1989 (b).

House of Lords Select Committee on Science and Technology, *Hazardous waste disposal. Report with evidence*, Fourth Report Session 1988-1989, London, 1989 (HL Paper 40).

Hudson, A., *Good riddance. Waste management law in Alberta*, Edmonton, 1992.

Huitema, D., Hazardous decisions. The siting of hazardous waste disposal facilities in Canada and the United States, in: Coenen, F.H.J.M., D. Huitema and L.J. O'Toole (eds.), *Participation and the quality of environmental decision-making*, Dordrecht, 1998, pp. 223-244.

Hunold, C. and I.M. Young, Justice, democracy and hazardous siting, in: *Political Studies* 1998 (vol. 46), pp. 82-95.

Hupe, P.L., Intenties en interacties; over postmoderne en andere instrumenten van overheidssturing, in: Hufen J.A.M. and A.B. Ringeling (eds.), *Beleidsnetwerken; overheids-, semi-overheids-, en particuliere organisaties in wisselwerking*, Den Haag, 1990.

Inhaber, H., *Slaying the NIMBY dragon*, New Brunswick, 1998.

Jackson, J., P. Weller and the Waterloo Public Interest Research Group, Chemical nature, reprinted in: Greenbaum, A., A. Wellington and E. Baar (eds.), *Social conflict and environmental law. Ethics, economics and equity, volume 1*, Toronto, 1994: pp. 141-151 (original text from 1982).

Johnson, N., The judicial dimension in British politics, in: *West European Politics*, 1998 (vol. 21), no. 1, pp. 148-166.

Johnston, B. and M. Gismondi, A forestry boom in Alberta?, in: *Probe Post*, Spring, 1989, pp. 16-19.

Jordan, A., Integrated pollution control and the evolving style and structure of environmental regulation in the UK, in: *Environmental Politics*, 1993 (vol. 2), pp. 405-427.

Jordan, A., The impact on UK legislation, in: Lowe, P. and St. Ward (eds.), *British environmental policy and Europe. Politics and policy in transition*, London, 1998, pp. 171-194.

Kiernan, M.J., Urban planning in Canada. A synopsis and some future directions, in: *Plan Canada*, 1990 (vol. 30), no. 1, pp. 11-22.

Kimber, R., J.J. Richardson and S.K. Brookes, The deposit of poisonous waste act 1972. A case of government by reaction?, in: *Parliamentary Affairs*, 1974 (vol. 27)., pp. 198-219.

Kiser, L.L. and E. Ostrom, The three worlds of action. A metatheoretical synthesis of institutional approaches, in: Ostrom, E. (ed.), *Strategies of political inquiry*, Beverly Hill, 1982, pp.179-222.

Kitson, T. and R. Harris, A burning issue? Planning controls, pollution controls and waste incineration, in: *Journal of Planning and Environmental Law*, 1994 (vol. 1), pp. 3-7.

Korsten, A.F.A., *Het spraakmakende bestuur. Een studie naar de effecten van participatie in relatie tot demokratiemodellen en sociale ongelijkheid*, Den Haag, 1979.

Krawetz, N., *Hazardous waste management. A review of special concerns and aspects of public involvement*, Edmonton, 1979.

Krawetz and MacDonald Research Management Consultants and Maureen Payne and Associates, *Hazardous waste management in Alberta. A chronology*, Edmonton, 1983.

Kruhlak, R.M., Who serves the public interest?, in: *News Brief Environmental Law Centre*, 1995, vol. 10, no. 4, pp 1-2.

Kucera, E., D. Labossiere, R. Pelser and C. Enns, *Compliance guide to Manitoba's hazardous waste legislation*, Winnipeg, 1993 (2nd edition).

Kuhn, R., Public participation in the hearings on the Canadian nuclear fuel waste disposal concept, in: Sinclair, A.J. (ed.), *Canadian environmental assessment in transition*, Waterloo, 1997, pp. 19-49.

Kwasniak, A.J., Municipal and land-use planning, in: Hughes, E.L., A.R. Lucas and W.A. Tilleman (eds.), *Environmental law and policy*, Toronto, 1998 (2nd edition), pp. 487-512.

Lafferty, W.M. and J. Meadowcroft, *Democracy and the environment. Problems and prospects*, Cheltenham, 1996.

Lafferty, W.M., *Democracy in an ecological state: problems and prospects*, paper prepared for the conference on 'The ecological state', Seville, 28 November - 1 December, 1996.

Lash, S., B. Szerszynski and B. Wynne, Introduction: Ecology, realism and the social sciences, in: Lash, S., B. Szerszynski and B. Wynne (eds.), *Risk, environment and modernity. Towards a new ecology*, London, 1996, pp. 1-26.

Latonas, G.P., *Siting hazardous waste management facilities in Alberta and Manitoba. The voluntary approach*, Pickering, 1996 (paper to the Nuclear fuel waste environmental assessment panel).

Latonas, G.P., *Site selection process for a hazardous waste management system*, 1988 (unpublished).

Latonas, G.P. and E. Kucera, *Proposed site selection criteria for the development of a hazardous waste management system*, Winnipeg, 1988.
Laux, F.A., *Planning law and practice in Alberta*, Toronto, 1990 (1st edition) and 1996 (2nd edition).
Leroy, P. De ontwikkeling van het milieubeleid en de milieubeleidstheorie, in: P. Glasbergen, *Milieubeleid. Een beleidswetenschappelijke inleiding*, Den Haag, 1994, pp. 35-58.
Leroy, P. , *Herrie om de heimat. Milieuproblemen, ruimtelijke organisatie en milieubeleid*, Antwerpen, 1983.
Lindblom, C. E., The market as prison, in: *Journal of Politics*, 1982 (vol. 44), pp. 324-336.
Lindblom, C.E. and D.K. Cohen, *Usable knowledge. Social science and social problem solving*, New Haven, 1979.
Lintsen, H., *Ingenieur van beroep. Historie, praktijk, macht en opvattingen van ingenieurs in Nederland*, Den Haag, 1985.
Lowe, P. and St. Ward (eds.), *British environmental policy and Europe. Politics and policy in transition*, London, 1998.
Lowe, P. and St. Ward, Lessons and prospects. The prospects for the UK environment in Europe, in: Lowe, P. and St. Ward (eds.), *British environmental policy and Europe. Politics and policy in transition*, London, 1998, pp. 285-299.
Lowi, T., *The end of liberalism*, New York, 1979.
MacDonald, D., *The politics of pollution*, Toronto, 1991.
MacGregor-Dawson, R., W.F. Dawson and N. Ward, *Democratic government in Canada*, Toronto, 1997 (5th edition).
MacLaren, V.W., Waste management. Moving beyond the crisis, in: Mitchell, B. (ed.), *Resource and environmental management in Canada. Addressing conflict and uncertainty*, Toronto, 1995, pp. 29-54.
MacWilliam, A.G., Highlights of Alberta environmental legislation, in: The Canadian Institute (ed.), *The changing face of environmental law and regulation in Alberta*, Toronto, 1996, no page numbers.
Manitoba Eco-Network, *A snapshot of public participation in the site selection process for the Manitoba Hazardous Waste Management Corporation*, Winnipeg, 1989.
Mathews, F., Introduction. Ecology and democracy, in: *Environmental Politics*, 1995 (a), vol. 4, no. 4, pp. 1-12.
Mathews, F., Community and the ecological self, in: *Environmental Politics*, 1995 (b), vol. 4, no. 4, pp. 66-100.
Mazmanian, D. and D. Morell, *Beyond superfailure. America's toxics policy for the 1990s*, Boulder, 1992.
McAuslan, P., *The ideologies of planning law*, Oxford, 1980.
McGrew, A.G. and M. J. Wilson (eds.), *Decision making. Approaches and analysis*, Manchester, 1982.
Michiels, F.M.C.A., *Hoofdzaken van het bestuursrecht*, Deventer, 1999.
Morell, D. and Magorian, C., *Siting hazardous waste facilities. Local opposition and the myth of pre-emption*, Cambridge, 1982.
Moore, V., *A practical approach to planning law*, London, 1997 (6th edition).
Munton, D., *Hazardous waste siting and democratic choice*, Washington D.C., 1996.
Nemetz, P.N., W.T. Stansbury and F. Thompson, Social regulation in Canada. An overview and comparison with the American model, in: *Policy Studies Journal*, 1986 (vol. 14), pp. 580-603.
Neufeld, R.A., The environmental assessment process in Alberta, in: The Canadian Institute (ed.), The changing face of environmental law and regulation in Alberta, Toronto, 1996, no page numbers.

Newton, K., *Second city politics. Democratic process and decision making in Birmingham*, London, 1976.

Nikiforuk, A., Ralph Klein's toxic folly, in: *Canadian Business*, February 1996, pp. 66-73.

Noor, L., *Een omstreden combinatie. Verwikkelingen rond de combinatie van specieberging en natuurontwikkeling in het project 'Waaier van Geulen'*, Amsterdam, 1999.

Northey, R. and W.A. Tilleman, Environmental Assessment, in: Hughes, E.L., A.R. Lucas and W.A. Tilleman (eds.), *Environmental law and policy*, Toronto, 1998 (2nd edition), pp. 189-234.

Northon, P., The judiciary, in: Jones, B., A. Gray, D. Kavanagh, M. Moran, P. Norton and A. Seldon (eds.), *Politics UK*, London, 1998, pp. 407-425.

NRCB (Natural Resources Conservation Board), *Guide to the Natural Resources Conservation Board process*, Edmonton, 1996 (a).

NRCB (Natural Resources Conservation Board), *The pre-hearing conference. Get involved - here's how*, Edmonton, 1996 (b).

NRCB (Natural Resources Conservation Board), *Decision report application #9101 - Chem-Security (Alberta) Ltd.. Incinerator expansion of the Swan Hills special waste treatment centre*, Edmonton, 1992.

OECD, *Environmental performance review of Canada*, Paris, 1995.

Ostrom, E., *Governing the commons. The evolution of institutions for collective action*, Cambridge, 1990.

Ostrom, E., An agenda for the study of institutions, in: *Public Choice* 48 (1986), pp. 3-25

Ostrom, E., L. Schroeder and S. Wynne, *Institutional incentives and sustainable development. Infrastructure policies in perspective*, Boulder, 1993.

Parenteau, R., *Public participation in environmental decision-making*, Montreal, 1988.

Petts, J., Public participation and environmental impact assessment, in: Petts, J. (ed.), *Handbook of environmental impact assessment*, vol. 1, London, 1999, pp. 145-177.

Petts, J., Effective waste management. Understanding and dealing with public concerns, in: *Waste Management and Research*, 1994 (vol. 12), pp. 207-222.

Petts, J., Incineration risk perceptions and public concern. Experience in the U.K. improving risk communication, in: *Waste Management and Research*, 1992 (vol. 10), pp. 169-182.

Petts, J. and G. Eduljee, *Environmental impact assessment for waste treatment and disposal facilities*, Chichester, 1994.

Pirsig, R.M., *Lila. An inquiry into morals*, New York, 1991.

Pollitt, C., *Managerialism and the public services. Cuts or cultural change in the 1990s?*, Oxford, 1993 (2nd ed.).

Portney, K.E., *Siting hazardous waste treatment facilities. The NIMBY syndrome*, Westport, 1991.

Press, D., *Democratic dilemmas in the age of ecology. Trees and toxics in the American West*, Durham, 1994.

Querido, A., *Een eeuw staatstoezicht op de volksgezondheid*, Den Haag, 1965.

Rabe, B.G., The politics of sustainable development. Impediments to pollution prevention and policy integration in Canada, in: *Canadian Public Administration*, vol. 40 (1998), no. 3 (Fall), pp. 415-435.

Rabe, B.G., *Beyond NIMBY. Hazardous waste facility siting in Canada and the United States*, Washington D.C., 1994.

Rawls, J., *A theory of justice*, Cambridge, 1971.

Renn, O., Th. Webler and P. Wiedemann, *Fairness and competence in citizen participation. Evaluating models for environmental discourse*, Boston, 1995.

Richards, A., *Implementing a voluntary, co-managed process to site a facility for Canada's nuclear fuel waste*, Winnipeg, 1997 (Presentation to the Canadian Environmental

Assessment Agency panel review of the concept for disposal of Canada's nuclear fuel waste).
Richards, A., *The riskiest step in a voluntary siting process. The initial community expression of interest*, Winnipeg, 1996 (paper to the International conference on deep geological disposal of radioactive wastes).
Richards, A., *Contrasting public participation and co-management experiences in hazardous waste management facility siting in rural and urban communities*, paper to the Twelfth annual meeting of the International Association for Impact management, Washington D.C., 1992.
Richards, S.T. and L. Hendrickson, *Guide to federal and provincial pollution control legislation in Manitoba*, Winnipeg, 1986.
Romanelli, T.A. and C. Marchand, *The delegation of planning responsibilities in Canada*, Toronto, 1991.
Rose, C., *The dirty man of Europe. The great British pollution scandal*, London, 1990.
Rowson, J., The appraisal of environmental impacts assessment processes. Towards a situated approach, in: Sinclair, A.J. (ed.), *Canadian environmental assessment in transition*, Waterloo, 1997, pp. 323-359.
Royal Commission on Environmental Pollution, *Seventeenth Report. Incineration of waste, London*, 1993 (Cm 2181).
Royal Commission on Environmental Pollution, *Managing waste. The duty of care*, London, 1985 (eleventh report of the RCEP, Cmnd. 9675).
Ruijter, P. de, *Voor volkshuisvesting en stedebouw. Over woninghervormers en de beweging voor een goede stedebouw 1890-1920*, Amsterdam, 1986.
Rydin, Y., *Urban and environmental planning in the UK*, Houndsmils, 1998.
Sabatier, P.A., Book review of Governing the commons: the evolution of institutions for collective action by E. Ostrom, in: *Public Administration Review*, vol. 86. no.1 (March 1992), pp. 248-249.
Schrecker, T., Of invisible beasts and the public interest. Environmental cases and the judicial system, in: Boardman, R. (ed.), *Canadian environmental policy. Ecosystems, politics, and process*, Toronto, 1992, pp. 83-105.
Sherbaniuk, R., *History project. Alberta Special Waste Management Corporation*, Edmonton, 1992.
Sherbaniuk, R., The price of protection, in: *Alberta Views*, summer 1998 (no. 3), pp. 26-33.
Simon, H.A., Why public administration?, in: *Journal of Public Administration research and Theory*, 1998 (vol. 8), no. 1, pp. 1-11.
Sims, G.H.J., *A history of the Atomic Energy Control Board*, Ottawa, 1980.
Siting Process Task Force, *Opting for co-operation. Report of the siting process task force on low-level radioactive waste disposal*, Ottawa, 1987.
Smith, A., *Integrated pollution control. Change and continuity in the UK industrial pollution policy network*, Aldershot, 1997.
Smith, P.J., Alberta. A province just like any other?, in: Brownsey, K. and M. Howlett (eds.), *The provincial state. Politics in Canada's provinces and territories*, pp. 243-264.
Smits, J., *Milieubeleid gestuurd? Een onderzoek naar de invloed van het BUGM op het gemeentelijk milieubeleid in Noord-Brabant*, Nijmegen, 1995.
STF (Siting Task Force), *The Deep River Initial Assessment Report*, Ottawa, 1995.
Study Group on EIA Hearing Procedures, *Public review. Neither judicial, nor political, but an essential forum for the future of the environment*, Hull, 1988.
Sunkin, M., Judicialization of politics in the United Kingdom, in: *International Political Science Review*, 1994 (vol. 15), no. 2, pp. 125-133.

Taft, K., *Shredding the public interest. Ralph Klein and 25 years of one-party government*, Edmonton, 1996.

Tang, S.Y., Institutional arrangements and the management of common pool resources, in: *Public Administration Review*, 1991 (vol. 51), no. 1, pp. 42-51.

Tewdwr-Jones, M., Policy implications of the 'plan-led' planning system, in: *Journal of Planning and Environmental Law*, 1994 (vol. 2), pp. 584-593.

Thompson, J., Towards a green world order: environment and world politics, in: *Environmental Politics*, 1995 (vol. 4), no. 4, pp. 31-48.

Thornley, A., *Urban planning under Thatcherism. The challenge of the market*, London, 1993 (2nd edition).

Tilleman, W.A., Public participation in the environmental impact assessment process. A comparative study of impact assessment in Canada, the United States and the European Community, in: *Columbia Journal of Transnational Law*, 1995 (vol. 33), pp. 337-439.

Tingley, D., *Public consultation and environmental law reform. Learning as we go*, Edmonton, 1994.

Tollefson, C., Public participation and judicial review, in: Hughes, E., A.R. Lucas and W.A. Tilleman (eds.), *Environmental law and policy*, Toronto, 1998 (2nd edition), pp. 235-276.

UK Royal Commission on Environmental Pollution, *Seventeenth Report. Incineration of waste* (Cm2181), London, 1993.

Valk, A.J.J. van der, *Amsterdam in aanleg. Planvorming en dagelijks handelen 1850-1900*, Amsterdam, 1989.

Vogel, D., *National styles of regulation. Environmental policy in Great Britain and the United States*, Ithaca, 1986.

Ward, H., State, association and community in a sustainable, democratic polity. Towards a green associationalism, in: Coenen, F.H.J.M., D. Huitema and L.J. O'Toole (eds.), *Participation and the quality of environmental decision-making*, Dordrecht, 1998, pp. 27-45.

Ward, S., Thinking global, acting local? British local authorities and their environmental plans, in: *Environmental Politics*, 1993 (vol. 2), no. 3, pp. 453-478.

Weale, A., United Kingdom, in: Jänicke, M. and H. Weidner (eds.), *National environmental policies. A comparative study of capacity building*, Berlin, 1997, pp. 89-108.

Weale, A., *The new politics of pollution*, Manchester, 1992.

Weale, A., *A tale of two strategies. Comparing Dutch and British approaches to environmental policy*, Norwich, 1991 (CSERGE Discussion Paper 8).

Weatherhead, P., Planning issues and clinical waste incineration, in: *Wastes Management*, 1994 (October), pp. 24-25.

Webler, Th., 'Right' discourse in citizen participation. An evaluative yardstick, in: Renn, O., Th. Webler and P. Wiedemann, *Fairness and competence in citizen participation*, Boston, 1995, pp. 35-86.

Webler, Th. and O. Renn, A brief primer on participation. Philosophy and practice, in: Renn, O., Th. Webler and P. Wiedemann, *Fairness and competence in citizen participation*, Boston, 1995, pp. 17-34.

Wetenschappelijke Raad voor het Regeringsbeleid, *Besluiten over grote projecten*, Den Haag, 1994.

Williams, B.A. and A.R. Matheny, *Democracy, dialogue, and environmental disputes. The contested languages of social regulation*, New Haven, 1995.

Wilson, J.Q., *The politics of regulation*, New York, 1978.

Wood, D.H., The Planning Act. A sourcebook, Toronto, 1996.

Wynne, B., *Risk management and hazardous waste. Implementation and the dialectics of credibility*, Berlin, 1987.

Wynne, B. and D. Hortensius, The rationalities of problem definition. The Netherlands and hazardous waste management, in: Wynne, B., *Risk management and hazardous waste. Implementation and the dialectics of credibility*, Berlin, 1987, pp. 84-113.

Yee, E., E. Kucera, and J.J. Keleher, *Hazardous waste management in Manitoba*, Winnipeg, 1985.

Zundert, C. van, Beleid betaald. Aansprakelijkheid en schadevergoeding, in: Heldeweg, M.A. (ed.), *Bestuursrecht met beleid*, Groningen, 2000, pp. 333-351.

Zwerver, S., *Schets van het bestemmingsplanrecht. Zwerftochten door een ruimtewet*, Zwolle, 1975.

APPENDIX 1: LIST OF INTERVIEWEES

In the UK - interviews for chapters 3 and 4

Interviewee	Affiliation	Place and date of interview
Mr. A. Anjum	Environment Agency	London, 29 April 1997.
Mr. M. Bates	Suffolk County, Waste Regulation Branch	Norwich, 17 March 1997.
Mr. A. Barker	University of Essex, Department of Government	Colchester, 23 April 1997.
Mr. K. Billington	Leader Rotherham Metropolitan Borough Council	Rotherham, 19 January 2000.
Mr. A. Blowers	Open University, Department of Geography	Milton Keynes, 24 March 1997.
Mr. J. Brigham	Suffolk County, Waste Regulation Branch	Norwich, 17 March 1997.
Mrs. C. Brimblecombe	Suffolk County, Waste Regulation Branch	Norwich, 17 March 1997.
Mr. P. Brimblecombe	University of East Anglia	Norwich, 17 March 1997.
Mr. T. Butterfield	Cranford Environmental Advisors, formerly with Browning Ferris Environmental Services Ltd.	Coventry, 5 January 2000.
Mr. K. Critchley	Councilor, Newport Borough Council	Newport, 12 January 2000.
Mrs. S. Critchley	Newport Against Hazardous Waste Plant	Newport, 12 January 2000.
Mrs. P. Elliot	Friends of the Earth Essex	Phone, 23 April 1997.
Mrs. P. Foster	Secretary Laughton District Conservation and Historical Society	Rotherham, 19 January 2000.
Mr. M. Gandy	University of Sussex	London, 17 April 1997.
Mr. S. Gerrard	University of East Anglia	Norwich, 25 April 1997 and 15 May 1997.
Mr. C.J. Heseltine	C.J. Heseltine and Associates, Planning and Development Consultants	Rotherham, 17 January 2000.

Mr. F. Jones	Stop Toxic Waste Incinerators in Cleveland (STINC)	Billingham, 15 December 1999.
Mr. M. Jacobs	London School of Economics	London, 10 March 1997.
Mr. A. Jordan	University of East Anglia	Norwich, 17 March 1997.
Mrs. J. Kirton-Darling	Secretary Stop Toxic Waste Incinerators in Cleveland (STINC)	Middlesbrough, 6 January 2000.
Mr. B. Kraus	Stop Toxic Waste Incinerators in Cleveland (STINC)	Middlesbrough, 6 January 2000.
Mr. R. Kuper	University of Hertfordshire	Phone, 14 May 1997.
Mrs. A. Layard	Oxford University	Oxford, 24 April 1997.
Mrs. J. Lennaghan	Institute for Public Policy Research	London, 5 March 1997.
Mr. R. Macrory	Royal Commission on Environmental Pollution/Imperial College	London, 15 May 1997.
Mr. B. Marfleet	Suffolk County, Waste Regulation Branch	Norwich, 18 March 1997.
Mr. B. Marker	Department of the Environment	Phone, May 1997.
Mr. A. Marsh	Stop Toxic Waste Incinerators in Cleveland (STINC)	Hartlepool, 7 January 2000
Mr. C. Mearcy	Institute of Waste Management	Phone, 22 April 1997.
Mr. B. Middleton	Cory Environmental Services, formerly Davy McKee Consultants	London, 26 January 2000.
Mr. R. Newman	Chair Laughton District Conservation and Historical Society	Rotherham, 21 January 2000.
Mr. Th. O'Riordan	University of East Anglia-CSERGE	Norwich, 25 April 1997.
Mr. R. Page	Greenpeace UK	Phone, 18 March 1997.
Mrs. J. Petts	University of Loughborough/University of Birmingham	Loughborough, 8 May 1997 and Birmingham December 1999.
Mr. R. Richards	Newport Borough Council, Planning department	Newport, 11 January 2000.
Mrs. R. Roll	Ruth Roll Environmental Communications	Long Buckby, 14 January 2000.
Mr. L. Searles	London Planning Advisory Committee	London, 29 April 1997.

Mr. D. Short	Department of the Environment,	London, 30 April 1997.
Mr. D . Stubbs	Coordinator Development Control, Rotherham Metropolitan Borough Council, Planning Service	Rotherham, 17 January 2000.
Mr. B Sulek	Deputy plant manager, PARC Environmental Services	Newport, 13 January 2000.
Mr. M. Sunkin	University of Essex, Department of Law	Colchester, 11 March 1997.
Mrs. A. Thomas	Friends of the Earth UK	Phone/e-mail, March 1997.
Mr. G. Tringham	Birmingham City Council	Phone, May, 1997.
Mr. D. Walsh	Leader Redcar and Cleveland Borough Council, former char of the Economic Development and Planning Committee Cleveland County Council.	Redcar, 6 January 2000.
Mr. H. Ward	University of Essex, Department of Government	Colchester, 24 February 1997, 13 March 1997, 23 April 1997.
Mr. A. Weale	University of Essex, Department of Government	Colchester, 25 February 1997 and 14 May 1997.
Mr. P. Whaley	Assistant Director of Planning, Stockton-on-Tees Borough Council.	Stockton-on-Tees, 14 December 1999.
Mr. R. Wight	University of Essex, Department of Law	Colchester, 13 March 1997.

In the Netherlands - interviews for chapter 5

Interviewee	Affiliation	Place and date of interview
Mr. J. van Dijk	Former member of the provincial executive (a.o. for environmental affairs) of the province of Groningen	Groningen, 9 June 2000.
Mr. J. van Heck and Mrs. R. Van Heck	Association for the Conservation of Environmental and Nature Maas and Waal	Beneden-Leeuwen, 9 May 2000.
Mr. H. Huisman	Waste Consultation Organ (Afval Overleg Orgaan)	Utrecht, 20 April 2000.
Mr. J. van Kerkhof	Municipality of Druten	Druten, 13 June 2000.
Mrs. E. van Lankveld-Meggelaars	Grontmij/Delgromij	Arnhem, 3 May 2000.
Mr. E. van der Meulen	Grontmij/Delgromij	Arnhem, 3 May 2000.
Mr. A. Op den Kamp	Op den Kamp Consultancy Group	Den Haag, 16 May 2000.
Mr. G. Penders	Province of Gelderland	Arnhem, 2 May 2000.
Mr. L. de Prieëlle	City of Dordrecht	Dordrecht, 6 June 2000.
Mrs. W. Remmerswaal	Inhabitant of Borgsweer, Municipality of Delfzijl	Delfzijl, 20 April 2000.
Mr. L. Slangen	Province of Groningen	Groningen, 17 April 2000.
Mr. P.G. van der Sleen	Province of Groningen	Groningen, 17 April 2000.
Mr. D. Stoppelenburg	Werkgroep Eemsmond, Association for the Wadden Sea	Groningen, 20 April 2000.
Mr. J. Toxopeus	Muncipality of Druten	Druten, 13 June 2000.
Mr. L. van Veen	Watco Nederland (Ecoservice)	Dordrecht, 17 May 2000.
Mrs. S. Vermeulen-Keijzer	Rijkswaterstaat	Arnhem, 26 April 2000.

In Canada - interviews for chapters 3 and 6

Interviewee	Affiliation	Place and date of interview
Mrs. A. Armour	York University, former member Low Level Radioactive Waste Siting Task Force	Toronto, 11 June 1999.
Mr. V. Bullough	Miller Environmental Corporation, formerly with BOVAR Waste Management	Winnipeg, 30 June 1999.
Mrs. J. Butler	Former councilor Town of Swan Hills, former member of Community Liaison Group Swan Hills	Swan Hills, 21 June 1999.
Mrs. B. Connell	Former employee of the Manitoba Hazardous Waste Management Corporation	Winnipeg, 6 July 1999.
Mr. D. Ediger	Manitoba Environment	Winnipeg, 28 June 1999.
Mr. R. Edmunds	Miller Environmental Corporation, formerly Manitoba Hazardous Waste Management Corporation	Winnipeg, 30 June 1999.
Mr. A. Fernandes	Alberta Environment	Edmonton, 17 June 1999.
Mr. D. Foidart	Community Liaison Group R.M. of Montcalm	Letellier, 5 July 1999.
Mr. G. Fontaine	Community Liaison Group R.M. of Montcalm	Letellier, 5 July 1999.
Mr. K. Froese	University of Alberta	Edmonton, 15 June 1999.
Mr. R. Gallant	Manitoba Hazardous Waste Management Corporation	Montcalm, 7 July 1999.
Mrs. M. Giroux	Former member Community Liaison Group Swan Hills	Swan Hills, 23 June 1999.
Mr. D. Gorber	SENES Consultants	Toronto, 4 June 1999.
Mr. W. Harris	University of Alberta, former member Hazardous Waste Management Committee and Hazardous Waste Implementation Committee province of Alberta	Edmonton, 15 June 1999.
Mr. P. Homenuck	York University	Toronto, 10 June 1999.

Mr. R. Huang	Alberta Environment	Edmonton, 17 June 1999.
Mrs. E. Hughes	University of Alberta	Edmonton, 9 July 1999.
Mr. H. Junck	Former councilor of Swan Hills	Swan Hills, 22 June 1999.
Mr. J. Jonasson	Manitoba Environment	Winnipeg, 30 June 1999.
Mrs. A Joyner	York University	Toronto, 10 June 1999.
Mr. M. Kitagawa	Toxics Watch	Edmonton, 26 June 1999.
Mrs. N. Krawetz	Formerly Alberta Environment, formerly Environment Council of Alberta	Edmonton, 9 July 1999.
Mr. H. Krenkel	BOVAR Waste Management	Swan Hills, 23 June 1999.
Mrs. N. Krogman	University of Alberta	Edmonton, 18 June 1999.
Mr. E. Kucera	Manitoba Environment	Winnipeg, 28 June 1999.
Mr. D. Kuhn	University of Guelph	Guelph, 3 June 1999.
Mr. G. Latonas	BOVAR Waste Management, formerly Manitoba Hazardous Waste Management Corporation	Calgary, 16 June 1999.
Mr. L. Leclair	Community Liaison Group R.M. of Montcalm	Letellier, 5 July 1999.
Mr. D. McCauley	Natural Resources Canada	Ottawa, 19 July 1999.
Mr. D. McRobert	Environmental Commissioner of Ontario	Toronto, 3 June 1999.
Mr. J. Murphy	Mayor Deep River	Deep River, 21 July 1999.
Mr. K O'Neill	Tire Recycling Management Association of Alberta, formerly Alberta Environment	Edmonton, 24 June 1999.
Mr. G Pollock	Major Town of Swan Hills	Swan Hills, 21 June 1999.
Mr. D. Powell	Natural Resources Conservation Board Alberta	Edmonton, 15 June 1999.
Mr. R. Pushchak	Ryerson Polytecnic University	Toronto, 1 June 1999.
Mr. B. Rabe	University of Michigan	Ann Arbor (USA), 7 June 1999.
Mr. P. Rennick	Former member Siting Task Force for Low Level Radioactive Waste	Toronto, 10 June 1999.

Mr. A. Richards	COGEMA Resources, formerly Manitoba Hazardous Waste Management Corporation	Saskatoon, 8 July 1999.
Mr. N. Rubin	Energy Probe	Toronto, 13 July 1999.
Mr. P. Sabourin	Montcalm Development Corporation	St. Jean.-le-Baptiste, 5 July 1999.
Mr. D. Schindler	University of Alberta	Phone, 14 June 1999
Mrs. A. Seifried	Alberta Environment	Edmonton, 17 June 1999
Mr. G. Sheng	York University	Toronto, 31 May 1999
Mr. R. Sherbaniuk	Formerly Alberta Special Waste Management Corporation	Edmonton, 14 June 1999.
Mr. J. Shires	Alberta Environment	Edmonton, 17 June 1999.
Mr. L. Simons	Town Clerk Deep River	Deep River, 20 July 1999.
Mr. L. Smith	Former major Deep River, former employee Atomic Canada Limited	Deep River, 21 July 1999.
Mrs. D. Spiller	Former member Citizen Liaison Group Swan Hills	Swan Hills, 22 June 1999.
Mr. L. Strachan	Manitoba Environment	Winnipeg, 30 June 1999.
Mr. D. Thompson	Former member Community Liaison Group Deep River	Ottawa, 21 July 1999.
Mrs. D. Tingley	Environmental Law Centre	Edmonton, 25 June 1999
Mrs. G. Underdown	Natural Resources Canada	Ottawa, 19 July 1999.
Mr. I. Urquhart	University of Alberta	Edmonton, 18 June 1999
Mrs. B. Vandersteen	Trevose Consultants, formerly Alberta Environment	Edmonton, 24 June 1999
Mr. P. Victor	York University	Toronto, 31 May 1999
Mrs. D. Walker	Councilor Deep River, Former Community Liaison Group Deep River	Deep River, 21 July 1999.
Mr. B. Watson	Town of Swan Hills	Swan Hills, 22 June 1999.
Mr. D. Wood	McCarthy Tétrault, former member Siting Task Force on Low Level Radioactive Waste	Toronto, 13 July 1999.
Mr. E. Yee	Manitoba Environment	Winnipeg, 28 June 1999.

APPENDIX 2: EXPLANATION OF CASE SELECTION

How have the three cases in each country been selected? The situation in Canada largely determined the way I went about case selection. The (for me) surprising finding that the community-based siting approach in that country was not based in decisions at the collective choice level posed a bit of a dilemma as it soon transpired that even in Canada, community-based decision-making is unique. I decided to go ahead in studying the three cases where attempts had been made at 'voluntary siting', considering that I was really interested in the effects of such an approach. The fact that the legislation does not really support such an approach made it perhaps even more interesting, because other institutions than the community could well be present in the process and possibly interfere and a process using argumentations related to the community's interest might invoke resistance.

There are only three Canadian cases where the observers agree that attempts to site a hazardous facility were voluntary; one of these cases actually involves low level radioactive waste instead of hazardous waste. These three cases are the Swan Hills facility in Alberta, the Montcalm facility in Manitoba (both hazardous waste), and the Deep River facility in Ontario (low level radioactive waste). There is no agreement on the question whether British Columbia attempted voluntary siting for its hazardous waste facility. While Munton (1996) argue that this is the case, Rabe (1994) implies it is not. Given the lack of agreement on the issue, I have decided to include the case of low level radioactive waste in my work instead of the British Columbia facility.

The Canadian case studies thus included a proposal for a incinerator, one for a treatment plant and one for landfill (perhaps not the right word for the low level radioactive waste repository, but still). I have consequently attempted one facility of each type in the UK and the Netherlands as well, which is relevant because these different types of plants tend to have different risk profiles in the eyes of the public. In each country I have looked explicitly for 'greenfield cases', that is proposals for hazardous waste facilities on locations that were not yet being used for waste treatment at the time of the proposals. This is mainly in order to prevent the history of waste treatment at that location influence the reaction of the local public. I do admit however, that in most cases a certain perception of the dangers of waste treatment was already present in the community, either by direct experiences or through the media. This was unfortunately unavoidable. I have aimed at the selection of cases that both had been completed (in the sense that some sort of 'final decision' had been made) but did not play before 1980 (because of the expected difficulties with fact finding). Finally, the reasons I had for selecting 'greenfield' cases also applied to in-company treatment facilities, that is facilities that are built by companies for their own wastes only. Thus, I looked for proposals pertaining to 'merchant facilities', as is the beautiful UK-term. These criteria were roughly attained, although sometimes the start of a case was earlier than was visible from the information I had at the outset of my work (especially in the Netherlands, where two really protracted cases have been studied). For all cases, I have chosen them regardless of outcome in the sense of approval or not.

The way in which I have identified case studies differs per country. Canadian case selection has already been explained. In the Netherlands, I have used the records of the Netherlands Commission for Environmental Impact Assessment in Utrecht to identify potentially suitable cases. Only for treatment facilities was there a certain amount of choice and from the several facilities that met the just specified criteria, I asked my supervisor to pick one at random. For hazardous waste incineration and landfill, it was more difficult to identify cases that met the criteria. As regards to incineration, the Commission's records reflected only one case (North Refinery). Unfortunately, this later turned out to have a long history leading to the incineration proposal, which centered on waste treatment, not incineration. I still decided to keep this case after I found out about this, because there is only one other 'merchant' incinerator in the Netherlands, but that one was already built in the early 1970s. In selecting proposals for landfills in the Netherlands, I ran into the problem that landfilling of hazardous wastes in the Netherlands is rare and proposals for greenfield sites do not exist. Therefore, I started looking at landfill proposals for dredging sludges (that also often qualify as hazardous waste). In that sphere, there were several proposals that met the criteria, and one again, my supervisor picked one case at random.

In the UK, it proved difficult to get a general overview of all proposals. My first source of information were the journals 'Mineral Planning' and 'Waste Planning'. In addition, I established contact with the Environmental Services Association (the waste sector representative), and spoke to various experts. Especially landfill proposals and incinerators were scarce, but with the help of the professor Judith Petts at Birmingham I have found a proposal for both types of development that met the criteria. Many examples of proposals for treatment plants were available, and I picked one at the advice of professor Petts.

ABOUT THE AUTHOR

Dave Huitema was born in Leeuwarden, The Netherlands on November 24, 1968. He studied Public Administration and Public Policy at the University of Twente from 1987 to 1993. His master's thesis received the prize for best thesis of the year 1993 from the Dutch Association for Public Administration (Vereniging voor Bestuurskunde). After graduating, Dave Huitema started working for the Center for Clean Technology and Environmental Policy (CSTM) at the University of Twente as a researcher, where he stayed until February 2002.

In his period at CSTM, Huitema specialized in governmental steering (economic instruments and network-steering) and environmental decision-making in general. For the research reported in this book, Huitema visited the University of Florida (USA), the University of Essex (UK – sponsored by the British Council), York University (Canada) and the University of Birmingham (UK).

Since February 2002, Huitema works as a Dutch Science Association (NWO) postdoctoral fellow at the Free University of Amsterdam. His work focuses on an assessment of the relations between the Dutch National Institute of Public Health and the Environment (RIVM) and the political system in the period 1975-2000. His work is part of a larger research effort under the name '*Rethinking political judgement and science-based expertise. Boundary work at the science-politics nexus of Dutch knowledge institutes*' (see: www.bsk.utwente.nl/Rethinking/).

Dave Huitema's publications include the following:

- Huitema, D. and A.H.L.M. van Snellenberg, Policy in Style, in: *Environmental Politics*, vol. 8 (1999), no. 3, pp.77-98.
- Bresssers, J.T.A. and D. Huitema, Economic Instruments for Environmental protection. Can we trust the magic carpet?, in: *International Political Science Review*, 1999 (vol. 20), no. 2, pp 175-196.
- Coenen F.H.J.M., D. Huitema and L.J. O'Toole, Jr. (eds.), *Participation and the quality of environmental decision-making*, Dordrecht, Kluwer Academic Publishers, 1998.

– Huitema, D. en J.P.P. Hinssen (eds.), *Natuurbeleid bestuurskundig bekeken. Een verkenning van sturingsstrategieën in het natuurbeleid en hun effectiviteit*, Enschede, Twente University Press, 1998 [A policy science perspective on nature management. An exploration of steering strategies and their effectiveness].

Address:
Institute for Environmental Studies (IVM),
Free University of Amsterdam,
De Boelelaan 1087,
1081 HV Amsterdam,
The Netherlands.
E-mail: dave.huitema@ivm.vu.nl

ENVIRONMENT & POLICY

1. Dutch Committee for Long-Term Environmental Policy: *The Environment: Towards a Sustainable Future*. 1994 ISBN 0-7923-2655-5; Pb 0-7923-2656-3
2. O. Kuik, P. Peters and N. Schrijver (eds.): *Joint Implementation to Curb Climate Change*. Legal and Economic Aspects. 1994 ISBN 0-7923-2825-6
3. C.J. Jepma (ed.): *The Feasibility of Joint Implementation*. 1995 ISBN 0-7923-3426-4
4. F.J. Dietz, H.R.J. Vollebergh and J.L. de Vries (eds.): *Environment, Incentives and the Common Market*. 1995 ISBN 0-7923-3602-X
5. J.F.Th. Schoute, P.A. Finke, F.R. Veeneklaas and H.P. Wolfert (eds.): *Scenario Studies for the Rural Environment*. 1995 ISBN 0-7923-3748-4
6. R.E. Munn, J.W.M. la Rivière and N. van Lookeren Campagne: *Policy Making in an Era of Global Environmental Change*. 1996 ISBN 0-7923-3872-3
7. F. Oosterhuis, F. Rubik and G. Scholl: *Product Policy in Europe: New Environmental Perspectives*. 1996 ISBN 0-7923-4078-7
8. J. Gupta: *The Climate Change Convention and Developing Countries: From Conflict to Consensus?* 1997 ISBN 0-7923-4577-0
9. M. Rolén, H. Sjöberg and U. Svedin (eds.): *International Governance on Environmental Issues*. 1997 ISBN 0-7923-4701-3
10. M.A. Ridley: *Lowering the Cost of Emission Reduction: Joint Implementation in the Framework Convention on Climate Change*. 1998 ISBN 0-7923-4914-8
11. G.J.I. Schrama (ed.): *Drinking Water Supply and Agricultural Pollution*. Preventive Action by the Water Supply Sector in the European Union and the United States. 1998 ISBN 0-7923-5104-5
12. P. Glasbergen: *Co-operative Environmental Governance: Public-Private Agreements as a Policy Strategy*. 1998 ISBN 0-7923-5148-7; Pb 0-7923-5149-5
13. P. Vellinga, F. Berkhout and J. Gupta (eds.): *Managing a Material World*. Perspectives in Industrial Ecology. 1998 ISBN 0-7923-5153-3; Pb 0-7923-5206-8
14. F.H.J.M. Coenen, D. Huitema and L.J. O'Toole, Jr. (eds.): *Participation and the Quality of Environmental Decision Making*. 1998 ISBN 0-7923-5264-5
15. D.M. Pugh and J.V. Tarazona (eds.): *Regulation for Chemical Safety in Europe: Analysis, Comment and Criticism*. 1998 ISBN 0-7923-5269-6
16. W. Østreng (ed.): *National Security and International Environmental Cooperation in the Arctic – the Case of the Northern Sea Route*. 1999 ISBN 0-7923-5528-8
17. S.V. Meijerink: *Conflict and Cooperation on the Scheldt River Basin*. A Case Study of Decision Making on International Scheldt Issues between 1967 and 1997. 1999 ISBN 0-7923-5650-0
18. M.A. Mohamed Salih: *Environmental Politics and Liberation in Contemporary Africa*. 1999 ISBN 0-7923-5650-0
19. C.J. Jepma and W. van der Gaast (eds.): *On the Compatibility of Flexible Instruments*. 1999 ISBN 0-7923-5728-0
20. M. Andersson: *Change and Continuity in Poland's Environmental Policy*. 1999 ISBN 0-7923-6051-6

ENVIRONMENT & POLICY

21. W. Kägi: *Economics of Climate Change: The Contribution of Forestry Projects.* 2000 ISBN 0-7923-6103-2
22. E. van der Voet, J.B. Guinée and H.A.U. de Haes (eds.): *Heavy Metals: A Problem Solved?* Methods and Models to Evaluate Policy Strategies for Heavy Metals. 2000 ISBN 0-7923-6192-X
23. G. Hønneland: *Coercive and Discursive Compliance Mechanisms in the Management of Natural Resourses.* A Case Study from the Barents Sea Fisheries. 2000 ISBN 0-7923-6243-8
24. J. van Tatenhove, B. Arts and P. Leroy (eds.): *Political Modernisation and the Environments.* The Renewal of Environmental Policy Arrangements. 2000 ISBN 0-7923-6312-4
25. G.K. Rosendal: *The Convention on Biological Diversity and Developing Countries.* 2000 ISBN 0-7923-6375-2
26. G.H. Vonkeman (ed.): *Sustainable Development of European Cities and Regions.* 2000 ISBN 0-7923-6423-6
27. J. Gupta and M. Grubb (eds.): *Climate Change and European Leadership.* A Sustainable Role for Europe? 2000 ISBN 0-7923-6466-X
28. D. Vidas (ed.): *Implementing the Environmental Protection Regime for the Antarctic.* 2000 ISBN 0-7923-6609-3; Pb 0-7923-6610-7
29. K. Eder and M. Kousis (eds.): *Environmental Politics in Southern Europe: Actors, Institutions and Discourses in a Europeanizing Society.* 2000 ISBN 0-7923-6753-7
30. R. Schwarze: *Law and Economics of International Climate Change Policy.* 2001 ISBN 0-7923-6800-2
31. M.J. Scoullos, G.H. Vonkeman, I. Thornton, and Z. Makuch: *Mercury - Cadmium-Lead: Handbook for Sustainable Heavy Metals Policy and Regulation.* 2001 ISBN 1-4020-0224-6
32. G. Sundqvist: *The Bedrock of Opinion.* Science, Technology and Society in the Siting of High-Level Nuclear Waste. 2002 ISBN 1-4020-0477-X
33. P.P.J. Driessen and P. Glasbergen (eds.): *Greening Society.* The Paradigm Shift in Dutch Environmental Politics. 2002 ISBN 1-4020-0652-7
34. D. Huitema: *Hazardous Decisions.* Hazardous waste siting in the UK, The Netherlands and Canada. Institutions and discourses. 2002 ISBN 1-4020-0969-0

For further information about the series and how to order, please visit our Website
http://www.wkap.nl/series.htm/ENPO

KLUWER ACADEMIC PUBLISHERS – DORDRECHT / BOSTON / LONDON